Pollutant Studies in Marine Animals

Editors

C. S. Giam, Ph.D.
Professor of Chemistry
Department of Industrial Environmental Health Sciences
Graduate School of Public Health
University of Pittsburgh
Pittsburgh, Pennsylvania

Lee E. Ray, Ph.D.
Senior Scientist
Finnsugar Biochemicals, Inc.
Schaumburg, Illinois

CRC Press, Inc.
Boca Raton, Florida

Library of Congress Cataloging-in-Publication Data

Pollutant studies in marine animals.

Bibliography: p.
Includes index.
1. Marine fauna--Effect of water pollution on.
I. Giam, C. S. II. Ray, Lee E., 1941-
QH545.W3P62 1987 591.5'2636 86-31690
ISBN 0-8493-5407-2

Direct all inquiries to CRC Press, Inc., 2000 Corporate Blvd., N.W., Boca Raton, Florida, 33431.

International Standard Book Number 0-8493-5407-2

Library of Congress Card Number 86-31690
Printed in the United States

PREFACE

Environmental pollution continues to be a problem of growing, world-wide concern. Due to their immense size, the oceans have been essentially viewed as "infinite, renewable" sinks. Because of this and a lack of earlier concern, there has been a progressive decline in certain regions, if not generally, in the oceans. While the quality of some regions (e.g., Galveston Bay, Chesapeake Bay, the Great Lakes) has improved during recent years, there is still great concern regarding the marine environment in general, including the life it supports. Although a great deal of effort has been devoted to studying the problem of pollution, most work has concentrated on freshwater and land-based systems. This volume attempts to summarize the joint efforts of a group of workers who developed analytical systems for use with marine animals and then used these systems to monitor select marine populations.

The approach of this group was somewhat analogous to that of a physician who orders a variety of clinical tests in attempting to diagnose a patient's problem. Thus, pollutant-induced changes were sought in metabolite (e.g., glutathione, ascorbate) concentrations, in hormone concentrations, in enzymes associated with pollutant metabolism (e.g., the cytochrome P-450 mixed-function oxidase system), in immunological capacity, and in cellular morphology. The work reported here details the specific methods used and summarizes the results of measurement of the parameters in both pristine and exposed animals. Not all systems exhibited changes in response to the pollutants tested; these systems are also discussed here in hopes of helping others avoid these potential, but insensitive, markers. It is also hoped that, by detailing methods and summarizing results in a single volume, future work along these lines will be encouraged and facilitated.

THE EDITORS

C.S. Giam, Ph.D., is a professor in the Graduate School of Public Health at the University of Pittsburgh where he also holds an appointment in the Department of Chemistry. He was a professor in the Department of Chemistry and the Department of Oceanography, Texas A&M University, College Station, for fifteen years. He has also served as Head of the Chemistry Department at Texas A & M, Dean of the College of Science at the University of Texas at El Paso, and Program Director for Environmental Quality at the National Science Foundation.

Dr. Giam received his B.S. degree (Honors) from the University of Singapore in 1955 and obtained his M.S. and Ph.D. degrees in Chemistry from the University of Saskatchewan in 1961 and 1963, respectively.

Dr. Giam's major research interests include heterocyclic chemistry (especially the reactions of pyridine and its derivatives) the distribution, transport, and fate of man-made organic pollutants in the atmosphere and oceans and the development of analytical procedures for studying ultratrace concentrations of pollutants, pesticides, and other compounds. He has been involved in numerous large scale studies of atmospheric and marine pollutants.

Dr. Giam has published more than 130 research papers, edited several books, and contributed to many other books. He has given numerous invited and contributed presentations at both national and international meetings. He proposed and helped organize the first International Symposium on the Responses of Marine Organisms to Pollutants which was held in Plymouth, England in 1981. The series has continued with meetings biennially.

Lee E. Ray, Ph.D., is Senior Scientist at Finnsugar Biochemicals Inc. (formerly Fermco Development, Inc.) in Schaumburg, Illinois. He was previously Assistant Professor in the Department of Biology, Research Scientist in the Department of Chemistry, and Research Associate in the Department of Pharmacology and Toxicology, all at Texas A & M University, College Station. He was also a guest scientist at Carlsberg Laboratory, Copenhagen, Denmark.

Dr. Ray received his B.S. degree (cum laude) in 1963 from The Colorado College, Colorado Springs, his M.S. degree (Chemistry) from the University of Colorado in 1969, and his Ph.D. degree (Biochemistry) from the University of Nebraska at Lincoln in 1971.

Dr. Ray's major research interests include protein biochemistry, including isolation, stabilization, and assay of enzymes, enzyme modification (covalent) and immobilization, applications of enzymes in food systems, preparation and application of monoclonal antibodies, and the effects of environmental pollutants.

Dr. Ray has published more than 20 research papers. He has given several invited and contributed presentations at both national and international meetings. He has also helped organize various national and international meetings.

CONTRIBUTORS

Robert S. Anderson, Ph.D.
Professor
Chesapeake Biological Laboratory
Center for Environmental and Estuarine Studies
University of Maryland
Solomons, Maryland

R. S. Carr, Ph.D.
Research Scientist
Battelle New England Marine Research Laboratory
Duxbury, Massachusetts

C. R. Fries, Ph.D.
Assistant Professor
University Honors Program and the School of Life and Health Sciences
University of Delaware
Newark, Delaware

Richard Lee, Ph.D.
Professor
Skidaway Institute of Oceanography
Savannah, Georgia

Jerry M. Neff, Ph.D.
Research Leader
Department of Ocean Sciences and Technology
Battelle New England Marine Research Laboratory
Duxbury, Massachusetts

L. Quattrochi
Research Associate
Department of Medicine
University of California at San Diego
La Jolla, California

John J. Stegeman, Ph.D.
Associate Scientist
Department of Biology
Woods Hole Oceanographic Institute
Woods Hole, Massachusetts

Peter Thomas, Ph.D.
Research Scientist
Marine Science Institute
University of Texas
Port Aransas, Texas

M. R. Tripp, Ph.D.
Director and Professor
School of Life and Health Sciences
University of Delaware
Newark, Delaware

ACKNOWLEDGMENTS

Many people were involved in planning, organizing, developing, implementing, and completing the Pollutant Responses in Marine Animals Program. While it is impossible to mention everyone who has contributed to this program, we would like to thank the following individuals: Dr. Roger Baier, Dr. Grant Gross, and Ms. Joan Mitchell, all of the National Science Foundation; Dr. Thomas Duke and Dr. Gilles LaRoche, who served on the PRIMA Scientific Advisory Committee; Dr. Charles Gates, who provided statistical assistance; and Dr. Feenan Jennings, who provided useful advice and support. Finally, we would like to thank Fermco Development Inc. (Finnsugar Biochemicals, Inc.) for providing financial support for preparation of the final drafts of the manuscript for this book.

C. S. Giam
Lee E. Ray

ABBREVIATIONS USED IN THIS BOOK

AHH	Aryl hydrocarbon hydroxylase (benzo(a)pyrene monooxygenase)
ANF	α-Naphthoflavone
BaP	Benzo(a)pyrene
BNF	β-Naphthoflavone
DBP	Dibutyl phthalate
DEHP	Di(2-ethylhexyl) phthalate
DEP	Diethyl phthalate
DMSO	Dimethylsulfoxide
DTT	Dithiothreitol
ECD	Electron capture detector
EDTA	Ethylenediamine tetraacetate
FID	Flame ionization detector
GC	Gas chromatography
HCB	Hexachlorobenzene
HPLC	High pressure liquid chromatography
MC	Methylcholanthrene
MFO	Mixed function oxygenase(s)
PAE	Phthalic acid esters
PAGE	Polyacrylamide gel electrophoresis
PAH	Polycyclic aromatic hydrocarbon(s)
PBB	Polybrominated biphenyls
PCB	Polychlorinated biphenyls
PCP	Pentachlorophenol
PE	Petroleum ether
PEG	Polyethylene glycol
ppb	Parts per billion (μg/ℓ or μg/kg)
PRIMA	Pollutant Responses in Marine Animals
SDS	Sodium dodecyl (lauryl) sulfate
TRIS	Tris (hydroxymethyl) aminomethane

TABLE OF CONTENTS

Chapter 1
Pollutant Responses in Marine Animals: The Program 1
C. S. Giam, L. E. Ray, R. S. Anderson, C. R. Fries, R. Lee, J.M. Neff, J. J. Stegeman, P. Thomas, and M. R. Tripp

Chapter 2
Bioaccumulation and Depuration of Selected Organic Compounds in Marine Fish 23
L. E. Ray and C. S. Giam

Chapter 3
Cytochrome P-450 and Mixed-Function Oxygenase Systems in Marine Invertebrates 51
R. Lee and L. Quattrochi

Chapter 4
Monooxygenase Systems in Marine Fish 65
J. J. Stegeman

Chapter 5
Immunocompetence in Invertebrates 93
R. S. Anderson

Chapter 6
Histopathology and Histochemistry 111
M. R. Tripp and C. R. Fries

Chapter 7
Biochemical Stress Responses and Alterations of Tissue Ascorbic Acid and Glutathione Content 155
P. Thomas, R. S. Carr, and J. M. Neff

Index 181

Chapter 1

POLLUTANT RESPONSES IN MARINE ANIMALS: THE PROGRAM*

C. S. Giam, L. E. Ray, R. S. Anderson, C. R. Fries, R. Lee, J. M. Neff, J. J. Stegeman, P. Thomas, and M. R. Tripp

TABLE OF CONTENTS

I. Introduction 2

II. Objectives 3

III. Selection of Model Organic Pollutants 4
- A. Polycyclic Aromatic Hydrocarbons 5
- B. Polychloroaromatic Hydrocarbons 5
- C. Chlorinated Phenols 5

IV. Selection of Test Animals 6
- A. Chordata 6
- B. Arthropoda 7
- C. Annelida 7
- D. Mollusca 8

V. Selection of Biological Effects 8
- A. Uptake and Release Kinetics 8
- B. Biotransformations 9
- C. Corticosteroid Stress Response and Ascorbic Acid Depletion 10
- D. Histopathological Analysis 11
- E. Immunocompetence 12
- F. Integration 12

References 13

* Abbreviations used in this book can be found following the Acknowledgments in the front of this book.

I. INTRODUCTION

The major objective of the Pollutant Responses in Marine Animals (PRIMA) program was to develop and evaluate biological indexes of impending pollution-induced damage to marine animals. Possible indexes included ecological, physiological, morphological, and biochemical changes. Ecological data (e.g., primary and secondary productivity, community dynamics, diversity indexes, etc.) tend to be highly variable from year to year so that many years of detailed data are necessary for predictive purposes.[1] Certain physiological indexes (e.g., respiration, osmoregulation, etc.) tend to be insensitive except at concentrations approaching lethality.[2-5] Differences in such factors as experimental design and conditions, choice of toxicant, species, and physiological parameter examined, as well as normal variability within species and populations and seasonal and sexual differences, have rendered the vast bulk of toxicological data obtained for marine organisms unsuitable for quantitative comparative purposes.[6,7] Therefore, the PRIMA program utilized an integrated multidisciplinary approach to the problem, measuring morphological and biochemical changes as early warning indexes of pollutant damage to marine animals.

Subtle or gross morphological changes may occur in tissues, cells, and individual cellular ultrastructure in response to various types of stress, including chemical stress. These morphological changes may reflect an activation of the body's basic homeostatic defense mechanisms against the stressor, such as proliferation of the endoplasmic reticulum in hepatic cells due to induction of the mixed function oxidase (MFO) system[8,9] or hypertrophy of the adrenal cortex.[10-13] They may reflect a rapid depletion of essential micronutrients, such as skin and skeletal lesions in fish and shrimp caused by the pollutant-mediated depletion of ascorbic acid reserves[14,15] or an alteration in cellular immune response increasing susceptibility to disease.[16,17] Alternatively, these morphological changes may be due to a direct interaction of the pollutant with cellular structure[18,19] or with essential biochemical pathways such as those involved in energy production or lipid metabolism.[20-23] Histopathology has been used effectively for diagnostic purposes in human medicine for over 100 years. Pollutant stress produces histopathological lesions in marine animals as well,[17,24] thus such changes could be useful for diagnosing the health of pollutant-impacted marine ecosystems.

As indicated above, in many cases, stress-induced histopathology is directly caused by or results in subtle biochemical changes in the tissues of pollutant-stressed animals. These biochemical changes may be homeostatic (e.g., induction of the MFO system or activation of the pituitary-adrenocortical axis) or they may represent an alteration or derangement of key biochemical or physiological pathways (e.g., alterations of key catabolic biochemical pathways, depressed immune response, or depletion of essential micronutrients). Whether the initial response is homeostatic or strictly pathological, chronic stress ultimately results in a reduction of energy available for basic physiological processes as well as a decrease in disease resistance and in the overall well-being of the organism.

The PRIMA program studied the effect of selected organic pollutants on various biological parameters in marine animals. The hypothesis was that basic chemical lesions induced at the molecular level would be reflected in biochemical and morphological changes which could be detected and quantified by examining the appropriate biochemical and histopathological parameters.

Organic pollutants in marine systems accumulate rapidly in the benthos,[25] therefore benthic biota are most likely to be affected by chronic pollution stress. Benthic or demersal species from each of four phyla (annelids, mollusks, arthropods, and chordates) were studied as indicator species. By limiting the number of species studied, it was possible to examine each one intensively and to learn how they respond to representative chemical pollutants.

Various specific biological, biochemical, or morphological systems have been proposed for monitoring the presence or impact of pollutants in marine or aquatic ecosystems. Best known of these is the "mussel watch" proposed by Goldberg[26] to monitor the tissues of the mussel *Mytilus* and several related bivalve mollusks for certain pollutants. Potential, but unproven, indexes include elevated aryl hydrocarbon hydroxylase in fish liver as an index of petroleum pollution,[27-29] tumors in flatfish as an index of pollution by chemical carcinogens,[17,30] hyperglycemia as a general index of pollutant stress in fish[31,32] and crustaceans,[33] and activation of the pituitary-adrenocortical axis as an index of environmental stress in fish.[11,34]

The PRIMA program incorporated some of these and also introduced several other indexes as it was felt that an integrated multiparameter approach would yield more useful diagnostic and predictive information. The factors studied comprised morphological changes, including tissue, cellular, and subcellular alterations, and biochemical and homeostatic changes, including accumulation and release kinetics, steroid hormones, enzymes and detoxification mechanisms, and cellular and humoral immune reactions.

The chemicals evaluated were chosen to be representative of the pollutant types of major concern in the marine environment,[35] including petroleum-derived aromatic hydrocarbons and related oxygenated and halogenated derivatives. The compounds used in this program were the polycyclic aromatic hydrocarbon, benzo(a)pyrene, the halogenated hydrocarbon, hexachlorobenzene, and the halogenated phenol, pentachlorophenol. They were chosen as prototypes also because of their toxicity, potential for bioaccumulation, and presence in the marine environment.

This type of analysis was expected to allow prediction of the biological effects of other related chemical pollutants. Describing the patterns of response in various species to a particular class of pollutant compounds should ideally provide an effective means to estimate the degree of hazard associated with a given compound and enable definition of the sites and possibly the mechanisms of action of those compounds shown to be hazardous, as well as provide a clear index of the physiological parameters most suited to use as indicators, both for environmental diagnosis and prognosis.

II. OBJECTIVES

The general objectives of PRIMA were to characterize the biochemical, immunological, and histological responses of a few representative species from each of the major marine metazoan phyla (Annelida, Mollusca, Arthropoda, Chordata) to representative classes of organic pollutants. As discussed above, the pollutants used included aromatic hydrocarbons and related oxygenated and halogenated derivatives. The study provided detailed knowledge of the normal histology and biochemical/physiological parameters of these organisms and attempted to identify those indexes which are most subject to pollutant-mediated modulation. Sensitive biochemical and histological early-warning indexes of pollutant-induced damage were evaluated in marine animals in integrated laboratory and field studies. The ultimate objective was to develop sensitive, reliable diagnostic tools to evaluate the health of marine organisms in pollutant-impacted marine ecosystems.

Acute exposures of marine animals to high sublethal concentrations of the pollutants were used to identify the target organs and biochemical systems of the particular pollutant. Then responses to chronic exposure to low, environmentally realistic concentrations were studied in detail. An important aspect of this study was the generation of comprehensive data on a spectrum of biochemical and histological responses of marine animals to organic pollutants. The spectrum of responses to a particular chemical pollutant was expected to provide the "symptoms" of poisoning for the species under investigation. These symptoms or responses could then be used diagnostically to evaluate the health of natural populations in pollutant-impacted marine environments. This integrated approach was expected to provide species- and pollutant-specific responses to low-level, chronic exposure to model environmental contaminants. For example, a chemical might stimulate the corticosteriod stress response and yet not cause histological lesions, or vice versa. The integrated multidisciplinary approach was used to detect any characteristic, specific responses. The factors studied include the following:

1. Rates of accumulation, distribution in the body, rates of metabolic transformation, and rates of release of selected organic pollutants by marine animals. Emphasis was placed on characterizing cytochrome P-450-dependent microsomal mixed function oxidase systems and their role in pollutant disposition and effects in the tissues of the experimental animals. Defining the induction or inhibition of MFO at both high and low concentrations of the different organic pollutants could permit adaptive systems to be distinguished.
2. Effects of high-level acute or low-level chronic exposure to the model pollutants on the histology of various tissues and organs of the experimental animals. Exposure to high doses helps permit identification of target organs. Light and electron microscopic methods were used to detect and evaluate cellular and tissue lesions and anomalies resulting from pollutant exposure.
3. Responses of the immune system of marine animals to high-level acute or low-level chronic pollutant stress. Immune competency of pollutant-stressed fish was determined by measuring lymphocyte rosette formation and agglutinin titers. Alterations in cellular and humoral immune reactions of pollutant-stressed marine invertebrates were assayed in terms of hemocyte phagocytic activity, lysosomal enzyme profiles, and agglutinin titers. Initial studies included high doses to help identify target organs.
4. Effects of acute high-level or chronic low-level pollutant exposure on the adrenocortical stress response in fish and the analogous biochemical stress responses in invertebrates. Plasma corticosteroids and rate of cortisol turnover were measured in pollutant-stressed fish. Alterations of cortisol dynamics were correlated to ascorbic acid depletion and changes in blood glucose, cholesterol, and osmolality. Pollutant-induced changes in blood glucose and tissue ascorbate and catecholamines in marine invertebrates were also investigated.
5. Correlation of the various indexes studied so that cause and effect relationships could be established and diagnostic profiles for each pollutant and species could be constructed.
6. Applicability of laboratory assays of biochemical and morphological parameters when used in stressed and unstressed field populations of benthic organisms.

III. SELECTION OF MODEL ORGANIC POLLUTANTS

In defining and describing sublethal biochemical responses which might serve as sensitive diagnostic indexes of impending environmental damage, it seemed important to take a comparative approach. Thus the responses in question were investigated in several representative species and several model chemical pollutants, representative of the major pollutant classes, were compared to determine whether the sublethal responses elicited were specific to a particular chemical toxicant or were a generalized reaction to pollutant stress. It is generally accepted that the pollutants requiring top priority in studies of marine biological effects are those which: (1) are produced in large amounts (25×10^6 kg/year or more); (2) have use and disposal patterns tending to favor their eventual entry into the oceans; (3) have high toxicity and/or bioaccumulation potential in marine organisms, and (4) are highly resistant to biodegradation.[35-38]

Three classes of chemical pollutants which meet these criteria are hydrocarbons and related petrochemicals, some industrial and pesticidal chlorinated hydrocarbons, and several heavy metals. In the PRIMA program, representative organic chemicals (model pollutants) of the first two types were evaluated and compared. The model pollutants included a polycyclic aromatic hydrocarbon (benzo(a)pyrene [BaP]), a polychloroaromatic hydrocarbon (hexachlorobenzene [HCB]), and a chlorinated phenol (pentachlorophenol [PCP]). All investigators in the program used the same model pollutants obtained from the same source and batch.

A. Polycyclic Aromatic Hydrocarbons

Polycyclic aromatic hydrocarbons (PAH) are nearly ubiquitous trace components in the marine environment.[39] Marine sediments are often a particularly rich source of PAH, both in terms of the concentrations present and numbers of different PAH encountered.[40-43] PAH have also been detected in the tissues of benthic marine invertebrates, especially mollusks.[39,44-48]

Marine PAH originate from a variety of sources. Crude oil and coal generally contain a wide variety of different PAH at low concentrations. However, pyrolysis or combustion of these fossil fuels and other organic materials produces large amounts of PAH in the form of coal tars, carbon blacks, and airborne particulates. For example, two samples of coal tar analyzed by Lijinsky et al.[49] contained a mean of 6.5 g/kg benz(a)anthracene, 2.0 g/kg benzo(a)pyrene, and 17.8 g/kg fluoranthene. These PAH may reach the marine environment in surface runoff from land or in atmospheric fallout.[42] Suess[50] estimated that 5000 tons of BaP are released annually to the environment from all anthropogenic sources. The best estimate of the annual influx of petroleum to the marine environment from all sources is 6.2 million metric tons/year.[51] Thus petroleum represents a significant source of PAH in the marine environment. The possibility that some marine bacteria and algae synthesize PAH is uncertain.[52] Neff[39] estimated the annual input of total PAH to the aquatic environment from all sources to be 230,000 metric tons. The largest single contribution to the total was from petroleum spillage.

Much of the research on environmental PAH has concentrated on BaP because of its well-established carcinogenicity in mammals[53,54] and its extremely wide distribution in the environment.[39,50] Neff[39] estimated that nearly 70 tons of BaP enter the aquatic environment each year from all sources. Much of this comes from fallout or rainout of airborne BaP derived from combustion effluents.

B. Polychloroaromatic Hydrocarbons

The persistence and toxicity of pesticidal and industrial polychloroaromatic hydrocarbons in the marine environment have been extensively documented (see reviews by Zitko and Choi[55] and Peakall[56]). The pesticides DDT, mirex, and toxaphene and the industrial polychlorinated biphenyls have received considerable attention.[25,57-61] In its report titled "Assessing Potential Ocean Pollutants", the National Academy of Sciences[35] recommended that increased attention be paid to HCB because of its detection in many marine samples, its persistence, and its high toxicity. Hexachlorobenzene is used as a seed fungicide and wood preservative. It is also a byproduct in the waste stream from the manufacture of a number of organochlorine compounds.[62,63] Approximately 1 million kg of HCB is manufactured each year in the United States alone.[35] HCB has been detected in the tissues of marine and freshwater animals[64-66] and has been shown to cause cancer in laboratory mammals.[67,68] HCB is also very stable and persistant in the environment. It is highly resistant to photolysis in the air[69] and is not readily degraded by soil microbes.[70]

C. Chlorinated Phenols

A number of chlorinated phenols, including pentachlorophenol, have been detected in the aquatic environment. Pentachlorophenol was selected as the model compound from this group because more PCP is manufactured and it is usually found at higher concentration in environmental samples than are other chlorinated phenols. It is commonly used as a herbicide, fungicide, insecticide, and as a wood perservative for the lumber and construction industry.[62,71] Annual production of PCP in the United States alone is estimated to be in excess of 23 million kg.[72] It has been detected in polluted coastal waters and organisms[73-75] and is relatively persistant in aquatic ecosystems.[76] PCP is extremely toxic to freshwater and marine invertebrates and fish. Ninety-six-hour LC_{50} values are usually well below 1 ppm and significant sublethal responses have been detected at concentrations of a few parts per billion in sensitive species (see papers in Rao[77]).

IV. SELECTION OF TEST ANIMALS

The suitability of a species for evaluating diagnostic and predictive models depends on its ability to reliably reflect the environment from which it is sampled. Alterations in environmental parameters should result in reproducible, quantifiable, and meaningful changes in physiological, biochemical, and morphological characteristics in the species under investigation. Because marine sediments are generally the major sinks for chemical contaminants entering the marine environment, all the species selected for investigation are representatives of the marine benthic or demersal fauna. Representative species from the four major phyla were chosen to ensure a broad comparative evaluation of environmental impact. The principal species chosen include the following:

Chordata:	*Fundulus heteroclitus* (mummichog)
	Pseudopleuronectes americanus (winter flounder)
	Mugil cephalus (striped mullet)
Arthropoda:	*Callinectes sapidus* (blue crab)
Annelida:	*Neanthes (Nereis) virens* (clam worm)
	Glycera sp. (blood worm)
Mollusca:	*Mercenaria mercenaria* (hard-shell clam)

A. Chordata

F. heteroclitus, P. americanus, and *M. cephalus* are representatives of estuarine and coastal marine environments, areas most likely to receive substantial inputs of contaminants. *F. heteroclitus* are generally available, *P. americanus* are abundant in New England, and *M. cephalus* are representative of marine demersal fish fauna of the southeastern U.S.

Although winter flounder do migrate, it is to a limited extent and they do not range extensively, at least in Cape Cod waters.[78] There is evidence that the mummichog in a given population return to the same areas of a given marsh or estuary with each tidal cycle.[79] Striped mullet make local alongshore migrations and seasonal offshore migrations for spawning, but the extent of these movements is not great. Broadhead and Mefford[80] reported that 90% of recaptures of taged mullet off Florida after 270 days were within 32 km of the tagging point. Thus each of these species may be exposed to its respective local environment on a regular, continuing basis, an important consideration in the selection of representative species for field studies.

F. heteroclitus is a hardy species noted for its adaptability, naturally tolerating temperatures from near freezing to 25°C, and salinities from 0 to 35‰.[81] They are summer spawners and are normally inactive at winter temperatures. *M. cephalus* is an extremely eurythermal and euryhaline species. It tolerates temperatures from 4.5 to 36°C and salinities from 0 to 81‰ and is commonly encountered from freshwater to hypersaline lagoons.[82] In the western Gulf of Mexico, peak spawning activity takes place in the spring and fall. *P. americanus,* on the other hand, is more restricted in its range of preferred temperature and salinity[78] and is a winter spawner. *F. heteroclitus* and *M. cephalus* were expected to exhibit responses of a hardy adaptable species, and *P. americanus,* those of a less hardly species. All three species adapt well to aquarium conditions.

The abundant background literature concerning the biology of these three teleosts includes information directly related to the studies proposed here. Several investigators have reported on pertinent metabolic processes in these species, including the cytochrome P-450 MFO system[28,83-87] and corticosteroid dynamics.[88-90] There have been extensive histological examinations of all three species.[91]

Although the above are the principal species used, other species were also employed. Mixed function oxygenase systems in *Stenotomus chrysops* (scup or porgy) were studied because of possible anomalies. *Micropogon undulatus* (croaker) were studied as they show responses to PAH inducers wholly typical of those in mammals.

B. Arthropoda

The blue crab, *C.sapidus,* is the most abundant, commercially important, and widely distributed of the Atlantic portunid crabs. Its range extends from Nova Scotia to northern Argentina, including Bermuda and the Antilles. It also occurs in Europe from Denmark to the Black and Mediterranean Seas.[92] Major fisheries for this species occur in the Chesapeake Bay, the Carolinas, and the coastal states of the Gulf of Mexico. It is a major predator and scavenger in the coastal and estuarine ecosystem.

Because of its economic and ecological importance, *C. sapidus* has been extensively studied. These studies have included larval development,[93] respiration and osmoregulation,[94,95] glucose metabolism,[96] sensitivity to environmental pollutants,[97-99] ecdysteroid metabolism,[100] and metabolism of xenobiotics.[101,102]

C. sapidus was chosen because of its wide geographic distribution, its economic importance, and the extensive literature about its biology and physiology.

C. Annelida

Polychaete worms are extremely important members of the benthic macrofauna. Thus, the worm *Neanthes (Nereis) virens* was chosen as a representative of the macrofauna in close contact with marine sediments. Interactions between benthic macrofauna, meiofauna, and microfauna appear to be responsible for the degradation of petroleum hydrocarbons and possibly other organic pollutants in marine sediments.[103-105] Polychaete worms are able to metabolize a variety of aromatic hydrocarbons.[103,105,106] Many of them are deposit feeders and process large volumes of sediment through their digestive tracts. Even the predatory benthic genera such as *Neanthes* tend to ingest significant quantities of sediment.

Certain polychaete species are the dominant animals in areas of oil spills, refinery effluents, brine effluents from oil fields, and natural oil seeps.[107-111] The ubiquitous species *Capitella capitata* dominated the benthos after the West Falmouth oil spill[110] and parts of Long Beach Harbor receiving refinery wastes.[109] *C. capitata* experimentally exposed for several weeks to crude oil had much higher MFO activity in third-generation worms than first- or second-generation worms.[105] Rossi and Anderson[106] exposed the polychaete *Neanthes arenaceodentata* to fuel oil and found that third-generation juvenile worms were significantly more resistant than the first generation. Possibly, the increased resistance of polychaetes to oil after three generations was due to an increase in MFO activity which acted to detoxify the aromatic hydrocarbons. Grassle and Grassle[112,113] have shown by a study of electrophoretic patterns that *C. capitata* is actually a complex of at least six sibling species. Thus, exposure to oil may result in selection for species or strains resistant to oil. In insects, pesticide-resistant strains have higher MFO activity than less tolerant strains.[114-116] This increase in enzyme activity appears to allow oxidation and detoxification of pesticides.

Species of the genus *Neanthes* are worldwide in distribution in near-shore marine and estuarine sediments and are easily maintained in laboratory culture. *N. virens* is the largest of the Atlantic nereid polychaetes, often attaining a length of 18 in.[117] Its large size allows easy dissection of various tissues for the analysis of biochemical transformation of xenobiotics and for other biochemical and morphological studies. *N. virens* was chosen because of its availability, its ecological importance in the benthic ecosystem, its large size and ease of handling, and because many aspects of its biology are well known.[118] Its ability to metabolize aromatic hydrocarbons has been investigated.[103,105]

Glycera sp., another polychaete with the same general physiological attributes as *Neanthes,* was used for studies of immunological competence in invertebrates. The use of *Neanthes,* for immune studies was discontinued for technical reasons (e.g., the relative difficulty in obtaining coelomic fluid and the rapid and irreversible clotting of the coelomocytes, which made studies of cellular immunology impossible). In contrast to *Neanthes, Glycera* has abundant, easily obtained coelomic fluid and coagulation of its coelomocytes is minimal.

D. Mollusca

The hard-shell clam *M. mercenaria* was selected as the representative mollusk because of its wide geographic distribution and its commercial importance. This species has been studied extensively and many aspects of its biology are known.[119,120] Bivalve mollusks have received considerable attention in marine pollution effects and monitoring studies because of their sedentary habits and tendency to accumulate pollutants in their tissues to higher concentrations and retain them longer than other marine species.[121-125] Most mollusks apparently lack the MFO cytochrome P-450 system and so are unable to metabolize aromatic hydrocarbons.[126] However, Anderson[127] was able to induce aryl hydrocarbon hydroxylase activity in the oyster *Crassostrea virginica* by exposure to aromatic hydrocarbons. There is some evidence that bivalves can metabolize other types of xenobiotics.[128-130]

V. SELECTION OF BIOLOGICAL EFFECTS

In clinical medicine, the diagnosis of a specific pathology usually depends on the detection of several different symptoms, each characteristic of that condition. The use of multiple symptoms increases the likelihood that a given pathology will be accurately identified and characterized, since, nearly always, no single symptom is uniquely characteristic of only one disease. Similarly in these studies, several related biological parameters were evaluated as possible diagnostic tools. These included accumulation/release kinetics, biotransformation of organic pollutants, corticosteroid and related stress responses, ascorbic acid metabolism, histopathology and histochemistry, and immunocompetence. This is the type of information that is needed to assess accurately the health of marine exosystems and the cause, from among many possibilities, of pathology in stressed ecosystems.

A. Uptake and Release Kinetics

Most environmental pollutants, with the exception of a few that have very high acute toxicity (e.g., polychlorinated dibenzodioxins and some organophosphate insecticides), rely for their chronic deleterious effects in marine animals upon their tendency to bioaccumulate in the tissues of exposed animals to concentrations orders of magnitude higher than those in the ambient medium. The majority of organic pollutants with biomagnification potential are lipophilic and poorly water soluble. Their accumulation and retention in the tissues of marine animals depend on a partitioning between the ambient medium and the tissue lipids.[131] Thus Neely et al.[132] showed a direct correlation between the octanol/water partition coefficients of several organic pollutants and their accumulation by fish. Neff et al.[124,133] showed a relationship between molecular size and alkyl substitution of several aromatic hydrocarbons and their accumulation by the clam *Rangia cuneata.* In many cases, the onset of pathological symptoms due to these lipophilic pollutants is related to the rate of accumulation of the pollutants or the attainment of a "critical" concentration in the tissues. Thus, it is extremely important to have accurate information about the rates of uptake, distribution in the body, and rates of release of the pollutants under investigation in the test animals. Correlation of uptake/release kinetics with the other biological parameters is especially valuable in evaluating the sensitivity and usefulness of these latter parameters as diagnostic tools.

B. Biotransformations

An important process in the evolution of marine animals was the development of detoxification systems which allow animals to metabolize foreign organic compounds. Aquatic animals are being exposed to increasing amounts of many foreign compounds of anthropogenic origin, including pesticides, industrial wastes, and fossil fuels. Thus, a logical and potentially very important component of any integrated toxicological program is the investigation of the metabolism or biotransformation of foreign organic compounds. These foreign organic compounds, many of which are lipophilic, are converted by reduction, oxidation, hydrolysis, or conjugation to more water-soluble metabolites which can more easily be disposed of by the animal. Although these metabolic pathways are called detoxification systems, the metabolites of some foreign compounds, such as those of BaP, have toxic or carcinogenic properties. The systems are therefore more correctly referred to as toxicogenic/detoxification systems.[134] In mammals, fish, and some invertebrates, the oxidation of foreign compounds by hydroxylation, O-dealkylation, N-dealkylation, or epoxidation is carried out by a group of hepatic microsomal enzymes, referred to as the cytochrome P-450 mixed function oxygenase (MFO) system.[39,83,135-138] The polar metabolites formed by these oxidative processes can be conjugated with sugars, sulfate, or glutathione and disposed of via the urine or feces.

The metabolism of PAH is of interest because of their persistence and toxicity. Certain aromatic hydrocarbons, e.g., BaP, have carcinogenic and mutagenic properties. In mammals, these aromatic hydrocarbons are metabolically activated by MFO, resulting in the production of reactive epoxides that damage DNA and initiate the carcinogenic process.[139,140] The unstable arene oxides formed by MFO can be nonenzymatically hydrated to phenols, enzymatically hydrated to diols by an epoxide hydrase, or conjugated to glutathione by the action of glutathione S-transferase.[141] These biotransformations may largely determine the biological half-life and hence the potential for effect of various organic compounds. They may also result in production of metabolites which are substantially more toxic than the parent compounds and which may be responsible for some pathologic effects.

MFO activity in fish can be induced by exposure of the animals to some chlorinated hydrocarbons or PAH.[39,142-146] Induction in fish liver is usually characterized by increases in the levels of cytochrome P-450 and BaP hydroxylase activity; this pattern is also typical of mammals treated with PAH. Studies by Payne,[27] Kurelec et al.,[147] Stegeman,[28] and others have shown that fish from petroleum-contaminated areas have higher levels of aryl hydrocarbon hydroxylase than fish from cleaner areas nearby. Among marine invertebrates, the MFO system has been identified in crustaceans and polychaetes[105,148,149] and in some mollusks.[127] Chronic exposure to low concentrations of foreign organic compounds might result in chronically elevated (induced) activity of these enzymes and perhaps serve as a basis for the development of resistance to these compounds.[150,151] However, a chronically induced MFO system may produce pathological side effects.

MFO systems in fish and invertebrates appear to be multicomponent systems similar to the microsomal electron transport systems in mammals. Principal components of the system include NADPH-cytochrome c reductase, cytochrome P-450, and phospholipid. The major MFO activity in fish, as in mammals, is associated with microsomes from hepatic tissues, although Stegeman et al.[152] observed BaP hydroxylase activity in virtually all tissues examined. For many of the tissues (e.g., neural, epidermal, and lenticular), this was the first such observation. In many of these fish tissues, they also identified cytochrome P-450 for the first time. In crabs, MFO activity also occurs in most tissues,[102] but the highest activity appears in microsomes of the green gland and stomach.[102,153] In polychaetes, MFO activity is greatest in intestine. Previous work suggested that the MFO system was absent in mollusks from "clean" areas, and was not inducible by exposure to oil extracts.[154] However, Anderson[127] was able to measure low levels of BaP hydroxylase activity in the oyster, *Crassostrea virginica,* by studying the conversion of ^{14}C-BaP to labeled metabolites in vitro. As part of the PRIMA work, the same technique has been used to quantify BaP hydroxylase activity in *Mercenaria;* the greatest MFO activity in *Mercenaria* was found in the digestive gland.

Defining the functions of MFO systems and linking a given characteristic of MFO systems to the action of environmental chemicals require knowing the features associated with varied biological or physiological conditions. In mammals, sex, diet, and age are among the factors known to influence MFO systems. Marked sex differences also occur in cytochrome P-450 and/or MFO activity of fish and crustaceans.[102,155] Moreover, in both fish and crustaceans, there are changes in MFO activity associated with season and gonadal status or with molting. The implication is that MFO of marine species might be regulated by hormonal factors, as well as participating in hormone metabolism. Developmental stage might also regulate the function of MFO systems. In *Fundulus,* MFO activity appears in embryos prior to hatching, but the distribution between hepatic and extrahepatic tissues changes with hatching.[156]

In mammals there are multiple cytochromes P-450, and differences in MFO activity in different tissues or sexes are presumably linked to qualitative or quantitative differences in the complement or set of cytochromes P-450.[157] The differences in MFO due to sex, season, or molting status offer presumptive evidence for the occurrence of multiple cytochromes P-450 in marine species also. Direct evidence has been obtained for multiple P-450s in *Raja erinacea.*[158]

Steroids, including sex hormones and corticosteroids, and several foreign organic compounds share common pathways of metabolism and give rise to important interactions.[159-161] Thus, as a result of the introduction of MFO-inducing foreign compounds into an animal, the rate of hormone production or destruction may be changed.[162,163] Several environmental pollutants have also been shown to have a direct stimulatory or inhibitory effect on adrenal and testicular steroidogenesis in fish.[164,165] Conversely, elevated cortisol titers in fish may result in induction of some enzymes of the MFO system.[166] Singer and Lee[102] reported very large changes in aryl hydrocarbon hydroxylase activity in the green gland of blue crabs *C. sapidus* during the molt cycle. These changes in MFO activity of a crustacean during the molt cycle strongly suggest that green gland MFO activity functions in controlling the concentration of steroid molting hormones (crustecdysones) in the hemolymph of the crab. These interactions could result in changes in such processes as vitellogenesis, molting, and reproduction. These changes should also correlate with histopathological and immune responses.

C. Corticosteroid Stress Response and Ascorbic Acid Depletion

The response of the pituitary-adrenocortical axis of fresh-water and salmonoid fish to environmental stress, including chemical pollution, has been well documented.[11,34,167] Release of adrenocorticotropic hormone from the pituitary in response to stress stimulates the synthesis and release of cortisol (the primary circulating glucocorticoid in teleosts) from the interrenal gland (homologous to the adrenocortex). The release of cortisol into the blood results in or is accompanied by changes in osmotic and ionic parameters of the body fluids, hyperglycemia, hypercholesterolemia, increased blood catecholamine titers, changes in hepatic and muscle catabolic activity, and depletion of interrenal ascorbic acid.[168,169] Stress-induced elevations in blood cortisol and catecholamine titers, if persistent for a long time, may eventually lead to depletion of nutrient reserves including catabolic substrates (glycogen and lipid) and micronutrients such as ascorbic acid. Exhaustion of the pituitary-adrenocortical axis results in circulatory collapse and other systemic changes characteristic of chronic stress. This condition is called the general adaptation syndrome.[170]

During chronic stress, metabolic requirements for ascorbic acid increase many fold.[171] In animals that have a dietary requirement for ascorbate (e.g., invertebrates, fish, and the mammalian primates), these increased requirements often exceed dietary intake, resulting in gradual ascorbate depletion. During stress, the elevated ascorbate requirements are due to ascorbate's role in catecholamine synthesis and stabilization,[171] antihistamine activity,[172] phagocytosis and γ-globulin synthesis in leukocytes,[171] collagen synthesis in wound repair and connective tissue maintenance[173] and synthesis and stabilization of components of the pollutant-metabolizing cytochrome P-450 MFO system.[174] Thus ascorbate depletion may result in increased susceptibility to infectious disease,[175,176] skeletal and growth anomalies,[14] characteristic tissue lesions,[15] and depressed MFO activity.[177]

Induction of the MFO system by pollutants may interfere with the corticosteroid stress response. Cortisol is an alternative substrate for the MFO system,[163] so that MFO induction by pollutants increases the metabolic turnover of cortisol, exacerbating the stress-induced systemic pathology. Although many aspects of this integrated biochemical response to stress have been explored in freshwater fish, little has been done in this area with marine species, especially in relation to pollutant stress.

A stress response, qualitatively similar in many respects to that seen in fish, has been described in crustaceans.[178] Induction of the MFO system may interfere with the metabolism of the steroid molting hormone, ecdysterone, producing alterations in molting physiology and carbohydrate metabolism.[102] Several aspects of this stress response, including blood cortisol, cholesterol, catecholamine, and glucose titers, cortisol turnover, and depletion of tissue ascorbate reserves, show substantial potential as sensitive indicators of early and chronic low-level pollutant stress in fish. Hemolymph glucose, catecholamine, and ecdysteroid levels, and tissue ascorbate concentration may have similar diagnostic utility in crustaceans. In polychaete worms, coelomic fluid glucose and tissue catecholamines and ascorbate concentration may be diagnostic of stress conditions.

Moore[179] has reported that stress in general causes labilization of lysosomes in *Mytilus edulis*. When anthracene is injected into *M. edulis*, a similar labilization of lysosomes occurs.[180] It is important to determine if this process is a general one among mollusks subjected to various environments.

D. Histopathological Analysis

Lipophilic organic pollutants produce in marine animals a variety of focal or generalized lesions, many of which can be detected and characterized by histopathological and histochemical analysis at the light or electron microscope level. Lipophilic pollutants tend to interact with cell membranes.[181] Possible consequences of this interaction include lysis resulting in cell death, or modification of membrane properties leading to a change in cell function (e.g., change in membrane permeability or ion transporting properties, malfunction or labilization of lysosomes, inhibition of membrane-bound enzyme systems, etc.) or cellular reproduction (e.g., neoplastic cells may be formed). A wide variety of histopathological lesions have been described in marine animals exposed to organic pollutants in the laboratory or field. These include fatty infiltration of the liver of fish exposed to oil,[19,182] labilization of lysosomes in mollusks in response to various types of pollutants,[183-185] fin rot in fish exposed to oil or other pollutants,[186,187] atrophy of digestive diverticular epithelium and degeneration of vascular connective tissues concomitant with leukocytic infiltration in oysters exposed to polychlorinated biphenyls,[188] and cytoplasmic and nuclear degeneration in epithelial cells from the gills of shrimp exposed to sodium pentachlorophenol.[189]

Many of the biochemical responses under investigation in these studies produce or are the product of histological changes in the tissues of the pollutant-stressed animal. For example, activation of the pituitary-adrenal axis in fish results in hypertrophy of interrenal tissues,[12,13] and induction of the hepatic MFO system results in proliferation of the endoplasmic reticulum.[19] Stress-induced depletion of ascorbic acid produces characteristic histopathologic lesions in fish[175] and crustaceans.[15] Histological examination has been used for many years as a valuable diagnostic tool in clinical medicine. Its utility in investigations of pollutant-mediated pathology in marine animals is gaining acceptance.[17,24,190]

E. Immunocompetence

Normal functioning of the immune system is essential to the general health of any organism. Any alteration, however subtle, in this system might ultimately exert profound effects on the animal in question. There are many reports that exposure to environmental pollutants or other stressors causes alterations in both the cellular and humoral immune responses in mammals,[16,191-193] fish,[194-196] and marine invertebrates.[197-199]

Environmental and pollutant stress have been shown to induce immunosuppression in fish as manifested by lower resistance to infection,[190,195] decreased phagocytic activity,[176] and decreased antibody synthesis.[200] In light of these data, it is apparent that the immunological parameters of fishes may be quite sensitive to environmental pollutants.

There is also direct and indirect evidence for impaired immune function in certain invertebrate species. Thus, phenol is known to decrease immunological competency of *M. mercenaria.*[198,201] Clams in polluted waters are susceptible to infestation by *Polydora* sp., whereas unstressed clams are only rarely infected.[202] Similarly, shrimp that are exposed to polychlorinated biphenyls have atypical viral infections.[197,203] Clearly, the details of the impact of pollutants on immune parameters of invertebrates need to be defined more completely.

F. Integration

Each of the various morphological, physiological, and biochemical parameters investigated is a significant indicator of the health and toxic responses of an organism and also bears a meaningful relationship to the others. In addition to metabolism of foreign compounds, the cytochrome P-450 MFO system is responsible for metabolism of a variety of important steroids.[160,163] In vitro studies of steroid metabolism by MFO in fish and invertebrates thus have a direct nexus with studies of serum and hemolymph levels of cortisol, corticosteroid, and ecdysteroids. Similarly, relationships between ascorbic acid and MFO activity are becoming well established.[174,177] The suppression of gamete formation in *N. virens* collected from an oil-polluted area in Portland, Maine may be due to an alteration in hormone production which is regulated by cytochrome P-450-dependent systems.

Histological status of tissues and immune competence are important indexes of the health of experimental animals and, when combined with biochemical parameters, provide a detailed picture of the etiology of pollutant-induced pathology in marine animals. Activation of the pituitary-adrenal axis and induction of the MFO system produce readily detectable histological changes in several tissues of fish and may alter several parameters of the immune response. Stimulation or induction of analogous systems in marine invertebrates also can be expected to result in histological changes in several tissues. In addition, primary pathological lesions produced by exposure to the model pollutants can be expected to produce a variety of secondary biochemical changes in the affected animals.

Immune competence of fish and invertebrates may be influenced directly by the pollutants or secondarily via alterations in histological or biochemical parameters in the affected animals. For example, corticosteroid status has a direct effect on the immune system,[204] and ascorbate depletion affects the ability of leukocytes to respond to antigenic challenge.[171]

Knowledge of the identity and quantity of pollutants and their rates of uptake and release and distribution in the animal is important for evaluating the sensitivity and applicability of the measured parameters as diagnostic tools. This information will also establish the pollutant status of these organisms and facilitate interpretation of the other data. Some of the interrelationships between a pollutant and various biological processes are summarized in Figure 1.

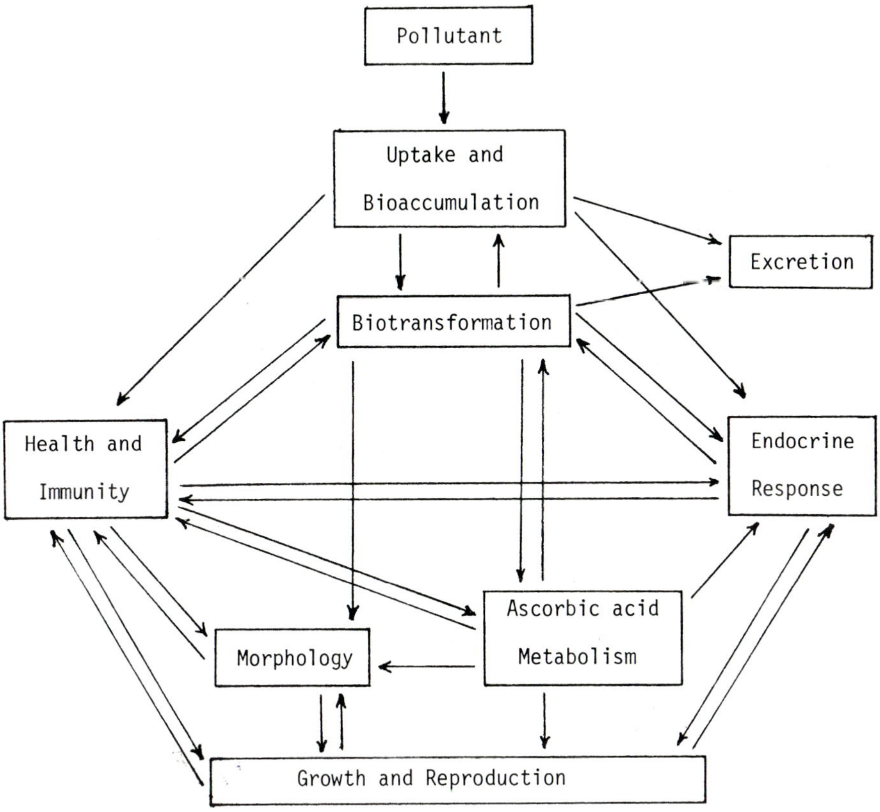

FIGURE 1. Some possible interactions between pollutants and various biological processes.

REFERENCES

1. **Saward, D.,** An experimental approach to the study of pollution effects in aquatic ecosystems, in *Ecological Toxicology Research. Effects of Heavy Metal and Organohalogen Compounds,* McIntyre, A. D. and Mills, C. F., Eds., Plenum Press, New York, 1975, 86.
2. **Waldichuk, M.,** Trends in methodology for evaluation of effects of pollutants on marine organisms and ecosystems, *CRC Crit. Rev. Environ. Control,* 3, 167, 1973.
3. **Wilson, K. W.,** The laboratory estimation of the biological effects of organic pollutants, *Proc. R. Soc. London, Ser. B,* 198, 459, 1975.
4. **Perkins, E. J.,** The evaluation of biological response by toxicity and water quality assessments, in *Marine Pollution,* Johnston, R., Ed., Academic Press, New York, 1976, 505.
5. **Perkins, E. J.,** The need for sublethal studies, in *The Assessment of Sublethal Effects of Pollutants in the Sea,* Cole, H. A., Ed., The Royal Society, London, 1979, 27.
6. **Anderson, J. W., Neff, J. M., and Petrocelli, S. R.,** Sublethal effects of oil, heavy metals and PCBs on marine organisms, in *Survival in Toxic Environments,* Khan, M. A. Q. and Bederka, J. P., Jr., Eds., Academic Press, New York, 1974, 83.
7. **Stegeman, J. J.,** Biological and ecological effects of petroleum in the marine environment. Proc. 2nd IOC/WMO Workshop on Marine Pollutant (Petroleum) Monitoring, Monaco, June 1976.
8. **Allen, J. R. and Abrahamson, L. J.,** Morphological and biochemical changes in the liver of rats fed polychlorinated biphenyls, *Arch. Environ. Contam. Toxicol.,* 1, 265, 1975.

9. **Sabo, D. J. and Stegeman, J. J.,** Some metabolic effects of petroleum hydrocarbons in marine fish, in *Physiological Responses of Marine Biota to Pollutants,* Vernberg, F. J., Calabrese, A., Thurberg, F. P., and Vernberg, W. B., Eds., Academic Press, New York, 1977, 279.
10. **Nussdorfer, G. G., Mazzocchi, G., and Rebonato, L.,** Long-term trophic effect of ACTH on rat adrenocortical cells, *Z. Zellforsch. Mikrosk. Anat.,* 115, 30, 1971.
11. **Chavin, W.,** Teleostean endocrine and paraendocrine alterations of utility in environmental studies, in *Responses of Fish to Environmental Changes,* Chaven, W., Ed., Charles C Thomas, Springfield, Ill., 1973, 199.
12. **Mudge, J. E., Dively, J. L., Neff, W. H., and Anthony, A.,** Interrenal histochemistry of acid-exposed brook trout *Salvelinus fontinalis* (Mitchell), *Gen. Comp. Endocrinol.,* 31, 208, 1977.
13. **Malo-Michele, M.,** Reactions cytologiques de l'axe hypophyso-interrenalien de la saupe, *Boops salpa* L. (Teleosteen marin), a la diminution de la salinite, aux injections de metropirone et de reserpine et a un stress neurogene, *Gen. Comp. Endocrinol.,* 38, 300, 1979.
14. **Mayer, F. L., Mehrle, P. M., and Crutcher, P. L.,** Interactions of toxaphene and vitamin C in channel catfish, *Trans. Am. Fish. Soc.,* 107, 326, 1978.
15. **Magarelli, P. C., Jr., Hunter, B., Lightner, D. V., and Colvin, L. B.,** Black death: an ascorbic acid deficiency disease in penaeid shrimp, *Comp. Biochem. Physiol.,* 63A, 103, 1979.
16. **Street, J. C. and Sharma, R. P.,** Alteration of induced cellular and humoral immune responses by pesticides and chemicals of environmental concern: quantitative studies of immunosuppression by DDT, Aroclor 1254, carbaryl, carbufuran, and methylparathion, *Toxicol. Appl. Pharmacol.,* 32, 587, 1975.
17. **Sinderman, C. J.,** Pollution-associated diseases and abnormalities of fish and shell-fish: a review, *Fish. Bull.,* 76, 717, 1978.
18. **Barry, M. and Yevich, P. P.,** The ecological, chemical and histopathological evaluation of an oil spill site. III. Histopathological studies, *Mar. Pollut. Bull.,* 6, 171, 1975.
19. **Hawkes, J. W.,** The effects of petroleum hydrocarbon exposure on the structure of fish tissues, in *Fate and Effects of Petroleum Hydrocarbons in Marine Organisms and Ecosystems,* Wolfe, D. A., Ed., Pergamon Press, New York, 1977, 115.
20. **Jackim, E., Hamlin, J. M., and Sonis, S.,** Effects of metal poisoning on five liver enzymes in the killfish *(Fundulus heteroclitus), J. Fish. Res. Board Can.,* 27, 383, 1970.
21. **Kasza, L., Weinberg, M. A., Carter, C., Hinton, D. E., Trump, B. F., and Brouwer, E. A.,** Acute, subacute and residual effects of polychlorinated biphenyl (PCB) in rats. II. Pathology and electron microscopy of liver and serum enzyme study, *J. Toxicol. Environ. Health,* 1, 689, 1976.
22. **Chambers, J. E., Heitz, J. R., McCorkle, F. M., and Yarbrough, J. D.,** Enzyme activities following chronic exposure to crude oil in a simulated ecosystem. I. American oysters and brown shrimp, *Environ. Res.,* 20, 133, 1979.
23. **Chambers, J. E., Heitz, J. R., McCorkle, F. M., and Yarbrough, J. D.,** Enzyme activities following chronic exposure to crude oil in a simulated ecosystem. II. Striped mullet, *Environ. Res.,* 20, 140, 1979.
24. **Couch, J., Gardner, G., Harshbarger, J. C., Tripp, M. R., and Yevich, P. P.,** Histological and physiological evaluation of some marine fauna, in *Marine Bioassays,* Marine Technology Society, Washington, D. C., 1974, 156.
25. **Giam, C. S., Chan, H. S., and Neff, G. S.,** Concentrations and fluxes of phthalates, DDTs and PCBs to the Gulf of Mexico, in *Marine Pollutant Transfer,* Windom, H. L. and Duce, R. A., Eds., D. C. Heath & Co., Lexington, Mass., 1976, 375.
26. **Goldberg, E. D.,** The mussel watch — a first step in global marine monitoring, *Mar. Pollut. Bull.,* 6, 111, 1975.
27. **Payne, J. F.,** Field evaluation of benzopyrene hydroxylase induction as a monitor for marine petroleum pollution, *Science,* 191, 945, 1976.
28. **Stegeman, J. J.,** Influence of environmental contamination on cytochrome P-450 mixed function oxygenases in fish: implications for recovery in the Wild Harbor Marsh, *J. Fish. Res. Board Can.,* 35, 668, 1978.
29. **Walton, D. G., Penrose, W. R., and Green, J. M.,** Petroleum-inducible mixed function oxidase of cunner *(Tautogolabrus adspersus* Wallbaum 1972): some characteristics relevant to hydrocarbon monitoring, *J. Fish. Res. Board Can.,* 35, 1547, 1978.
30. **Stich, H. F. and Acton, A. B.,** The possible use of fish tumors in monitoring for carcinogens in the marine environment, *Prog. Exp. Tumor Res.,* 20, 44, 1976.
31. **Silbergeld, E. K.,** Blood glucose: a sensitive indicator of environmental stress in fish, *Bull. Environ. Contam. Toxicol.,* 11, 20, 1974.
32. **Hattingh, J.,** Blood sugar as an indicator of stress in freshwater fish *Labeo capensis, J. Fish. Biol.,* 10, 191, 1977.
33. **Telford, M.,** Blood glucose in crayfish. II. Variations induced by artificial stress, *Comp. Biochem. Physiol.,* 48A, 555, 1974.

34. **Shreck, C. B. and Lorz, H. W.,** Stress response of coho salmon *(Oncorhynchus kisutch)* elicited by cadmium and copper and potential use of cortisol as an indicator of stress, *J. Fish. Res. Board Can.,* 35, 1124, 1978.
35. National Academy of Sciences, *Assessing Potential Ocean Pollutants,* NAS, Washington, D.C., 1975.
36. **Flinn, F. E., Thomas, T. J., and Bishop, M. D.,** Identification Systems for Selecting Chemical Classes as Candidates for Evaluation, EPA-650/1-74-001, Washington, D.C., 1974.
37. **Zitko, V.,** Comments on the status and direction of environmental research, in *Biological Effects of Pollutants on Marine Organisms,* Giam, C. S., Ed., D.C. Heath & Co., Lexington, Mass., 1977, 191.
38. **Hutzinger, O., Tulp, M. Th. M., and Zitko, V.,** Chemicals with pollution potential, in *Aquatic Pollutants: Transformation and Biological Effects,* Hutzinger, O., van Lelyveld, I. H., and Zoetesman, B. D. J., Eds., Pergamon Press, New York, 1978, 13.
39. **Neff, J. M.,** *Polycyclic Aromatic Hydrocarbons in the Aquatic Environment, Sources, Fates and Biological Effects,* Applied Science Publ., Barking, Essex, England. 1979.
40. **Giger, W. and Blumer, M.,** Polycyclic aromatic hydrocarbons in the environment: isolation and characterization by chromatography, visible, ultraviolet and mass spectrometry, *Anal. Chem.,* 46, 1663, 1974.
41. **Blumer, M. and Youngblood, W. W.,** Polycyclic aromatic hydrocarbons in soils and recent sediments, *Science,* 188, 53, 1975.
42. **Hites, R. A.,** Sources of polycyclic aromatic hydrocarbons in the aquatic environment, in *Sources, Effects and Sinks of Hydrocarbons in the Aquatic Environment,* American Institute of Biological Sciences, Washington, D.C., 1976, 325.
43. **Hites, R. A., Laflame, R. E., and Farrington, J. W.,** Sedimentary polycyclic aromatic hydrocarbons: the historical record, *Science,* 198, 829, 1977.
44. **Dunn, B. P. and Stich, H. F.,** Monitoring procedures for chemical carcinogens in coastal waters, *J. Fish. Res. Board Can.,* 33, 2040, 1976.
45. **Dunn, B. P. and Young, D. R.,** Baseline levels of benzo(a)pyrene in southern California mussels, *Mar. Pollut. Bull.,* 7, 231, 1976.
46. **Ehrhardt, M.,** Petroleum hydrocarbons in oysters from Galveston Bay, *Environ. Pollut.,* 3, 257, 1972.
47. **Mix, M. C., Riley, R. T., King, K. I., Trenhold, S. R., and Schaffer, R. L.,** Chemical carcinogens in the marine environment. Benzo[a]pyrene in economically-important bivalve molluscs from Oregon estuaries, in *Fate and Effects of Petroleum Hydrocarbons in Marine Ecosystems and Organisms,* Wolfe, D. A., Ed., Pergamon Press, New York, 1977, 421.
48. **Mix, M. C. and Schaffer, R. L.,** Benzo[a]pyrene concentrated in mussels *(Mytilus edulis)* from Yaquina Bay, Oregon, during June 1976 to May 1978, *Bull. Environ. Contam. Toxicol.,* 23, 677, 1979.
49. **Lijinsky, W., Domsky, I., Mason, G., Romaki, H. Y., and Safavi, T.,** The chromatographic determination of trace amounts of polynuclear hydrocarbons in petroleum, mineral oil and coal-tar, *Anal. Chem.,* 35, 952, 1963.
50. **Suess, J. M.,** The environmental load and cycle of polycyclic aromatic hydrocarbons, *Sci. Total Environ.,* 6, 239, 1976.
51. National Academy of Sciences, *Petroleum in the Marine Environment,* Ocean Affairs Board, Washington, D.C., 1975.
52. **Hase, A. and Hites, R. A.,** On the origin of polycyclic aromatic hydrocarbons in recent sediments: biosynthesis by anaerobic bacteria, *Geochim. Cosmochim. Acta,* 40, 1141, 1976.
53. **Heidelberger, C.,** Studies on the mechanisms of carcinogenesis by polycyclic aromatic hydrocarbons and their derivatives, in *Polynuclear Aromatic Hydrocarbons: Chemistry, Metabolism and Carcinogenesis,* Freudenthal, R. I. and Jones, P. W., Eds., Raven Press, New York, 1976, 1.
54. **Gelboin, H. V.,** Studies on the mechanism of microsomal hydroxylase induction and its role in carcinogen action, *Rev. Can. Biol.,* 31, 31, 1972.
55. **Zitko, V. and Choi, P. M. K.,** PCB and other industrial halogenated hydrocarbons in the environment, *Fish. Res. Board Can. Tech. Rep. No. 272,* 1971.
56. **Peakall, D. B.,** PCBs and their environmental effects, *CRC Crit. Rev. Environ. Control,* 5, 469, 1975.
57. **Giam, C. S.,** Trace analyses of phthalates and chlorinated hydrocarbons in marine samples, in *Strategies for Marine Pollution Monitoring,* Goldberg, E. D., Ed., Wiley-Interscience, New York, 1976, 61.
58. **Giam, C. S., Chan, H. S., and Neff, G. S.,** Phthalate ester plasticizers, DDT, DDE, and polychlorinated biphenyls in biota from the Gulf of Mexico, *Mar. Pollut. Bull.,* 9, 249, 1978.
59. **Giam, C. S., Richardson, R. L., Taylor, D., and Wong, M. K.,** DDT, DDE and PCBs in the tissue of reef dwelling groupers (Serranidae) in the Gulf of Mexico and the Grand Bahamas, *Bull. Environ. Contam. Toxicol.,* 11, 189, 1974.
60. **Giam, C. S., Richardson, R. L., Wong, M. K., and Sackett, W. M.,** Polychlorinated biphenyls in Antarctic biota, *Antarctic J. U.S.,* 8, 303, 1974.

61. **Giam, C. S., Wong, M. K., Hanks, A. R., Sackett, W. M., and Richardson, R. L.,** Chlorinated hydrocarbons in plankton from the Gulf of Mexico and Northern Caribbean, *Bull. Environ. Contam. Toxicol.*, 9, 376, 1973.
62. **Green, M. B.,** Polychloroaromatics and heteroaromatics of industrial importance, in *Polychloroaromatic Compounds*, Suschitzky, H., Ed., Plenum Press, New York, 1974, 403.
63. **Quinlivan, S. C., Ghassemi, M., and Leshendok, T. V.,** Sources, characteristics and treatment and disposal of industrial wastes containing hexachlorobenzene, *J. Hazard. Mater.*, 1, 343, 1977.
64. **Johnson, J. L., Stalling, D. L., and Hogan, J. W.,** Hexachlorobenzene (HCB) residues in fish, *Bull. Environ. Contam. Toxicol.*, 11, 393, 1974.
65. **Ofstad, E. B., Lunde, G., and Martinsen, K.,** Chlorinated aromatic hydrocarbons in fish from an area polluted by industrial effluents, *Sci. Total Environ.*, 10, 219, 1978.
66. **Niimi, A. J.,** Hexachlorobenzene (HCB) levels in Lake Ontario salmonids, *Bull. Environ. Contam. Toxicol.*, 23, 20, 1979.
67. **Cabral, J. R. P., Mollner, T., Raitano, F., and Shubile, P.,** Carcinogenesis of hexachlorobenzene in mice, *Int. J. Cancer*, 23, 47, 1979.
68. **Cabral, J. R. P., Shubik, P., Mollner T., and Raitano, F.,** Carcinogenic activity of hexachlorobenzene in hamsters, *Nature (London)*, 269, 510, 1977.
69. **Plimmer, J. R. and Klingebiel, U. V.,** Photolysis of hexachlorobenzene, *J. Agric. Food Chem.*, 24, 721, 1976.
70. **Isensee, A. R., Holden, E. R., Woolson, E. A., and Jones, G. E.,** Soil persistence and aquatic bioaccumulation potential of hexachlorobenzene, *J. Agric. Food Chem.*, 24, 1210, 1976.
71. **Cirelli, D. P.,** Patterns of pentachlorophenol usage in the United States of America — an overview, in *Pentachlorophenol*, Rao, K. R., Ed., Plenum Press, New York, 1978, 13.
72. **von Rumker, R., Lawless, W., and Meiners, A. F.,** Production, distribution, use and environmental impact potential of selected pesticides, Office of Pesticide Programs, U.S. Environmental Protection Agency, Washington, D.C., 1974.
73. **Ernst, W. and Weber, K.,** Chlorinated phenols in selected estuarine bottom fauna, *Chemosphere*, 11, 867, 1978.
74. **Weber, K. and Ernst, W.,** Levels and patterns of chlorophenols in water of the Weser Estuary and German Bight, *Chemosphere*, 11, 873, 1978.
75. **Wegman, R. C. C. and Hofstee, A. W. M.,** Chlorophenols in surface waters of the Netherlands (1976 to 1977), *Water Res.*, 13, 651, 1979.
76. **Pierce, R. H. and Victor, D. M.,** The fate of pentachlorophenol in an aquatic ecosystem, in *Pentachlorophenol*, Rao, K. R., Ed., Plenum Press, New York, 1978, 41.
77. **Rao, K. R., Ed.,** *Pentachlorophenol Chemistry, Pharmacology and Environmental Toxicology*, Plenum Press, New York, 1978.
78. **Bigelow, H. B. and Schroeder, W.C.,** *Fishes of the Gulf of Maine. Fishery Bulletin of the Fish and Wildlife Service*, Vol. 53, U.S. Government Printing Office, Washington, D.C., 1953.
79. **Lotrich, V. A.,** Summer home range and movements of *Fundulus heteroclitus* (species: *Cyprinodentidae)* in a tidal creek, *Ecology*, 56, 191, 1975.
80. **Broadhead, G. C. and Mefford, H. P.,** The migration and exploitation of the black mullet *Mugil cephalus* L. in Florida as determined from tagging during 1949 to 1953, *Fla. Board Conserv. Mar. Res. Lab. Tech. Ser.*, 18, 1956.
81. **Griffith, R. W.,** Environment and salinity tolerance in the genus *Fundulus*, *Copeia*, 2, 319, 1974.
82. **Moore, R. H.,** General ecology, distribution and relative abundance of *Mugil cephalus* and *Mugil curema* on the South Texas coast, *Contrib. Mar. Sci. Univ. Tex.*, 18, 241, 1974.
83. **Pohl, R. J. Bend, J. R., Guarino, A. M., and Fouts, J. R.,** Hepatic microsomal mixed-function oxidase activity of several marine species from coastal Maine, *Drug. Metab. Dispos.*, 2, 545, 1974.
84. **Stegeman, J. J. and Sabo, D. J.,** Aspects of the effects of petroleum hydrocarbons on intermediary metabolism and xenobiotic metabolism in marine fish, in *Sources, Effects and Sinks of Hydrocarbons in the Aquatic Environment*, American Institute of Biological Sciences, Washington, D.C., 1976, 423.
85. **Bend, J. R., James, M. O., and Dansette, P. M.,** *In vitro* metabolism of xenobiotics in some marine animals, *Ann. N.Y. Acad. Sci.*, 298, 505, 1977.
86. **Yarbrough, J. D. and Chambers, J. E.,** Crude oil effects on microsomal mixed-function oxidase system components in the striped mullet *(Mugil cephalus)*, *Life Sci.*, 21, 1095, 1977.
87. **Chambers, J. E.,** Induction of microsomal mixed-function oxidase system components in striped mullet by short-term exposure to crude oil, *Toxicol. Lett.*, 4, 227, 1979.
88. **Campbell, C. M., Walsh, J. M., and Idler, D. R.,** Steroids in the plasma of the winter flounder *(Pseudopleuronectes americanus* Walbaum). A seasonal study and investigation of steroid involvement in oocyte maturation, *Gen. Comp. Endocrinol.*, 29, 14, 1976.

89. **Leach, G. J. and Taylor, M. H.,** Seasonal measurements of serum glucose and serum cortisol in a natural population of *Fundulus heteroclitus* L., *Comp. Biochem. Physiol.*, 56A, 217, 1977.
90. **DiMichele, L. and Taylor, M. H.,** Histopathological and physiological responses of *Fundulus heteroclitus* L. to naphthalene exposure, *J. Fish. Res. Board Can.*, 35, 1060, 1978.
91. **Wolke, R. E.,** personal communication.
92. **Williams, A. B.,** The swimming crabs of the genus *Callinectes* (Decapoda: Portunidae), *Fish. Bull.*, 72, 685, 1974.
93. **Costlow, J. D., Jr. and Bookhout, C. G.,** Larval development of *Callinectes sapidus* Rathbun reared in the laboratory, *Biol. Bull.*, 116, 373, 1959.
94. **Leffler, C. W.,** Some effects of temperature on the growth and metabolic rate of juvenile blue crabs, *Callinectes sapidus*, in the laboratory, *Mar. Biol.*, 14, 104, 1972.
95. **Leffler, C. W.,** Ionic and osmotic regulation and metabolic response to salinity of the juvenile *Callinectes sapidus* Rathbun, *Comp. Biochem. Physiol.*, 52A, 545, 1975.
96. **Lynch, M. P. and Webb, K. L.,** Variations in serum constituents of the blue crab, *Callinectes sapidus:* glucose, *Comp. Biochem. Physiol.*, 45A, 127, 1973.
97. **Bookhout, C. G. and Costlow, J. D., Jr.,** Effects of mirex on larval development of blue crabs, *Water Air Soil Pollut.*, 4, 113, 1975.
98. **Bookhout, C. G. and Monroe, R. J.,** Effects of malathion on the development of crabs, in *Physiological Responses of Marine Biota to Pollutants,* Vernberg, F. J., Calabrese, A., Thurberg, F. P., and Vernberg, W. B., Eds., Academic Press, New York, 1976, 3.
99. **Fox, F. R. and Rao, K. R.,** Effects of sodium pentachlorophenate and 2,4-dinitrophenol on hepatopancreatic enzymes in the blue crab *Callinectes sapidus*, in *Pentachlorophenol,* Rao, K. R., Ed., Plenum Press, New York, 1978, 265.
100. **Faux, A., Horn, D. H. S., Middleton, E. J., Fales, H. M., and Lowe, M. E.,** Moulting hormones of a crab during ecdysis, *Chem. Commun.*, 1969, 175, 1969.
101. **Lee, R. F., Ryan, C., and Neuhauser, M. L.,** Fate of petroleum hydrocarbons taken up from food and water by the blue crab, *Callinectes sapidus, Mar. Biol.*, 37, 363, 1976.
102. **Singer, S. C. and Lee, R. F.,** Mixed function oxygenase activity in blue crab, *Callinectes sapidus:* tissue distribution and correlation with changes during molting and development, *Biol. Bull.*, 153, 377, 1977.
103. **Lee, R. F.,** Metabolism of petroleum hydrocarbons in marine sediments, in *Sources, Effects and Sinks of Hydrocarbons in the Aquatic Environment,* American Institute of Biological Sciences, Washington, D. C., 1976, 334.
104. **Prouse, N. J. and Gordon, D. C., Jr.,** Interactions between the deposit-feeding polychaete, *Arenicola marina* and oiled sediment, in *Sources, Effects and Sinks of Hydrocarbons in the Aquatic Environment,* American Institute of Biological Sciences, Washington, D.C., 1976, 407.
105. **Lee, R. F., Singer, S. C., Tenore, K. R., Gardner, W. S., and Philpot, R. M.,** Detoxification system in polychaete worms: importance in the degradation of sediment hydrocarbons, in *Marine Pollution: Functional Responses,* Vernberg, W. B., Calabrese, A., Thurberg, F. P., and Vernberg, F. J., Eds., Academic Press, New York, 1979, 23.
106. **Rossi, S. S. and Anderson, J. W.,** Petroleum hydrocarbon resistance in the marine worm *Neanthes arenaceodentata* (polychaeta: Annelida), induced by chronic exposure to No. 2 fuel oil, *Bull. Environ. Contam. Toxicol.*, 20, 513, 1978.
107. **Armstrong, H. W., Fucik, K., Anderson, J. W., and Neff, J. M.,** Effects of oil field brine effluent on sediments and benthic organisms in Trinity Bay, Texas, *Mar. Environ. Res.*, 2, 55, 1979.
108. **Baker, J. M.,** Investigation of refinery effluent effects through field surveys, in *Marine Ecology and Oil Pollution,* Baker, J. M., Ed., John Wiley & Sons, New York, 1976, 201.
109. **Reish, D. J.,** Effect of pollution abatement in Los Angeles harbours, *Mar. Pollut. Bull.*, 2, 71, 1971.
110. **Sanders, H. L., Grassle, J. F., and Hampson, G. R.,** The West Falmouth oil spill. I. Biology, Tech. Rep. No. 72-20, Woods Hole Oceanographic Institution, Woods Hole, Mass, 1972.
111. **Spies, R. B., Davis, P. H., and Stuermer, D. H.,** The infaunal benthos of petroleum-contaminated sediments: study of a community at a natural oil seep, in *Ecological Impact of Oil Spills,* American Institute of Biological Sciences, Washington, D.C., 1979, 735.
112. **Grassle, J. F. and Grassle, J. P.,** Temporal adaptations in sibling species of *Capitella,* in *Ecology of Marine Benthos,* Coull, B. C., Ed., University of South Carolina Press, Columbia, 1977, 177.
113. **Grassle, J. P. and Grassle, J. F.,** Sibling species in the marine pollution indicator *Capitella* (polychaete), *Science,* 192, 567, 1976.
114. **Brooks, G. T.,** Chlorinated insecticides: retrospect and prospect, in *Pesticide Chemistry in the 20th Century,* Plimmer, J. R., Ed., American Chemical Society, Washington, D.C., 1977, 1.
115. **Walker, C. R. and Terriere, L. C.,** Induction of microsomal oxidases by dieldrin in *Musca domestica, Ent. Ex. Appl.*, 13, 260, 1970.

116. **Wilkinson, C. F. and Brattsten, L. B.,** Microsomal drug metabolizing enzymes in insects, *Drug Metab. Rev.*, 1, 153, 1972.
117. **Miner, R. W.,** *Field Book of Seashore Life,* G. P. Putnam & Sons, New York, 1950.
118. **Bass, N. R. and Brafield, A. E.,** The life-cycle of the polychaete *Nereis virens, J. Mar. Biol. Assoc. U.K.*, 52, 701, 1971.
119. **Carriker, M. R.,** Interrelation of functional morphology, behavior and autecology in early stages of the bivalve, *Mercenaria mercenaria, J. Elisha Mitchell Sci. Soc.*, 77, 168, 1961.
120. **Ansell, A. D.,** The rate of growth of the hard clam *Mercenaria mercenaria* throughout the geographic range, *J. Conserv. Perm. Int. Explor. Merc.*, 31, 364, 1968.
121. **Stegeman, J. J. and Teal, J. M.,** Accumulation, release and retention of petroleum hydrocarbons by the oyster *Crassostrea virginica, Mar. Biol.*, 22, 27, 1973.
122. **DiSalvo, L. H., Guard, H. E., and Hunter, L.,** Tissue hydrocarbon burden of mussels as potential monitor of environmental hydrocarbon insult, *Environ. Sci. Technol.*, 9, 247, 1975.
123. **Neff, J. M. and Anderson, J. W.,** Accumulation, release and distribution of benzo[a]pyrene-^{14}C in the clam, *Rangia cuneata, in Proc. 1975 Jt. Conf. Prev. Control Oil Pollut.*, American Petroleum Institute, Washington, D.C., 1975, 469.
124. **Neff, J. M., Cox, B. A., Dixit, D., and Anderson, J. W.,** Accumulation and release of petroleum-derived aromatic hydrocarbons by four species of marine animals, *Mar. Biol.*, 38, 279, 1976.
125. **Boehm, P. D. and Quinn, J. G.,** The persistence of chronically accumulated hydrocarbons in the hard shell clam, *Mercenaria mercenaria, Mar. Biol.*, 44, 227, 1977.
126. **Vandermuelen, J. H. and Penrose, W. R.,** Absence of aryl hydrocarbon hydroxylase (AHH) in three marine bivalves, *J. Fish Res. Board Can.*, 35, 643, 1978.
127. **Anderson, R. S.,** Benzo(a)pyrene metabolism in the American oyster *Crassostrea virginica,* Ecological Research Series EPA-600/3-78-009, U.S. Environmental Protection Agency, Washington, D.C., 1978.
128. **Carlson, G. P.,** Detoxification of foreign compounds by the quahaug, *Mercenaria mercenaria, Comp. Biochem. Physiol.*, 43B, 295, 1972.
129. **Hanzel, M. E. and Carlson, G. P.,** Azoreductase activity in the hard clam *Mercenaria mercenaria* (L.), *J. Exp. Mar. Biol. Ecol.*, 14, 225, 1974.
130. **Khan, M. A. Q., Stanton, R. H., and Reddy, G.,** Detoxification of foreign compounds by invertebrates, in *Survival in Toxic Environments,* Khan, M. A. Q. and Berderka, J. P., Jr., Eds., Academic Press, New York, 1974, 177.
131. **Hamelink, J. L. and Spacie, A.,** Fish and chemicals: the process of accumulation, *Ann. Rev. Pharmacol. Toxicol.*, 17, 167, 1977.
132. **Neely, W. B., Branson, D. R., and Blau, G. E.,** Partition coefficient to measure bioconcentration potential of organic chemicals in fish, *Environ. Sci. Technol.*, 8, 1113, 1974.
133. **Neff, J. M., Anderson, J. W., Cox, B. A., Laughlin, R. B., Jr., Rossi, S. S., and Tatem, H. E.,** Effects of petroleum on survival, respiration and growth of marine animals, in *Sources, Effects and Sinks of Hydrocarbons in the Aquatic Environment,* American Institute of Biological Sciences, Washington, D.C., 1976, 515.
134. **Safe, S., Wyndham, C., Crawford, A., and Kohli, J.,** Metabolism: detoxification or toxification, in *Aquatic Pollutants: Transformation and Biological Effects,* Hutzinger, O., van Lelyveld, I. H., and Zoetemann, B. C. J., Eds., Pergamon Press, New York, 1978, 299.
135. **Hutson, D. H.,** Mechanism of biotransformation, in *Foreign Compound Metabolism in Mammals,* Hathway, D. E., Ed., The Chemical Society, London, 1975, 449.
136. **Philpot, R. M., James, M. O., and Bend, J. R.,** Metabolism of benzo[a]pyrene and other xenobiotics by microsomal mixed-function oxidases in marine species, in *Sources, Effects and Sinks of Hydrocarbons in the Aquatic Environment,* American Institute of Biological Sciences, Washington, D.C., 1976, 184.
137. **Malins, D. C.,** Metabolism of aromatic hydrocarbons in marine organisms, *Ann. N.Y. Acad. Sci.*, 298, 482, 1977.
138. **Kobayashi, K.,** Metabolism of pentachlorophenol in fish, in *Pentachlorophenol,* Rao, K. R., Ed., Plenum Press, New York, 1978, 89.
139. **Ames, B. N., Sims, P., and Grover, P. L.,** Epoxides of carcinogenic polycyclic hydrocarbons are frameshift mutagens, *Science,* 176, 47, 1972.
140. **Miller, J. A. and Miller, E. C.,** Chemical carcinogenesis: mechanisms and approaches to its control, *J. Natl. Cancer Inst.*, 47, 5, 1971.
141. **Bend, J. R., Ben-Zui, Z., Van Anda, J., Dansette, P. M., and Jerina, D. M.,** Hepatic and extrahepatic glutathione S-transferase activity toward several arene oxides and epoxides in the rat, in *Carcinogensis, Vol. 1, Polynuclear Aromatic Hydrocarbons: Chemistry, Metabolism and Carcinogenesis,* Freudenthal, R. I. and Jones, P. W., Eds., Raven Press, New York, 1976, 63.
142. **Gruger, E. H., Jr., Wekell, M. M., Numoto, P. T., and Craddock, D. R.,** Induction of aryl hydrocarbon hydroxylase in salmon exposed to petroleum dissolved in seawater and to petroleum and polychlorinated biphenyls, separate and together, in food, *Bull. Environ. Contam. Toxicol.*, 17, 512, 1977.

143. **Gruger, E. H., Jr., Wekell, M. M., and Robisch, P. A.,** Effects of chlorinated biphenyls and petroleum hydrocarbons on the activity of hepatic aryl hydrocarbon hydroxylase of coho salmon *(Oncorhynchus kisutch)* and chinook salmon *(O. tshawytscha)*, in *Fate and Effects of Petroleum Hydrocarbons in Marine Ecosystems and Organisms,* Wolfe, D. A., Ed., Pergamon Press, New York, 1977, 323.
144. **Varanasi, U. and Malins, D. C.,** Metabolism of petroleum hydrocarbons: accumulation and biotransformation in marine organisms, in *Effects of Petroleum on Artic and Subarctic Marine Environments and Organisms. Vol. II. Biological Effects,* Malins, D.C., Ed.,Academic Press, New York, 1977, 175.
145. **Carlson, G. P.,** Induction of cytochrome P-450 by halogenated benzenes, *Biochem. Pharmacol.,* 27, 361, 1978.
146. **Addison, R. F., Zinck, M. E., Willis, D. E., and Darrow, D. C.,** Induction of hepatic mixed function oxidases in trout by polychlorinated biphenyl and butylated monochlorodiphenyl ethers, *Toxicol. Appl. Pharmacol.,* 49, 245, 1979.
147. **Kurelec, B., Britvic, S., Rijavec, M., Muller, W. E. G., and Zahn, R. K.,** Benzo[a]pyrene monooxygenase induction in marine fish — molecular response to oil pollution, *Mar. Biol.,* 44, 211, 1977.
148. **Payne, J. F.,** Mixed function oxidases in marine organisms in relation to petroleum hydrocarbon metabolism and detection, *Mar. Pollut. Bull.,* 8, 112, 1977.
149. **Lee, R. F., Furlong, E., and Singer, S.,** Metabolism of hydrocarbons in marine invertebrates: aryl hydrocarbon hydroxylase from the tissues of the blue crab, *Callinectes sapidus,* and the polychaete worm *Nereis* sp., in *Pollutant Effects on Marine Organisms,* Giam, C. S., Ed., D.C. Heath & Co., Lexington, Mass., 1977, 111.
150. **Webb, R. E., Randolph, W. C., and Horsefall, F., Jr.,** Hepatic benzpyrene hydroxylase activity in endrin-susceptible and resistant pine mice, *Life Sci.,* 11, 477, 1972.
151. **Bend, J. R., Foureman, G. L., and James, M. O.,** Partially induced hepatic mixed-function oxidase systems in individual members of certain marine species from coastal Maine and Florida, in *Aquatic Pollutants: Transformation and Biological Effects,* Hutzinger, O., von Lelyveld, I. H., and Zoeteman, B. C. J., Eds., Pergamon Press, New York, 1978, 483.
152. **Stegeman, J. J., Binder, R. J., and Orren, A.,** Hepatic and extra-hepatic microsomal electron transport components and mixed-function oxygenases in the marine fish *Stenotomus versicolor, Biochem. Pharmacol.,* 28, 3431, 1979.
153. **Burns, K.,** Hydrocarbon metabolism in the intertidal fiddler crab *Uca pugnax, Mar. Biol.,* 36, 5, 1976.
154. **Vandermeulen, J. H., Keizer, P. D., and Penrose, W. R.,** Persistence of non-alkane components of Bunker C oil in beach sediments of Chedablucto Bay, and lack of their metabolism by molluscs, in *Proc. 1977 Oil Spill Conf.,* American Petroleum Institute Publ. No. 4284, American Petroleum Institute, Washington, D.C., 1977, 469.
155. **Stegeman, J. J. and Chevion, M.,** Sex differences in cytochrome P-450 and mixed function oxygenase activity in gonadally mature trout, *Biochem. Pharmacol.,* 29, 553, 1980.
156. **Binder, R. L. and Stegeman, J. J.,** Induction of aryl hydrocarbon hydroxylase activity in embryos of an estuarine fish, *Biochem. Pharmacol.,* 29, 949, 1980.
157. **Lu, A. Y. H. and Levin, W.,** The resolution and reconstitution of the liver microsomal hydroxylation system, *Biochim. Biophys. Acta,* 344, 205, 1974.
158. **Elmamlouk, T. H., Philpot, R. M., and Bend, J. R.,** Separation of two forms of cytochrome P-450 from hepatic microsomes of 1,2,3,4-dibenzanthracene pretreated little skates *(Raja erinacea), Pharmacologist,* 19, 160, 1977.
159. **Conney, A. H. and Klutch, A.,** Increased activity of androgen hydroxylases in liver microsomes of rats pretreated with phenobarbitol and other drugs, *J. Biol. Chem.,* 238, 1611, 1963.
160. **Brown, S. S.,** Metabolic interactions involving drugs or foreign compounds, in *Foreign Compound Metabolism in Mammals,* Hathway, D. E., Ed., The Chemical Society, London, 1975, 591.
161. **Gessner, T.,** A comparison of steroid glucuronyltransferases of rat liver and hepatoma, and effect of phenobarbitone and 3-methylcholanthrene induction, *Biochem. Soc. Trans.,* 4, 522, 1976.
162. **Levin, W., Welsh, R. M., and Conney, A. H.,** Increased liver microsomal androgen metabolism by phenobarbitol: correlation with decreased androgen action on the seminal vesicles of the rat, *J. Pharmacol. Exp. Ther.,* 188, 287, 1974.
163. **Kupfer, D. and Bulger, W. H.,** Interactions of chlorinated hydrocarbons with steroid hormones, *Fed. Proc.,* 35, 2603, 1976.
164. **Freeman, H. C. and Idler, D. R.,** The effect of polychlorinated biphenyl (PCB) on steroidogenesis and reproduction in the brook trout *(Salvelinus fontinalis), Can. J. Biochem.,* 53, 666, 1975.
165. **Freeman, H. C. and Sangolang, G. B.,** A study of the effects of methyl mercury, cadmium, arsenic, selenium, and a PCB (Aroclor 1254) on adrenal and testicular steroidogenesis *in vitro,* by the gray seal *Halichocrus grypus, Arch. Environ. Contam. Toxicol.,* 5, 369, 1977.
166. **Hansson, T. and Lidman, U.,** Effects of cortisol administration on components of the hepatic microsomal mixed function oxidase system (MFO) in immature rainbow trout *(Salmo gairdnerii* Rich.), *Acta Pharmacol. Toxicol.,* 43, 6, 1978.

167. **Singley, J. A. and Chavin, W.,** The adrenocortical-hypophyseal response to saline stress in the goldfish *Carassius auratus* L., *Comp. Biochem. Physiol.*, 51A, 749, 1975.
168. **Wedemeyer, G.,** Stress-induced ascorbic acid depletion and cortisol production in two salmonid fishes, *Comp. Biochem. Physiol.*, 29, 1247, 1969.
169. **Mazeaud, M. M., Mazeaud, F., and Donaldson, E. M.,** Primary and secondary effects of stress in fish: some new data with a general review, *Trans. Am. Fish. Soc.*, 106, 201, 1977.
170. **Selye, H.,** The evolution of the stress concept, *Am. Sci.*, 61, 692, 1973.
171. **Lewin, S.,** *Vitamin C: Its Molecular Biology and Medical Potential,* Academic Press, New York, 1976.
172. **Subramanian, W., Nandi, B. K., Majumder, A. K., and Chatterjee, I. B.,** Role of L-ascorbic acid on detoxification of histamine, *Biochem. Pharmacol.*, 22, 1671, 1973.
173. **Barnes, M. J. and Kodicek, E.,** Biological hydroxylations and ascorbic acid with special regard to collagen metabolism, in *Vitamins and Hormones,* Vol. 30, Harris, R. S., Munson, P. L., Glover, J., and Dkzfalusy, E., Eds., Academic Press, New York, 1972, 1.
174. **Street, J. C. and Chadwick, R. W.,** Ascorbic acid requirements and metabolism in relation to organochlorine pesticides, *Ann. N.Y. Acad. Sci.*, 258, 132, 1975.
175. **Halver, J. E., Smith, R. R., Tolbert, B. M., and Baker, E. M.,** Utilization of ascorbic acid in fish, *Ann. N.Y. Acad. Sci.*, 258, 81, 1975.
176. **Wedemeyer, G.,** The role of stress in the disease resistance of fishes. Diseases of fishes and shellfishes, *Am. Fish. Soc. Spec. Publ.*, 5, 30, 1970.
177. **Sikic, I., Mimnaugh, G., Litterst, C. L., and Gram, T. E.,** The effects of ascorbic acid deficiency and repletion on pulmonary, renal and hepatic drug metabolism in the guinea pig, *Arch. Biochem. Biophys.*, 179, 663, 1977.
178. **Telford, M.,** The effects of stress on blood sugar composition of the lobster, *Homarus americanus, Can. J. Zool.*, 46, 819, 1968.
179. **Moore, M. N.,** Cytochemical demonstration of latency of lysosomal hydrolases in digestive cells of the common mussel, *Mytilus edulis,* and changes induced by thermal stress, *Cell. Tissue Res.*, 175, 279, 1976.
180. **Moore, M. N.,** Cytochemical determination of cellular responses to environmental stressors in marine organisms, *Rapp. P.-V. Reun. Cons. Int. Explor. Mes.*, 179, 7, 1980.
181. **Roubal, W. T.,** Spin-labeling of living tissue — a method for investigating pollutant-host interaction, in *Pollution and Physiology of Marine Organisms,* Vernberg, F. J. and Vernberg, W. B., Eds., Academic Press, New York, 1974, 367.
182. **Ernst, V. V., Neff, J. M., and Anderson, J. W.,** Effects of the water-soluble fraction of no. 2 fuel oil on the development of the estuarine fish, *Fundulus grandis* Baird and Girard, *Environ. Pollut.*, 14, 25, 1977.
183. **Bayne, B. L., Livingstone, D. R., Moore, M. N., and Widdows, J.,** A cytochemical index and a biochemical index of stress in *Mytilus edulis* L., *Mar. Pollut. Bull.*, 7, 221, 1976.
184. **Bayne, B. L., Moore, M. N., Widdows, J., Livingstone, D. R., and Salkeld, P.,** Measurement of the responses of individuals to environmental stress and pollution: studies with bivalve molluscs, in *The Assessment of Sublethal Effects of Pollutants in the Sea, Cole, H. A., Ed.*, The Royal Society, London, 1979, 165.
185. **Moore, M. N.,** Cellular responses to polycyclic aromatic hydrocarbons and phenobarbitol in *Mytilus edulis, Mar. Environ. Res.*, 2, 255, 1979.
186. **Mearns, A. J. and Sherwood, M. J.,** Distribution of neoplasms and other diseases in marine fishes relative to the discharge of waste water, *Ann. N.Y. Acad. Sci.*, 298, 210, 1977.
187. **Minchew, C. D. and Yarbrough, J. D.,** The occurrence of fin rot in mullet *(Mugil cephalus)* associated with crude oil contamination of an estuarine pond-ecosystem, *J. Fish. Biol.*, 10, 319, 1977.
188. **Lowe, J. I., Parrish, P. R., Patrick, J. M., Jr., and Forester, J.,** Effects of the polychlorinated biphenyl Aroclor 1254 on the American oyster *Crassostrea virginica, Mar. Biol.*, 17, 209, 1972.
189. **Doughtie, D. G. and Rao, K. R.,** Ultrastructural changes induced by sodium pentachlorophenate in the grass shrimp, *Palaemonetes pugio,* in relation to the molt cycle, in *Pentachlorophenol,* Rao, K. R., Ed., Plenum Press, New York, 1978, 213.
190. **Hodgins, H. O., McCain, B. B., and Hawkes, J. W.,** Marine fish and invertebrate diseases, host disease resistance, and pathological effects of petroleum, in *Effects of Petroleum on Arctic and Subarctic Marine Environments and Organisms Vol. 2 Biological Effects,* Malins, D. C., Ed., Academic Press, New York, 1977, 95.
191. **Koehler, L. D.,** Immunosuppression produced by lead, cadmium and mercury, *Am. J. Vet. Res.*, 34, 1457, 1973.
192. **Adkinson, N. F.,** Environmental influences on the immune system and allergic reactions, *Environ. Health Perspect.*, 20, 97, 1977.
193. **Monjan, A. A. and Collector, M. I.,** Stress-induced modulation of the immune response, *Science,* 196, 307, 1977.

194. **Robohm, R. A. and Nitkowski, M. F.,** Physiological response of the cunner *Tautogolabrus adspersus*, to cadmium. IV. Effects on the immune system, NOAA Tech. Rep. NMFS SSRF-681, U.S. Department of Commerce, Washington, D.C., 1974, 15.

195. **Snieszko, S. F.,** The effects of environmental stress on outbreaks of infectious diseases of fish, *J. Fish. Biol.*, 6, 197, 1974.

196. **Roales, R. R. and Perlmutter, A.,** The effects of sublethal doses of methylmercury and copper, applied singly and jointly, on the immune response of the blue gourami *(Trichogaster trichopterus)* to viral and bacterial antigens, *Arch. Environ. Contam. Toxicol.*, p. 325, 1977.

197. **Couch, J. A. and Courtney, L.,** Interaction of chemical pollutants and virus in a crustacean: a novel bioassay system, *Ann. N.Y. Acad. Sci.*, 298, 497, 1977.

198. **Fries, C. R. and Tripp, M. R.,** Cytological damage in *Mercenaria mercenaria* exposed to phenol, in *Fate and Effects of Petroleum Hydrocarbons in Marine Organisms and Ecosystems*, Wolfe, D. A., Ed., Pergamon Press, New York, 1977, 174.

199. **Fries, C. R. and Tripp, M. R.,** Depression of phagocytosis in *Mercenaria* following chemical stress, *Dev. Comp. Immunol.*, 4, 233, 1980.

200. **Goncharov, G. D. and Mikyakov, V. R.,** The effects of low concentrations of phenol on the antibody formation in carp, *Cyprinus carpio*, Probl. Aquat. Toxicol. 1970 (Transl. U.S. Bur. Sport Fish. Wildl.), Washington, D.C., 1971, 171.

201. **Fries, C. R. and Tripp, M. R.,** Effects of phenol on clams, *Mar. Fish. Rev.*, 38, 10, 1976.

202. **Jeffries, H. P.,** A stress syndrome in the hard clam, *Mercenaria mercenaria, J. Invert. Pathol.*, 20, 242, 1972.

203. **Couch, J. and Nimmo, D.,** Ultrastructural studies on shrimp exposed to the pollutant chemical, polychlorinated biphenyl (Aroclor 1254), *Bull. Pharmacol. Environ. Pathol.*, 11, 17, 1974.

204. **Liebovitz, B. and Siegel, B. V.,** Ascorbic acid, neutrophil function, and the immune response, *Int. J. Vit. Nutr. Res.*, 48, 159, 1978.

Chapter 2

BIOACCUMULATION AND DEPURATION OF SELECTED ORGANIC COMPOUNDS IN MARINE FISH

L. E. Ray and C. S. Giam

TABLE OF CONTENTS

I. Analytical Methods for Determination of Benzo(a)pyrene, Hexachlorobenzene, and Pentachlorophenol 24
- A. Equipment 24
- B. Extraction Procedures 24
  1. Simplified Procedures for Water 24
  2. Detailed High Sensitivity/High Resolution Procedures for Water 25
  3. Detailed High Sensitivity/High Resolution Procedures for Tissue 26
  4. Detailed High Sensitivity/High Resolution Procedures for Sediment 26

II. Sample Collection and Handling 27
- A. Water 27
- B. Sediment 27
- C. Animals 28
- D. Sand Column Dosing System 28

III. Results 28
- A. Validation of Methods 28
  1. Simplified Procedures 28
  2. High Sensitivity/High Resolution Procedures 29
- B. Bioaccumulation Studies 30
  1. Hexachlorobenzene 30
  2. Pentachlorophenol 32
- C. Field Studies 35
  1. Portland, Maine 35
  2. San Luis Pass, Texas 40
  3. Corpus Christi Bay, Texas 43

Acknowledgments 47

References 47

I. ANALYTICAL METHODS FOR DETERMINATION OF BENZO(A)PYRENE, HEXACHLOROBENZENE, AND PENTACHLOROPHENOL

All organic solvents should be ''distilled in glass'', pesticide or nanograde quality, e.g., Burdick and Jackson or MCB. They should be concentrated 50- to 100-fold and analyzed by gas chromatography (GC) to demonstrate the absence of interfering peaks. Other chemicals were reagent-grade or the best available quality. Distilled water used for preparing NaCl solutions and for other aspects of this work was continuously extracted with petroleum ether (PE) in a 20-ℓ extractor until a 50-fold concentrate of the solvent showed no interfering peaks when analyzed by GC using both electron capture detector (ECD) and flame ionization detector (FID). The PE was changed every 24 hr. Chromatography columns (10 × 400 mm; Chromaflex) with a 50-mℓ reservior are convenient and were used for the work described below. Glassware and other equipment were thoroughly washed with warm Micro cleaning solution (International Products Corp.), rinsed with distilled water, and heated at 320°C overnight. Just prior to use, glassware was rinsed with PE. These and other salient points related to analysis of samples for very low concentrations of organic compounds have been discussed in greater detail in other publications from this group.[1-3]

A. Equipment

Benzo(a)pyrene was quantitated by high pressure liquid chromatography (HPLC) using a Waters C-18 μBondapak column (4.0 × 300 mm). The mobile phase was H_2O:CH_3OH (10:90) at a flow rate of 1.5 mℓ/min. Fluorescence of BaP was measured at 403 nm with excitation at 365 nm.

Hexachlorobenzene was quantitated using a gas chromatograph (GC) equipped with a ^{63}Ni electron-capture detector (ECD), and a 6-ft glass column (4 mm i.d.) packed with 5% QF-1 on 100 to 120 mesh Supelcoport.[4] Temperatures were 160, 250, and 300°C for the column, injection port, and detector, respectively. The carrier gas was 5% methane/argon at a flow of 30 mℓ/min. Alternatively, a 15-m fused-silica capillary column of SP-2100 (SGE) or DB-1 (J & W) was used instead of the packed column. The injection port and detector temperatures were 225 and 325°C, respectively; the oven temperature increased from 100 to 250°C at a rate of 10°C/min, starting 2 min after injection. The carrier gas was H_2 at 15 psi and the make-up gas was 5% methane/argon at a flow of about 30 mℓ/min.

At high concentrations, pentachlorophenol was quantitated using a HPLC with a variable wavelength UV detector at 254 nm. A Waters C-18 μBondapak column (4.0 × 300 mm) was used for separation. The mobile phase was CH_3CN:H_2O) (60:40) with 1% acetic acid at a flow rate of 2.0 mℓ/min. For lower concentrations, PCP was converted to its methyl ether (pentachloroanisole; see Section I.B.2), which was then determined by GC-ECD as described for HCB.

B. Extraction Procedures

1. Simplified Procedures for Water

The following procedures were used for analysis of relatively high concentrations of BaP, HCB, or PCP in water, e.g., water samples from laboratory studies of bioaccumulation.[5] They are useful for higher concentrations since they are faster than and require less material (sample and reagents) than the more sensitive procedures described in the next section. These procedures and those in the following section are based on longer methods reported in the literature.[6-8]

Water samples (5 mℓ) for BaP were extracted twice by shaking for 30 sec with 1.0-mℓ portions of PE in a 2- or 3-dram vial with a Teflon®-lined cap. The PE extracts were removed with a Pasteur pipet and placed in a second vial. The combined PE extracts were then evaporated with a gentle stream of dry nitrogen filtered through a molecular sieve. The residue was dissolved in 200 μℓ CH_3OH and 50 to 100 μℓ were injected into the HPLC for quantitation as described above.

For HCB, water samples (5 mℓ) were extracted by shaking for 30 sec with 1.0 mℓ of isooctane in a 2- or 3-dram vial with a Teflon®-lined cap. A microsyringe was used to withdraw 2 to 10 μℓ of the isooctane layer. The sample was immediately injected into a GC for quantitation as described above.

For PCP, water samples (5 mℓ) were first acidified with 3 to 4 drops of 6 *N* H_2SO_4 delivered by Pasteur pipet and then extracted twice by shaking for 30 sec with 1.0-mℓ portions of PE:diethyl ether (1:1) in a 2- or 3-dram vial with a Teflon®-lined cap. The extracts were combined in a second vial and the solvent was evaporated with dry N_2 as described for BaP. The residue was dissolved in 200 μℓ CH_3CN and 50 to 100 μℓ were analyzed by HPLC as described above.

2. Detailed High Sensitivity/High Resolution Procedures for Water

The following procedures were used for analysis of low concentrations of BaP, HCB, or PCB in water, e.g., field samples.

For BaP, a 1-ℓ sample was extracted by shaking vigorously for 30 sec with 50 mℓ PE in a 2-ℓ separatory funnel. After the phases separated, the aqueous layer was drained off into a clean flask and the organic layer was transferred to a 250-mℓ Kuderna-Danish (K-D) evaporator. The water layer was then reextracted twice. The organic extracts were combined and were concentrated to 5 to 6 mℓ on a steam bath. The concentrated PE extract was then chromatographed through 10 g of 5% deactivated silica gel (type 40, 70 to 230 mesh; EM reagents) overlaid with ~1 cm (~1.6 g) anhydrous Na_2SO_4. Silica gel was heated at least 16 hr at 150°C (fully activated), cooled and deactivated by adding 5 g PE-extracted distilled water per 95 g gel and mixing for 1 to 2 hr. The sample extract was eluted with 50 mℓ hexane (which was discarded) and then with 200 mℓ 2% CH_3CN in PE. The effluent was concentrated to 5 to 6 mℓ in a K-D evaporator on a steam bath, dryed with Na_2SO_4, and then evaporated to dryness with a gentle stream of dry N_2. The residue was dissolved in 200 μℓ CH_3OH and an aliquot (50 to 100 μℓ) was analyzed for BaP using HPLC and spectrofluorometry as described above.

For HCB, a 1-ℓ sample was extracted 3 times with PE and the extracts were concentrated as described above. The concentrated PE extract was chromatographed through 10 g activated silica gel (prepared by heating at 110 to 120°C for at least 16 hr) overlaid with ~1 cm anhydrous Na_2SO_4. The extract was eluted with 50 mℓ PE; the first 15 mℓ was discarded. The subsequent effluent was concentrated as necessary in a K-D evaporator on a steam bath, dried with Na_2SO_4, and analyzed for HCB using GC-ECD as described above.

For PCP, the water sample (1 ℓ) was adjusted to pH $\geq$ 12 (checked with pH paper) with NaOH pellets and extracted 3 times by shaking vigorously for 30 sec with 50 mℓ hexane in a 2-ℓ separatory funnel. After the phases separated, the aqueous layer was drained off and then the organic layer was removed. The hexane extract can be either discarded or saved for the quantitation of base/neutrals including pesticides. The aqueous phase was acidified to pH $\leq$ 2 with concentrated H_2SO_4 and extracted 3 times with 50 mℓ PE:diethyl ether (1:1) as just described. The extracted aqueous phase was discarded and the combined PE:diethyl ether phases were concentrated in a K-D evaporator to 5 to 6 mℓ and dried with Na_2SO_4.

Samples with a low concentration of PCP (e.g., field samples) were derivatized with CH_2N_2 (generated from N-methyl-N′-nitro-N-nitrosoguanidine and trapped in diethyl ether) for 30 to 60 min at room temperature in a tightly capped 2- or 3-dram vial. The methylated sample (pentachloroanisole) was chromatographed through 4 g acidic alumina (80 to 200 mesh, Brockman activity I; heated at 150°C for a least 12 hr) overlaid with ~1 cm anhydrous Na_2SO_4. It was washed with 5 mℓ hexane (discarded) and eluted with 20 mℓ 10% benzene in hexane. The benzene/hexane fraction was concentrated in a K-D evaporator and analyzed by GC-ECD. For samples (e.g., laboratory exposure studies) where higher concentrations of PCP are expected, the dried PE:diethyl ether extract can be injected directly into the HPLC and eluted with $CH_3CN:H_2O:CH_3CO_2H$ (60:40:1) as described above.

3. Detailed High Sensitivity/High Resolution Procedures for Tissue

The following procedures were used for analysis of BaP, HCB, or PCP in tissue samples.

For BaP, tissue (50 g) was homogenized twice using a Polytron (Brinkmann Instruments, Inc.) tissue homogenizer with 100 mℓ 20% acetone in CH_3CN; the tissue was filtered (Whatman #1) after each homogenization. The filtrates were added to 700 mℓ 5% aqueous NaCl in a 1-ℓ separatory funnel and extracted with 3 × 100 mℓ PE. The PE layers were placed into a K-D evaporator and concentrated to 5 to 6 mℓ. If smaller amounts (5 to 10 g) of tissue are used, solvent volumes should be adjusted accordingly (see procedures for HCB and PCP). The concentrated PE extracts were next chromatographed through 10 g 5% deactivated silica gel overlaid with ~1 cm anhydrous Na_2SO_4. The sample was washed with 50 mℓ hexane (which was discarded) and eluted with 200 mℓ 2% CH_3CN in pentane; the second fraction was concentrated to 5 to 6 mℓ in a K-D evaporator. The concentrated sample was transferred to a 2- or 3-dram vial and evaporated to dryness with dry N_2. The residue was redissolved in 200 μℓ of CH_3OH and 50 to 100 μℓ were analyzed for BaP by HPLC as described above.

If interfering or overlapping peaks are obtained, the CH_3CN/pentane is evaporated off and the sample is then redissolved in and hydrolyzed in 1 mℓ 5% KOH/CH_3OH. After 30 min hydrolysis on a steam bath, the sample is diluted with 7 mℓ PE-extracted water and extracted with 3 × 1 mℓ PE. The combined extracts are evaporated to dryness, redissolved in CH_3OH, and analyzed by HPLC as just described.

For HCB, tissue (10 g) was homogenized in 2 × 30 mℓ 20% acetone in CH_3CN, filtered, diluted into 150 mℓ 5% aqueous NaCl, extracted with 3 × 10 mℓ PE, and concentrated as just described for BaP. For larger samples (up to 50 g), the volumes were increased to 2 × 50 mℓ 20% acetone/CH_3CN, 250 mℓ 5% NaCl, and 3 × 50 mℓ PE. The concentrated PE extract was chromatographed on 10 g activated silica gel (70 to 230 mesh, Woelm, heated at 150°C at least 16 hr) overlaid with ~1 cm anhydrous Na_2SO_4. The sample was eluted with 50 mℓ PE; the first 15 mℓ was discarded. The effluent was concentrated as necessary using a K-D evaporator on a stream bath followed by a gentle stream of dry N_2. To avoid loss of HCB, the sample should not be evaporated to dryness. As described above, the sample was injected into a GC-ECD for final quantitation of HCB.

For PCP, tissue (10 g) was homogenized with 2 × 30 mℓ 20% acetone/CH_3CN and filtered as described for HCB. The filtrate was added to 150 mℓ 5% aqueous NaCl, adjusted to pH $\geqslant$ 11 with NaOH pellets, and extracted with 3 × 30 mℓ hexane. The hexane layers may be discarded or saved for the analysis of base/neutrals. For larger samples (up to 50 g), the volumes used would be the same as just described for HCB. The aqueous phase was acidified to pH $\leq$ 2 with concentrated H_2SO_4 and extracted with 3 × 30 mℓ hexane. The hexane extracts were concentrated in a K-D evaporator to 3 to 5 mℓ, dried with anhydrous Na_2SO_4, transferred to a 3-dram vial, and evaporated to dryness with a gentle stream of dry N_2. The sample was derivatized with diazomethane in the vial as described in Section I.B.2. The derivatized sample was chromatographed through 4 g acidic alumina overlaid with ~1 cm anhydrous Na_2SO_4. The sample as washed with 5 mℓ hexane (which was discarded) and eluted with 20 mℓ 10% benzene in hexane which was collected, concentrated, and injected into a GC-ECD to quantitate pentachloroanisole.

4. Detailed High Sensitivity/High Resolution Procedures for Sediment

For BaP, wet sediment (50 g) was refluxed in 100 mℓ 5% KOH in CH_3OH for 2 hr. The refluxed sediment was then filtered (Whatman GF/C) and rinsed with 2 × 50 mℓ CH_3OH.

For maximum recovery, the condensor and flask should also be rinsed. The filtrate and rinses were added to 700 mℓ 5% aqueous NaCl in a separatory funnel and extracted with 3 × 100 mℓ pentane. The combined pentane layers were concentrated in a K-D evaporator to 5 to 6 mℓ and chromatographed through 10 g activated silica gel overlaid with ~1 cm Na_2SO_4. The sample was washed with 50 mℓ hexane (discarded) and then eluted with 200 mℓ 2% CH_3CN in PE. This was concentrated to dryness, dissolved in 200 μℓ CH_3OH, and 50 to 100 μℓ were analyzed for BaP by HPLC as detailed for tissue (Section I.B.3).

For HCB, sediment (50 g) was refluxed in 100 mℓ 20% acetone/CH_3CN for 2 hr. The refluxed sediment was then filtered (Whatman GF/C) and rinsed with 2 × 50 mℓ 20% acetone in CH_3CN. The condensor and reflux flask were also rinsed. The filtrate and rinses were added to 700 mℓ 5% aqueous NaCl in a separatory funnel and extracted with 3 × 100 mℓ PE. The PE extracts were concentrated to 5 to 6 mℓ in a K-D evaporator and chromatographed through 10 g activated silica gel. The sample was eluted with 50 mℓ PE; the first 15 mℓ was discarded and the remainder was concentrated using a K-D evaporator and/or a stream of N_2.

Sulfur and/or sulfur compounds are frequently present in these samples and must be removed to eliminate interference. This is accomplished by adding mercury (1 or 2 drops) to the concentrated sample in a 2- or 3-dram vial and shaking vigorously until no more black precipitate is formed. HCB in the sample was then quantitated using GC-ECD.

For PCP, sediment (50 g) was refluxed in 100 mℓ 20% acetone in CH_3CN for 2 hr, filtered, and rinsed with 2 × 50 mℓ 20% acetone in CH_3CN. The filtrate and rinses were added to 500 mℓ 5% aqueous NaCl, adjusted to pH $\geq$ 11 with NaOH pellets, and extracted with 3 × 50 mℓ hexane. The hexane layer may be discarded or used for the analysis of base/neutral compounds. The aqueous layer was acidified to pH $\leq$ 2 with concentrated H_2SO_4 and extracted with 3 × 50 mℓ hexane. The hexane extracts were concentrated to 5 to 6 mℓ in a K-D evaporator, dried with anhydrous Na_2SO_4, transferred to a 2- or 3-dram vial, and evaporated to dryness with dry N_2. The sample was then derivatized with CH_2N_2 as described above for PCP in water and tissue (Section I.B.2).

The derivatized sample was purified by chromatography through 4 g acidic alumina overlaid with ~1 cm anhydrous Na_2SO_4. The sample was washed with 5 mℓ hexane (which was discarded) and eluted with 20 mℓ 10% benzene in hexane which was collected, concentrated, and analyzed by GC-ECD.

II. SAMPLE COLLECTION AND HANDLING

Water, biota, and sediment samples were collected at various estuaries and harbors on the Texas Gulf coast, at New Bedford, Plymouth, and Boston Harbors, Massachusetts, and at Portland, Maine. Procedures for sampling and analyzing marine organisms have been discussed in detail in other publications from this group.[9-12]

A. Water

Surface samples were collected directly in clean solvent bottles by submerging the capped bottle, removing the cap, and then recapping after it was filled. Deep samples were obtained using appropriate samplers which could be opened and closed while submerged. Samples were transferred to clean solvent bottles after bringing them to the surface. At least 1 gal was obtained at each site. Samples were tightly capped and were refrigerated without preservatives until analyzed.

B. Sediment

Surface samples in shallow water were collected in solvent-cleaned, wide-mouth glass jars at low tide under 15 to 20 cm of water. Any water was decanted from the sediment sample. Sediment-filled jars were covered with solvent-rinsed aluminum foil, tightly capped, and frozen or refrigerated without preservatives until analyzed. In deep water, samples were obtained using a grab sampler. At the surface, glass jars were filled and the samples handled as just described.

C. Animals

Clams and *Neanthes (Nereis)* were collected by hand from the intertidal zone and from under the water's edge at low tide. Whole *Nereis* were wrapped in solvent-rinsed aluminum foil, frozen in liquid N_2 at the collection site, and stored frozen ($-20°C$) until analyzed. Clams were frozen as just described either whole or after dissection at the collection site.

Fish were collected from shallow inlets and marsh areas by seining with a 15-m hand-seine. In deeper water, a trawl was used. The trawl was emptied after about 5 min. For analytical work, large fish were dissected on site, wrapped in foil, and frozen in liquid N_2. Smaller fish were frozen whole. The material was stored frozen until it was analyzed.

For bioaccumulation studies, the fish were placed in large ice chests containing natural sea water, chilled slightly, and aerated during transport to the laboratory. Fish were acclimated at least 2 weeks in the laboratory in aquaria with Instant Ocean which was continuously aerated and filtered through activated charcoal. The fish were fed Oregon Moist pellets daily.

To measure bioaccumulation, fish were placed in exposure aquaria and continuously dosed with BCP, HCB, or PCP. BaP and HCB were supplied using the sand column dosing system described in Section II.D; PCP was supplied from a stock solution of PCP in Instant Ocean. The concentration of each chemical was controlled by adjusting the rate of flow through the sand column or the rate of addition from the stock reservoir. During exposure to HCB or PCP, activated charcoal and/or crushed oyster shell was used in the filter pump to remove particulates and help regulate pollutant concentration.

To measure depuration of any of the compounds, exposed fish were transferred to another aquarium containing fresh Instant Ocean. Water was circulated through a filter pump containing activated charcoal in all cases. During exposure and depuration, salinity was measured daily and maintained at 30 to 35‰ by addition of distilled water as necessary.

D. Sand Column Dosing System

To administer compounds (such as BaP and HCB) poorly soluble in water at a uniform rate during laboratory experiments, a sand column dosing system was developed in conjunction with Drs. Neff and Thomas.[13] Sand (100 g) was coated with BaP or HCB by dissolving the pollutant (100 mg) in benzene or diethyl ether (150 mℓ), adding the sand to the solution and allowing the solvent to evaporate in a fume hood. The coated sand was then placed between two layers of clean sand in a glass chromatography column. Both ends of the column were packed with glass wool to prevent loss of the sand. Water was percolated through the column in an upward direction using a peristaltic pump. The concentration of pollutant in the water was regulated by adjusting the flow rate of the peristaltic pump.

III. RESULTS

A. Validation of Methods

1. Simplified Procedures

Known amounts of BaP, HCB, or PCP were dissolved in CH_3CN and added to 5-mℓ aliquots of Instant Ocean to determine the recovery and precision of these procedures. All methods gave good precision and high recoveries using spiked water samples.[5] Values reported below are based on three or more replicate determinations.

Analysis of water samples containing BaP at 0.04 to 2.0 ppb gave recoveries of 82 to 90%. The mean and standard deviation were 87 ± 3%. The relative standard deviation for 10 replicate samples containing 1.0 ppb BaP was 6.4%.

For water samples spiked with HCB, the recovery was 77 to 82% at 0.5 to 1.0 ppb HCB, with a mean and standard deviation of 80 ± 4%. At 0.1 to 0.2 ppb HCB, the recovery was somewhat lower (50 to 62%). For 10 replicate samples at 1.0 ppb, the relative standard deviation was 3.2%.

The recovery of PCP was greater than 90% for water samples containing 80 to 200 ppb PCP; the average was 91 ± 11%. At 40 ppb PCP, the recovery was 74%. At 200 ppb, the relative standard deviation was 2.6% for 10 samples.

The minimum amounts necessary to give detector peak heights twice the baseline noise were 200 pg, 0.5 pg, and 50 ng for BaP, HCB, and PCP, respectively. Based on recovery, reproducibiltiy, normal sample size, and proportion analyzed, the practical detection limits in water for the simplified procedures were considered to be 0.04 ppb BaP, 0.01 ppb HCB, and 40 ppb PCP. The limit for PCP is based on UV detection; using GC-ECD its limit is the same as for HCB. Although these limits (especially for PCP) are somewhat higher than one would expect in open ocean samples, the procedures offer several advantages. For example, only a small sample volume is needed; the amount of solvent required for extraction is greatly reduced (typically less than 5% of that for the longer procedures); and the time for preparation and analysis is greatly reduced (e.g., 8 to 12 samples can be analyzed per hour). Thus, these procedures are excellent for monitoring routine laboratory experiments on toxicity, bioaccumulation, depuration, etc.

2. High Sensitivity/High Resolution Procedures

Known amounts of BaP, HCB, or PCP were dissolved in CH_3CN and added to Instant Ocean (50 to 1000 mℓ), tissue (3 to 50 g), or sediment (50 g) before the first extraction step for that sample type. Again, all methods gave good precision and high recovery using the spiked samples. Values reported below are based on at least three replicate determinations; most are from five or more.

Analysis of water samples containing BaP at 0.001 to 0.01 ppb gave recoveries of 81 to 108% with a mean and standard deviation of 91 ± 11%. For tissue spiked with BaP to give 0.2 to 34 ppb, the recovery was 69 to 102% with an average and standard deviation of 81 ± 10%. For sediment spiked to give 0.2 to 20 ppb, the recoveries were 76 to 95% and the average was 81 ± 12%.

For water samples spiked with HCB at 0.005 to 1 ppb, the recoveries were 71 to 98% with an average of 84 ± 9%. For tissue samples containing 2 to 100 ppb, recoveries were 83 to 93%, with an average of 90 ± 4%. For sediment samples containing 2 to 10 ppb, the recoveries were 71 to 85%; the average was 79 ± 5%.

The recovery of PCP from water samples containing 0.2 to 10 ppb was 71 to 112% with an average of 94 ± 13%. For tissue samples containing 33 to 167 ppb, the recovery was 97 to 109%, with an average of 103 ± 6%. For sediment samples containing 2 to 4 ppb, the recovery of PCP was 77 to 110%, with an average of 93 ± 14%.

Based on the considerations mentioned above, the practical limits of detection in water for these procedures were considered to be 0.001 ppb BaP, 0.0005 ppb HCB, and 0.2 ppb PCP. Again the limit for PCP is based on UV detection; with analysis by GC-ECD, the limit is 0.0005 ppb.

The recoveries and practical limits of detection in water, tissue, and sediment are summarized in Table 1. Of course the limits for both the shortened and long procedures can be improved by increasing the amount of sample processed and the proportion analyzed.

Table 1
SUMMARY OF RECOVERIES AND LIMITS OF DETECTION FOR BaP, HCB, AND PCP IN VARIOUS SAMPLES

Compound	Sample	Sample size	Average recovery[a] (%)	Detection limit[b] (ppb)
BaP	Water[c]	5 mℓ	87 ± 3	0.04
	Water[d]	1000 mℓ	91 ± 11	0.001
	Tissue	50 g	81 ± 10	0.004
	Sediment	50 g	81 ± 12	0.004
HCB	Water[c]	5 mℓ	80 ± 4	0.01
	Water[d]	1000 mℓ	84 ± 9	0.0005
	Tissue	50 g	90 ± 4	0.001
	Sediment	50 g	79 ± 5	0.001
PCP	Water[c]	5 mℓ	91 ± 11	0.01; 40[e]
	Water[d]	1000 mℓ	94 ± 13	0.0005; 0.2
	Tissue	50 g	103 ± 6	0.001; 4.0
	Sediment	50 g	93 ± 14	0.001; 4.0

[a] Average ± standard deviation for 3 or more determinations.
[b] Detection limits are based on sample size indicated and analysis as described in the text. Sensitivity can be improved by using more sample material or analyzing a larger proportion of the final material.
[c] Analyzed using the simplified procedures.
[d] Analyzed using the longer, higher sensitivity procedures.
[e] For PCP, the lower limit represents analysis by GC-ECD and the higher limit represents analysis by HPLC with a UV monitor at 254 nm.

B. Bioaccumulation Studies

The use of bioaccumulation factors as an estimate of potentially harmful effects of organic chemicals is important in studies of biological effects in aquatic systems,[14] however, it has not been extensively evaluated or used with marine organisms. In these studies, the rates and extents of both bioaccumulation and depuration in killifish continuously exposed to low concentrations of BaP, HCB, and PCP were determined under controlled experimental conditions. Low concentrations were used to identify threshold levels and to approximate more closely typical environmental exposures.

1. Hexachlorobenzene

Killifish *(Fundulus similis)* were exposed to HCB for up to 11 days. Periodically during this time, water samples and 12 fish were removed from the exposure aquarium and analyzed for HCB. The fish were analyzed in pooled lots containing four fish each. Concentrations of HCB in water ranged from 0.04 to 0.45 μg/ℓ (ppb) during the studies. The bioaccumulation factors (BF) measured in two experiments are summarized in Table 2. The data were also analyzed by reiterative approximations to the equation $BF = K(1 - e^{-\alpha t})$, where K represents the theoretical maximum (equilibrium) bioaccumulation factor when the time of exposure (t) goes to infinity, and α represents a constant dependent on the rates of accumulation, excretion, and metabolic turnover.[15] The theoretical maximum bioaccumulation factor was 375 according to this analysis.[13] The data of Table 2 and the theoretical curve calculated from the equation above are plotted in Figure 1. Approximately 4.8 hr was required to reach half the maximum BF, indicating that accumulation occurred relatively rapidly. These values were confirmed in a third experiment which gave a maximum BF of 420 and

Table 2
BIOACCUMULATION FACTORS FOR HEXACHLOROBENZENE IN KILLIFISH *(FUNDULUS SIMILIS)*

Exposure (hr)	Experiment 1	Experiment 2
1	ND[b]	65 ± 2[a]
2	68 ± 5	112 ± 6
4	ND	155 ± 20
6	215 ± 28	255 ± 71
8	ND	214 ± 40
10	ND	365 ± 101
12	206 ± 41	282 ± 39
24	236 ± 33	ND
48	ND	607 ± 196
72	128 ± 20	ND
96	216 ± 30	667 ± 47
168	139 ± 47	709 ± 74
264	235 ± 43	ND

Note: The bioaccumulation factor is the ratio of the concentration of HCB in killifish to that in the surrounding water. Fish were exposed to HCB, sacrificed at various times, and analyzed for HCB as described in the text. The concentrations of HCB in the water were determined at corresponding times and ranged from 0.04 to 0.45 μg/ℓ in experiment 1 and from 0.16 to 0.38 μg/ℓ in experiment 2.

[a] Values are averages of 3 pooled samples ± the standard error; ND, not determined.

From Giam, C. S., Murray, H. E., Ray, L. E., and Kira, S., *Bull. Environ. Contam. Toxicol.*, 25, 891, 1980. With permission.

took about 5.8 hr to reach half of that value. These values are substantially less than those reported for a variety of freshwater fish. For example, Veith et al.[16] reported bioaccumulation factors of 5,500 for HCB in rainbow trout *(Salmo gairdneri),* 16,200 in fathead minnow *(Pimephales promelas),* and 21,900 for green sunfish *(Lepomis cyanellus).* Isensee et al.,[7] on the other hand, have reported lower but variable bioaccumulation in catfish *(Ictalurus punctatus).* They reported factors of 160, 680, and 4170, respectively, after exposure to 0.01, 0.09, and 0.61 μg ^{14}C-HCB per gram sediment for 12 days.

After 11 days exposure, fish were removed from the aquarium with HCB and placed into one with fresh Instant Ocean. The HCB concentration in tissue decreased exponentially for 2 days, from about 70 to 1.8 ng/g (Figure 2). After 1 week, the HCB concentration in the depurated fish had leveled off at 0.8 ng/g, the same as measured in control fish. The half-life ($t_{1/2}$) for loss of HCB from the whole body was calculated to be 9.3 hr. In duplicate depuration experiments in this laboratory, exponential loss of the pollutant was also observed, but at a slightly slower rate, i.e., with a half-life of about 12 hr. The $t_{1/2}$ values reported in this study using *Fundulus similis* are considerably shorter than have been reported in other

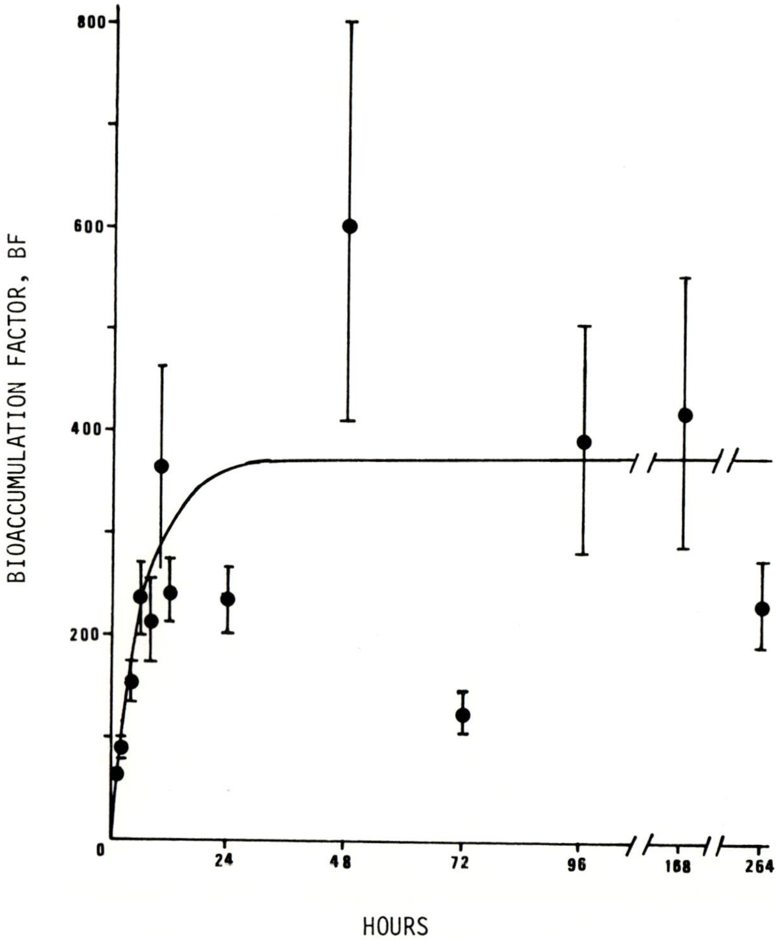

FIGURE 1. Bioaccumulation of hexachlorobenzene by killifish. The curve represents the best fit of the equation $BF = K(1 - e^{-\alpha t})$ to the experimental data (see text). Sample averages for two experiments are plotted with their standard errors. Each point represents the average of three to six pooled samples. (From Giam, C. S., Murray, H. E., Ray, L. E., and Kira, S., *Bull. Environ. Contam. Toxicol.*, 25, 891, 1980. With permission.)

studies. For example, Sanborn et al.[17] reported $t_{1/2}$ values ranging from 8 days in stomach to 19.6 days in liver of green sunfish. Other studies[7,16] have suggested that the bioaccumulation factor and $t_{1/2}$ increase with increasing concentration. Thus, it is possible that the low bioaccumulation factor and $t_{1/2}$ for depuration obtained in this work were related to the low HCB concentrations used for exposure.

This study showed that HCB can be bioaccumulated by marine fish, but to a lesser extent than in freshwater animals. Under the conditions tested, the maximum average bioaccumulation factor for HCB in *F. similis* was calculated as 375. When placed in clean water, *F. similis* rapidly eliminated the accumulated HCB, the concentration in tissue falling exponentially with a half-life of 9 to 12 hr.

2. *Pentachlorophenol*

Killifish *(F. similis)* were experimentally exposed to 57 to 610 μg PCP per liter for 7 to 10 days. Periodically, a water sample and two or three fish were removed and analyzed for PCP. Individual whole fish were analyzed. The average bioaccumulation factors as deter-

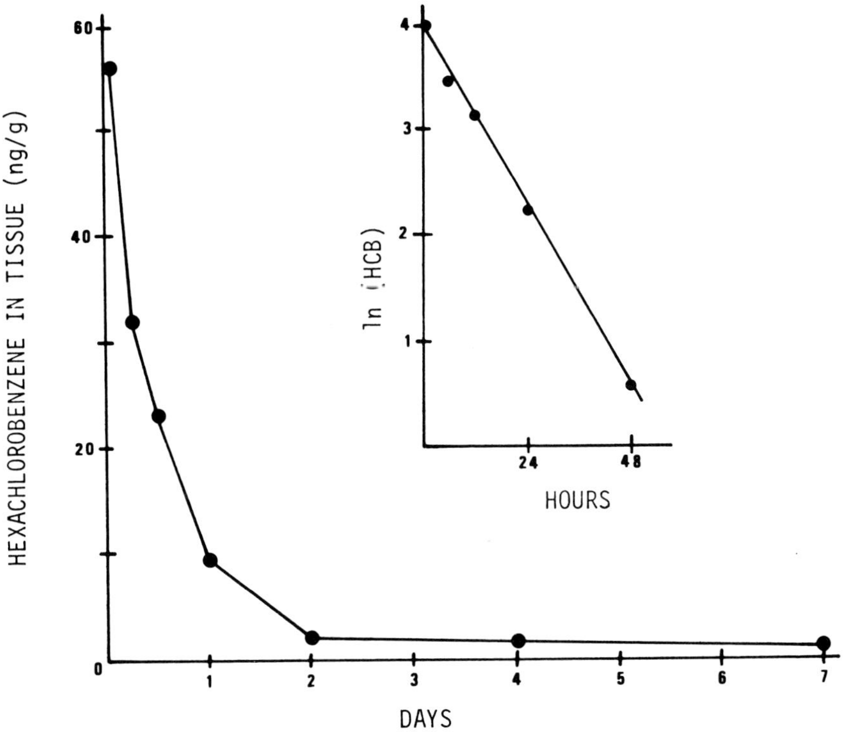

FIGURE 2. Loss of hexachlorobenzene from killifish. Killifish exposed to HCB for 11 days were placed in HCB-free Instant Ocean and sacrificed at the times indicated. The concentrations of HCB in their tissues decreased as shown (inset: semilog plot of the same data). Values are averages of three fish. (From Giam, C. S., Murray, H. E., Ray, L. E., and Kira, S., *Bull. Environ. Contam. Toxicol.*, 25, 891, 1980. With permission.)

mined in two experiments are summarized in Table 3. Analysis of the experimental data by reiterative approximations to the equation $BF = K(1 - e^{-\alpha t})$, as discussed above, gave the theoretical curve shown in Figure 3. By this analysis, the theoretical maximum bioaccumulation factor was 53, and approximately a 25 hr period was required to reach half this value.[18] Although this BF is comparable to those reported for killifish (BF = 30) and striped mullet *(Mugil cephalus;* BF = 38) after 96 hr exposure,[8] all of these values are considerably lower than for freshwater species. Kobayashi and Akitake[19] investigated the absorption of PCP by goldfish *(Carassius auratus)* at 3 concentrations (100, 200, and 400 μg/ℓ) and found that accumulation increased with time. They reported a bioaccumulation factor of 1000 after exposure to 100 μg PCP per liter for 120 hr and a maximum concentration of 116 μg PCP per gram body weight in fish exposed to 200 μg/ℓ. The maximum body concentration of PCP measured in our study was 33 μg/g; the concentration in water was 580 μg/ℓ. Furthermore, the bioaccumulation factors reported therein are relatively constant between 6 and 10 days. In contrast to Kobayashi and Akitake's finding that the concentration in fish increases with time, Pruitt et al.[20] have reported that PCP in bluegills *(Lepomis macrochirus)* was higher after 8 days than it was after 16 days. Based on this and the results of their depuration studies, Pruitt and co-workers[20] suggested that it took about 8 days to induce the enzyme systems necessary for the metabolism of PCP in bluegills. Hattula et al.[21] reported a bioaccumulation factor of 100 for trout *(Salmo trutta)* from Finland. They attributed the low bioaccumulation to possible differences in metabolism since this species normally lives at 5°C. Similarly, the low body burden of PCP (33 μg/g) measured in our work suggests that killifish may be better adapted to eliminate PCP.

Table 3
BIOACCUMULATION FACTORS FOR PENTACHLOROPHENOL IN KILLIFISH *(FUNDULUS SIMILIS)*

Exposure (hr)	Experiment 1	Experiment 2
4	10 ± 0.88[a]	ND[a]
8	19 ± 1.7	ND
12	16 ± 1.2	ND
24	49 ± 2.1	7.9 ± 0.60
48	36 ± 4.2	30 ± 1.8
96	49 ± 1.0	38 ± 6.9
168	64 ± 13	47 ± 4.4
240	ND	50 ± 3.0

Note: The bioaccumulation factor is the ratio of the concentration of PCP in killifish to that in the surrounding water. Killifish were exposed to PCP, sacrificed at times indicated, and analyzed for PCP as described in the text. Water samples were taken for analysis at corresponding times. The concentrations of PCP ranged from 57 to 120 μg/ℓ in experiment 1 and from 100 to 610 μg/ℓ in experiment 2.

[a] Values are averages of 2 or 3 individual fish ± the standard error; ND, not determined.

The kinetics of depuration were determined in fish exposed to PCP for 7 to 10 days and then transferred to aquaria with fresh Instant Ocean. The PCP concentration decreased slowly for several days, reaching a value of about 320 ng/g after 18 days (Figure 4). This corresponds to a half-life of about 4.7 days for the loss of PCP.[18] In contrast, control values in nonexposed fish were about 3 ng/g. The very early time periods were suggestive of rapid loss ($t_{1/2} \sim$ 4.2 hr) of a small amount of PCP; however, additional work is needed to confirm this. The value of 4.7 days agrees relatively well with the data of Holmberg et al.[22] who reported that PCP in the muscle of yellow eel *(Anguilla anguilla* L.) decreased from 9.4 to 3.6 μg/g in 8 days. Assuming exponential loss, this corresponds to $t_{1/2} \sim$ 5.8 days. Depuration is much faster in some species; e.g. in goldfish, a 50% loss of PCP occurs in about 10 hr,[19] while $t_{1/2}$ values of 6.2 to 23 hr have been reported for different tissues of rainbow trout.[23] If the labile species suggested by our work is confirmed, its $t_{1/2}$ would be very similar to those in rainbow trout. Glickman et al.[23] showed a much longer retention of pentachloroanisole (PCA) in rainbow trout ($t_{1/2} \sim$ 6 to 23 days). In conjunction with the long $t_{1/2}$ values we measured, this might indicate some conversion of PCP to PCA and subsequent accumulation by killifish. The methylation could be carried out by bacteria in the exposure system, by flora in the gut of the killifish, or by the killifish. Pierce and Victor[24] have reported the accumulation of PCA by fish following environmental methylation of PCP.

This work has shown that PCP can be rapidly accumulated by marine fish, but to a lesser extent than in freshwater species. The average maximum bioaccumulation factor (BF = 53) for PCP in killifish was reached by about 168 hr and persisted through at least 240 hr. Half

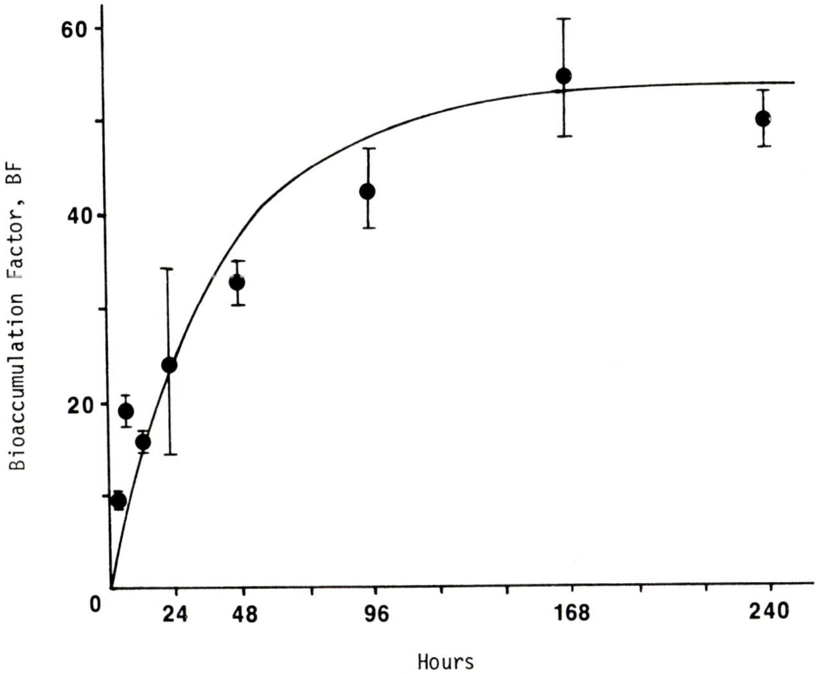

FIGURE 3. Bioaccumulation of pentachlorophenol by killifish. The curve represents the best fit of the equation BF = $(1 - e^{-\alpha t})$ to the experimental data. Sample averages for two experiments are plotted with standard errors. Each point represents the average value for three to six fish. (Reprinted with permission from *Chemosphere,* 11, Trujillo, D. A., Ray, L. E., Murray, H. E., and Giam, C. S., Bioaccumulation of pentachlorophenol by Killifish, Copyright 1982, Pergamon Press, Ltd.)

the plateau value was attained in about 25 hr. When transferred to clean water, exposed fish slowly ($t_{1/2} \sim 4.7$ days) eliminated accumulated PCP. A more labile pool of PCP ($t_{1/2} \sim 4.2$ hr) was also suggested. These data, with those of others, suggest differences in metabolism or in responses of enzyme systems in different organisms exposed to environmental pollutants.

C. Field Studies

1. Portland, Maine

Various organochlorine residues and two phthalate esters were measured in samples of sediment and biota from eight sites in the Portland, Maine area (Figure 5). The concentrations of these pollutants in sediment are listed in Table 4. Total DDT (including its degradation products, e.g., DDE) ranged from <0.03 to 42 ng/g and averaged 8.0 ng/g (Table 4). Total DDT was higher in the Back Cove region (sites 5 to 8) than along the Fore River (sites 1 to 4). This distribution was observed for most of the pollutants measured. Concentrations were lower along the Mississippi Delta and Gulf Coast where the average values were 4.2 and 1.3 ng/g, respectively.[25] Differences were most marked in comparison to the Back Cove.

PCB concentrations varied from 2.1 to 850 ng/g sediment (as Aroclor 1260) and averaged 120 ng/g sediment (Table 4). Again, the average values for the Mississippi Delta (19 ng/g) and Gulf Coast (2.0 ng/g) were lower,[25] particularly in comparison to the Back Cove area. Much higher values (0.1 to 100 μg/g) have been reported for Aroclor 1242 in sediment from the highly industrialized Lower Fox River, Wisconsin.[26]

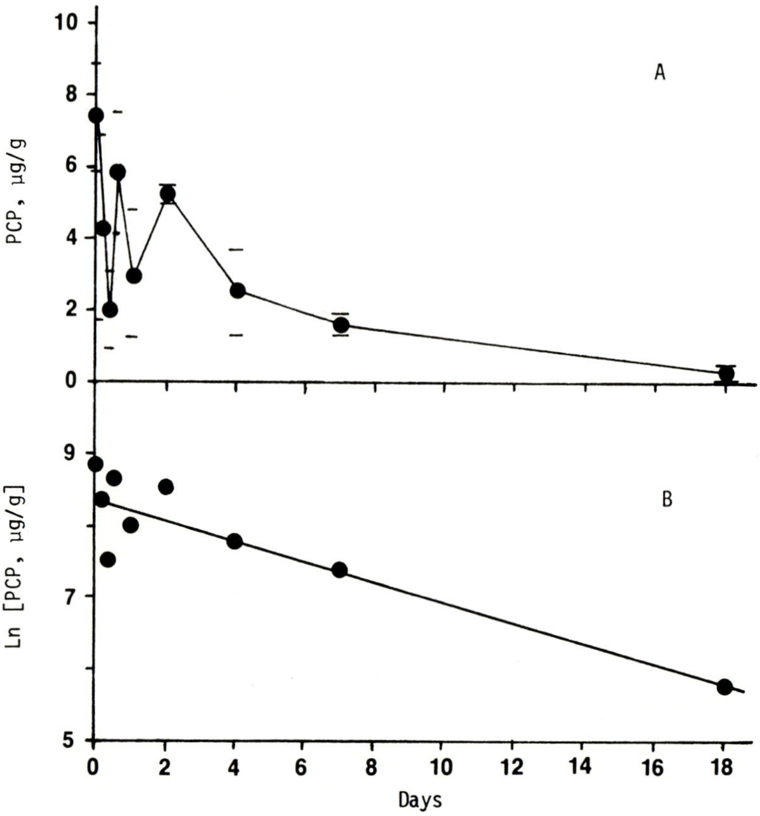

FIGURE 4. (A) Loss of pentachlorophenol from killifish. Killifish exposed to PCP for 7 days were transferred to PCP-free Instant Ocean and sacrificed at the times indicated. The tissue concentrations of PCP decreased as shown. Each point is the average for two fish; standard error limits are indicated. (B) Logarithmic plot of the data shown in (A). (Reprinted with permission from *Chemosphere,* 11, Trujillo, D. A., Ray, L. E., Murray, H. E., and Giam, C. S., Bioaccumulation of pentachlorophenol by Killifish, Copyright 1982, Pergamon Press, Ltd.)

The average value for chlordane (1.8 ng/g) was much higher than for α-hexachlorocyclohexane ([α-HCH], 0.14 ng/g). Both compounds were highest in the Back Cove, particularly at site 5.

Di(n-butyl)phthalate (DBP) varied from 40 to 280 ng/g (Table 4), and was slightly higher along the Fore River than in Back Cove. The average concentration of DBP in these samples (160 ng/g) was greater than that for the Mississippi Delta (13 ng/g) or Gulf Coast (7.6 ng/g).[25] The values for di(2-ethylhexyl)phthalate (DEHP) showed the same trend. Thus, the average values for DEHP were 1500, 69, and 6.6 ng/g for Portland, the Mississippi Delta, and the Gulf Coast, respectively. DEHP concentrations in sediment from Portland ranged from 60 to 7800 ng/g (Table 4). Like most other pollutants, DEHP was higher in Back Cove than along the river.

In these sediment samples, PCP averaged 0.83 ng/g and HCB averaged 0.14 ng/g. These values are lower than reported for other sites.[24,27]

Sediments from site 5 contained the highest concentrations of six of the eight compounds measured in this study. Samples from site 3 contained the lowest concentrations of five of the compounds. Comparison of the average concentrations of each of the compounds in sediment revealed the following trend: DEHP >> DBP > PCBs >> ΣDDT > chlordane

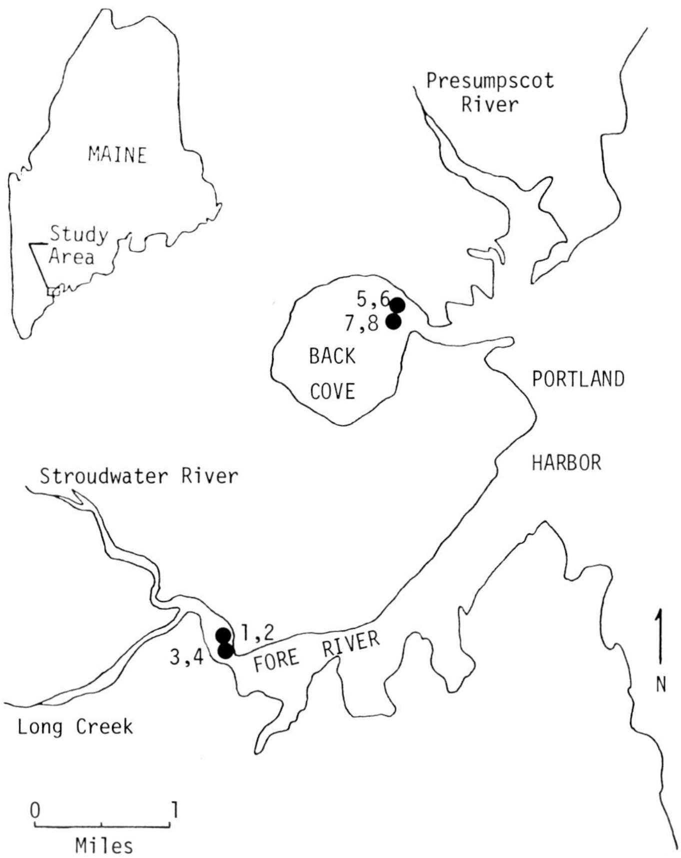

FIGURE 5. Portland, Maine. Samples were collected at the sites indicated. (Reprinted with permission from *Chemosphere*, 12, Ray, L. E., Murray, H. E., and Giam, C. S., Organic pollutants in marine samples from Portland, Maine, Copyright 1983, Pergamon Press, Ltd.)

> PCP > αHCH ~ HCB. Portland was relatively high in DDT, PCBs, DBP, and DEHP, and low in HCB, compared to other regions.

The concentrations of organic pollutants in *Nereis virens* and clams from Portland are summarized in Table 5. Clams from Site 2 were highest in total DDT which ranged from 4.0 to 11 ng/g. Total DDT for all biota averaged 6.2 ng/g, a value lower than reported for many other species. Total DDT averaged 19 ng/g in grouper from the Gulf of Mexico and the Bahamas;[28] 44 ng/g in fish from estuaries of Texas, Mississippi, Alabama, and Florida;[29] 15 ng/g in oysters from Mexican coastal lagoons[30] and 11 ng/g in biota from the Gulf of Mexico.[25] The Portland value is also lower than for species from England and the United States, but similar to or slightly higher than in species from Germany and Mexico.[31] White croaker from southern California were higher in total DDT and PCBs than the animals from Portland.[32]

PCB (as Aroclor 1260) in these samples ranged from 20 to 79 ng/g; again, clams from Site 2 were highest. In comparison, PCB averaged 44 ng/g in our studies (Table 5[33]); 33 ng/g in grouper from the Gulf of Mexico and the Bahamas;[28] 203 ng/g in fish from the Gulf of Mexico and along the Mississippi Delta;[34] 55 ng/g in oysters from Mexican coastal lagoons;[30] and 30 ng/g in other biota from the Gulf of Mexico.[25] Lobsters from four sites

Table 4
CONCENTRATIONS OF SELECTED ORGANIC POLLUTANTS IN SEDIMENT FROM PORTLAND, MAINE

Site	ΣDDT[a]	PCB	αHCH	CHLO	DBP	DEHP	PCP	HCB
1	11	32	0.19	0.52	120	320	2.4	0.09
2	5.1	19	0.03	3.5	170	740	1.7	0.37
3	<0.03	2.1	<0.03	<0.03	200	60	0.02	0.22
4	0.23	6.2	<0.07	<0.03	270	770	0.51	<0.03
5	42	850	0.70	9.8	280	7800	0.03	0.14
6	1.9	3.3	0.06	0.18	150	840	0.45	<0.03
7	1.8	6.1	<0.03	0.37	40	80	0.01	0.05
8	2.2	7.5	<0.03	0.22	50	1400	1.5	0.17
Average								
Sites 1—8	8.0	120	0.14	1.8	160	1500	0.83	0.14
Sites 1—4	4.1	15	0.08	1.0	190	470	1.2	0.18
Sites 5—8	12	220	0.21	2.6	130	2530	0.50	0.10

Note: Concentrations in ng/g; sample size, 50 g. Samples were prepared and analyzed as described in the text.

[a] Abbreviations: ΣDDT, various isomers of DDT and DDE; PCB, polychlorinated biphenyls (reported as Aroclor 1260); αHCH, α-hexachlorocyclohexane; CHLO, chlordane; DBP, di(*n*-butyl)phthalate; DEHP, di(2-ethylhexyl)phthalate; PCP, pentachlorophenol; HCB, hexachlorobenzene.

Table 5
CONCENTRATIONS OF SELECTED ORGANIC POLLUTANTS IN BIOTA FROM PORTLAND, MAINE

Species	Site	ΣDDT[a]	PCB	αHCH	CHLO	DBP	DEHP	PCP	HCB
Nereis virens	2	5.2	20	0.37	2.0	180	490	6.2	0.15
	5	4.5	50	0.65	5.4	70	380	4.5	0.26
Clams	2	11	79	0.23	<0.03	100	170	1.7	<0.10
	5	4.0	25	0.22	1.8	40	110	3.4	0.64
Average		6.2	44	0.37	3.1	100	290	4.0	0.35

Note: Concentrations in ng/g; average sample weight, 30 g. Samples were prepared and analyzed as described in the text.

[a] Abbreviations: ΣDDT, various isomers of DDT and DDE; PCB, polychlorinated biphenyls (reported as Aroclor 1260); αHCH, α-hexachlorocyclohexane; CHLO, chlordane; DBP, di(*n*-butyl)phthalate; DEHP, di(2-ethylhexyl)phthalate; PCP, pentachlorophenol; HCB, hexachlorobenzene.

in the New York Bight averaged 78 to 233 ng/g.[35] The total Aroclor content (range, 0.35 to 6.93 μg/g; average, 2.86 μg/g) of *M. mercenaria* from New Bedford Harbor, Massachusetts, was much higher than any of these values.[36] Aroclor 1242 in fish from the Lower Fox River was even higher (<0.2 to 90 μg/g).[26]

α-Hexachlorocyclohexane and chlordane ranged from 0.22 to 0.65 ng/g and from <0.03 to 5.4 ng/g, respectively. They were highest in *N. virens* from Site 5. Data on these compounds in the marine environment are relatively sparse, however, chlordane was undetectable in oysters from Mexican coastal lagoons.[30] The average concentration (0.37 ng/g) of αHCH at Portland was lower than for a variety of animals worldwide.[31]

DBP ranged from 40 to 180 ng/g and was highest in *N. virens* at Site 2. DEHP ranged from 110 to 490 ng/g and was also highest in *N. virens* at Site 2. For comparison, the following DEHP concentrations have been reported for other animals: 135 ng/g in starfish *(Luidia clathrata);* <1.0 ng/g in croaker *(Micropogon undulatus),* catfish *(Galeichthys felis),* and others; 9 ng/g in sand trout *(Cynoscion nothus),* and 20 ng/g in spade fish *(Chaetodipterus faber)* and blue crab *(Callinectes sapidus)* from the Gulf of Mexico.[37]

PCP ranged from 1.7 to 6.2 ng/g with the highest concentration found in *N. virens* at Site 2. These values are somewhat higher than found in jellyfish (0.1 to 1.0 ng/g) from the Gulf of Mexico[38] or in fish (<0.5 to 4.0 ng/g) from St Croix and St. John's estuaries in Canada.[39] However, they compare closely with values (3.4 to 8.3 ng/g) reported for oysters *(Crassostrea virginica)* from Galveston Bay, Texas.[12]

HCB varied from <0.10 to 0.64 ng/g (average, 0.35 ng/g) in the organisms examined. Clams from Site 5 were highest in HCB. In oysters from Galveston Bay, Texas, HCB ranged from 0.31 to 1.4 ng/g and averaged 0.63 ng/g.[12] Mosquitofish from the Mississippi River between Baton Rouge and New Orleans had HCB concentrations in the range of 72 to 380 ng/g, while crayfish from the same area contained 22 to 194 ng/g.[27] Fish from Lake Ontario were also found to have higher amounts of HCB (25 to 100 ng/g)[40] than the organisms from this study. Goldberg[41] has reported a wide range (1 to 560 ng/g) of HCB in marine fish.

Based on average values, the relative concentrations for these compounds in biota were DEHP > DBP > PCBs > ΣDDT > PCP > chlordane >αHCH ~ HCB. Except for the reversal of PCP and chlordane, this is the same order as observed in sediment. This supports our previous suggestion that, in an organism dwelling in sediment, the order of pollutant concentrations would be the same as in the sediment, whereas in an animal principally exposed through water, the order of pollutant concentrations would be the same as the water solubilities.[42] Generally, the biota samples from Portland were low in DDT and HCB and high in DEHP and PCP when compared to other regions. Biota at Site 2 were highest in five of the eight pollutants.

Although most industrial and petroleum-shipping activity is along the Fore River (near Sites 1 through 4), the most polluted sediment samples were those obtained from Back Cove (Sites 5 through 8). This suggests that the flushing action provided by the Fore River and its tributaries is important, and relatively efficient, in washing the pollutants out before they are sorbed onto the sediment. In the relatively poorly flushed Back Cove, the sediment seems more important for sorption of pollutants. The differences in pollutants in the biota were less marked (i.e., five of the most contaminated species were found at Site 2, whereas, three were at Site 5), making it difficult to draw any firm conclusions. It is certainly clear, however, that the flushing action of the river is not enough to protect the biota at Site 2 from concentrating the pollutants.

When the concentrations of pollutants in biota are compared to those in the sediment (i.e., when bioconcentration factors are calculated; Table 6), it is observed that five of the compounds are concentrated. Thus, for ΣDDTs, PCB, and HCH at Site 2, the average ratios are 1.6, 2.7, and 9.9, respectively. For PCP and HCB at Site 5, the average ratios are 130 and 3.3, respectively. This indicates that these compounds are preferentially partitioned into the tissues from the sediment. The fact that these ratios are very low compared to the usual bioaccumulation factors of a few hundred to several thousand when measured using water solutions of the pollutants suggests that sorption onto sediment serves as an important protective mechanism. Studies by Rubinstein et al.[43] with various species support this notion and suggest that a high organic content in the sediment offers additional protection. This protection is even more complete in the case of the other pollutants (chlordane, DBP, and

Table 6
BIOCONCENTRATION OF POLLUTANTS IN BIOTA FROM PORTLAND, MAINE

Species	Site	ΣDDT[a]	PCB	αHCH	CHLO	DBP	DEHP	PCP	HCB
Nereis virens	2	1.0	1.1	12	0.57	1.1	0.66	3.6	0.41
	5	0.11	0.06	0.93	0.55	0.25	0.05	150	1.9
Clams	2	2.2	4.2	7.7	0.01	0.59	0.23	1.0	0.27
	5	0.10	0.03	0.31	0.18	0.14	0.01	110	4.6
Average									
Site 2		1.6	2.7	9.9	0.29	0.85	0.45	2.3	0.34
Site 5		0.11	0.05	0.62	0.37	0.20	0.03	130	3.3

Note: Bioconcentration factors were calculated by dividing the tissue concentration of each compound by the concentration of the same compound in sediment at the appropriate site.

[a] Abbreviations: ΣDDT, various isomers of DDT and DDE; PCB, polychlorinated biphenyls (reported as Aroclor 1260); αHCH, α-hexachlorocyclohexane; CHLO, chlordane; DBP di(*n*-butyl)phthalate; DEHP, di(2-ethylhexyl)phthalate; PCP, pentachlorophenol; HCB, hexachlorobenzene.

Reprinted with permission from *Chemosphere,* 12, Ray, L. E., Murray, H. E., and Giam, C. S., Organic pollutants in marine samples from Portland, Maine, Copyright 1983, Pergamon Press, Ltd.

DEHP) where the tissue concentrations were less than those of the sediment. Neither species was markedly superior to the other as an indicator of pollution, since one species had higher bioconcentration factors half of the time, the other species, the other half.

2. San Luis Pass, Texas

The concentrations of HCB, BaP, PCP, PAEs, and PCBs in sediment from San Luis Pass, located at the west end of Galveston Island, Texas, are listed in Table 7. Phthalates, which averaged 159 ng/g in total, were by far the most prevalent pollutant. This is substantially higher than measured in sediment from the Mississippi Delta (82 ng/g) or the Gulf Coast (14 ng/g).[25] The most abundant phthalate in these samples was DEHP, which ranged from 13 to 170 ng/g and averaged 94 ng/g. This is in accord with its relatively high production compared to other phthalates,[44] but is much lower than measured in the Portland, Maine, region (Table 4). The other two phthalates, DBP and DEP, ranged from undetectable to 93 and 9 ng/g, respectively.

Second to the phthalates, the most prevelant compound measured in these sediments was BaP, which ranged from 0.01 to 6 ng/g. This is comparable to values (0.3 to 3.2 ng/g) in sediment samples from the outer Atlantic Shelf[45] and those (0.5 to 2.4 ng/g) in various lake and river samples.[46] Samples from Casco Bay, Maine, were much higher (average, 110 ng/g)[47] than at San Luis Pass.

Despite its environment stability, PCB concentrations were rather low (0.25 to 0.78 ng/g; average, 0.52 ng/g). This may reflect its restricted use and more carefully controlled disposal.

Earlier studies showed higher average concentrations in samples from the Mississippi Delta (19 ng/g) and Gulf Coast (2.0 ng/g), but lower in open waters of the Gulf (0.2 ng/g).[25] In contrast, contemporary samples from Portland, Maine, were much higher (Table 4) as were *M. mercenaria* (0.35 to 6.83 μg/g) from New Bedford Harbor, Massachusetts.[36]

HCB concentrations, which ranged from 0.05 to 1.5 ng/g (average, 0.49 ng/g), were only slightly lower than those for PCB. This is lower than reported for sediment from the lower Mississippi River,[27] but higher than for those from Portland (Table 4).

PCP was present in the lowest amounts (0.18 to 0.26 ng/g). This is in accord with its

Table 7
CONCENTRATIONS OF ORGANIC POLLUTANTS IN SEDIMENT FROM SAN LUIS PASS, TEXAS

Compound	Site 1	Site 2	Site 3	Average[a]
HCB	1.5	0.05	0.29	0.49
BaP	6.0	0.01	1.8	2.2
PCP	0.23	0.18	0.26	0.22
DBP[b]	87	<15	93	60
DEHP[b]	98	13	170	94
DEP[b]	9	<2	7	5
PCB[b]	0.78	0.25	0.54	0.52

Note: Concentrations (ng/g dry wt) are the average of duplicate samples; 50-g aliquots were analyzed as described in the text. Values are corrected for sample blanks. Dieldrin, chlordane (*cis* and *trans*), DDT, and DDE were undetectable (i.e., <0.03 ng/g) in these samples.

[a] Average of all values for all sites.
[b] DBP, di(*n*-butyl)phthalate; DEHP, di(2-ethylhexyl)phthalate; DEP, diethylphthalate; PCB, polychlorinated biphenyls (reported as Aroclor 1260).

Reprinted with permission from *Chemosphere*, 10, Murray, H. E., Ray, L. E., and Giam, C. S., Analysis of marine sediment, water and biota for selected organic pollutants, Copyright 1981, Pergamon Press, Ltd.

relatively high water solubility and facile degradation compared to other pollutants. The concentrations in these samples were somewhat lower than in samples from Portland (Table 4) and much lower than in some lake sediment samples (3.3 ng/g).[24]

Water samples from San Luis Pass did not contain detectable amounts (≤1 ng/ℓ) of BaP or HCB. PCP in water from three sites at San Luis Pass varied from 4.3 to 11 ng/ℓ (average, 8.2 ng/ℓ).

The concentrations of BaP, HCB, and PCP in various marine animals from San Luis Pass are shown in Table 8. BaP was undetectable (<0.01 ng/g) in these samples. Similarly, Heit et al.[48] were unable to detect BaP in mussels from Lake George, New York, but the limit of detection in their study was 10 ng/g. Other studies have found much higher concentrations of BaP (e.g., 0.12 ng/g in oysters from Galveston Bay;[12] 10 to 20 ng/g in oysters from Norfolk, Virginia;[49] 0.4 to 11 ng/g in mussels from Yaquina Bay, Oregon;[50] 0.1 to 32.5 ng/g in softshell clams from Coos Bay, Oregon;[51] and 0.6 to 49 ng/g in mussels from Vancouver Harbor[52]).

HCB concentrations varied from 0.29 to 14 ng/g in individual animals. The highest concentrations were found in blue crabs *(C. sapidus)*. This is higher than reported (0.31 to 1.4 ng/g) for oysters from Galveston Bay,[12] but much lower than reported for many other species. Thus, mosquitofish from the Mississippi River between Baton Rouge and New Orleans contained 72 to 380 ng/g,[27] crayfish from the same region contained 22 to 190 ng/g,[27] and fish from Lake Ontario contained 25 to 100 ng/g.[40] A wide range of HCB concentrations (1 to 560 ng/g) has been reported in marine fish.[41] There was a marked difference (> 10-fold) between the average HCB concentration (9.6 ng/g) in blue crabs and those (0.49 to 0.88 ng/g) for the other species in this study. This suggests that the benthic detritus contains more HCB than the water column, consistent with the relatively low

Table 8
CONCENTRATIONS OF BaP, HCB, AND PCP IN ANIMALS FROM SAN LUIS PASS, TEXAS

Species	Weight (g)	BaP	HCB	PCP
Blue crab	4.7	<0.01	14	1.9
(Callinectes sapidus)	2.6	<0.01	—	2.4
	6.5	<0.01	5.6	4.1
Average	—	<0.01	9.6	2.8
Brown shrimp	2.5	<0.01	1.1	17
(Penaeus aztecus)	2.3	<0.01	0.65	11
	5.4	<0.01	0.93	4
Average	—	<0.01	0.88	11
Dwarf squid	8.8	<0.01	0.58	4.3
(Lollingnucula brevis)	4.2	<0.01	0.83	3.0
	6.9	<0.01	0.70	1.4
Average	—	<0.01	0.71	2.9
Longnose killifish	5.1	<0.01	1.3	5.5
(Fundulus similis)	4.1	<0.01	0.35	5.6
	4.2	<0.01	0.29	4.7
Average	—	<0.01	0.65	5.3
Flounder	4.6	<0.01	0.35	2.7
(Unidentified	12.6	<0.01	0.32	1.6
species)	2.3	<0.01	0.80	3.5
Average	—	<0.01	0.49	2.6

Note: Concentrations (ng/g wet wt.) have been corrected for blanks. Individual animals were prepared and analyzed as discussed in the text.

Reprinted with permission from *Chemosphere,* 10, Murray, H. E., Ray, L. E., and Giam, C. S., Analysis of marine sediment, water and biota for selected organic pollutants, Copyright 1981, Pergamon Press, Ltd.

solubility of HCB in water. Thus, crabs, which are bottom-dwelling detritovores, ingest and accumulate greater amounts of HCB.

The concentrations of PCP varied from 1.4 to 17 ng/g in individual animals and was highest in brown shrimp *(Penaeus aztecus)* which averaged 7.5 ng/g (Table 8). This is higher than in oysters (average, 5.3 ng/g) from Galveston Bay,[12] jellyfish (0.1 to 1.0 ng/g) from the Gulf of Mexico,[38] and fish (<0.5 to 4.0 ng/g) from Canadian estuaries.[39] It appears that organisms from Galveston Bay may be higher in PCP than those from other locations. Consistent with its high water solubility, differences in PCP between species were fairly small in this study. Two of the shrimp contained much higher concentrations of PCP than the other animals, suggesting that they might have been from a population recently exposed to PCP.

In all animals except the blue crab, the relative concentrations were PCP > HCB > BaP. In sediment from Site 1, where the animals were collected, the relative concentrations were just the reverse, i.e., BaP > HCB > PCP. These animals contained relatively high concentrations of PCP, which is the most water soluble of these compounds. Thus, in animals exposed to pollutants mainly via water rather than sediment, body burden (and hence exposure history) seems related to the pollutant's solubility in water, not its concentration on sediment.[42] Sorption of pollutants onto sediment may thus decrease the exposure of animals that do not ingest sediment. Again, no single species in these studies seemed unambiguously better than the others as an indicator species to monitor marine pollution.

3. Corpus Christi Bay, Texas

The concentrations of HCB, PCP, PCBs, PAEs, and various chlorinated hydrocarbon pesticides (CHP) were measured in water and sediment from several sites in the Nueces Estuary (Corpus Christi Bay system), Texas (Figure 6).[53] This system is the location of many industrial plants, including oil refineries, petrochemical plants, and power-generating stations, which are potential sources of pollution. Industry is particularly abundant along the Tule Lake Channel and around Corpus Christi Bay. A general description of each site is included in Table 9 which summarizes the pollutant concentrations in water samples from the region.

The concentrations of HCB ranged from <0.01 to 0.61 ng/ℓ, with an average value of 0.24 ng/ℓ. The highest concentration occurred at the sewage outfall (Site 3). HCB concentrations (<700 to 2200 ng/ℓ)[27] reported for the Mississippi River are greater by a factor of 1000 or more.

The concentrations of PCP varied from 1.2 to 68 ng/ℓ with an average of 22 ng/ℓ (Table 9). These values are higher than those measured at San Luis Pass (see previous section) which is more remote from industrial activity. Pierce and Victor[24] reported a value (280 ng/ℓ) far higher than either the average or maximum in the Corpus Christi Bay region.

PCBs (as Aroclor 1260) averaged 4.7 ng/ℓ with a range of 1.6 to 9.8 ng/ℓ; Site 4 was highest. These values are somewhat higher than those reported for the Mississippi Delta (1.7 to 3.3 ng/ℓ; average, 2.5 ng/ℓ) or Gulf Coast (0.1 to 3.1 ng/ℓ; average, 1.6 ng/ℓ).[25]

As reported for Portland and San Luis Pass in the sections above, PAEs were the most abundant pollutant measured in these studies. Total PAEs averaged 480 ng/ℓ with a range of 290 to 880 ng/ℓ. DEHP, which averaged 350 ng/ℓ (range 210 to 770 ng/ℓ), was more prevalent than DBP which averaged 130 ng/ℓ (range 70 to 230 ng/ℓ; Table 9). These values are higher than those reported for the Mississippi Delta, where DEHP averaged 70 ng/ℓ (range, 23 to 230 ng/ℓ) and DBP averaged 95 ng/ℓ (range, 6.5 to 470 ng/ℓ),[25] or for other portions of the Gulf Coast, where DEHP averaged 130 ng/ℓ (range, 6 to 320 ng/ℓ) and DBP averaged 74 ng/ℓ (range, 3 to 270 ng/ℓ).[25] Site 2 was highest in DEHP, whereas Site 6 was highest in DBP. Interestingly, two of the lowest total PAE values were measured very near industries (sites 7 and 8).

Total DDT in the Corpus Christi Bay area ranged from 0.21 to 3.1 ng/ℓ with an average of 1.3 ng/ℓ (Table 9). In comparison, values along the Mississippi Delta ranged from 0.01 to 2.9 ng/ℓ (average 1.7 ng/ℓ); along the Gulf Coast they ranged from 0.01 to 0.6 ng/ℓ (average, 0.35 ng/ℓ).[25] Total CHPs ranged from 0.40 to 3.8 ng/ℓ with an average of 1.9 ng/ℓ (Table 9). The highest concentrations (3.8 ng/ℓ) was found at the sewage outfall (Site 3) and consisted of total DDT, 2.5 ng/ℓ; αHCH, 0.16 ng/ℓ; lindane, 0.11 ng/ℓ; chlordane, 0.93 ng/ℓ, and dieldrin, 0.09 ng/ℓ.

The concentrations of these pollutants in sediment from the Corpus Christi region are given in Table 10. HCB concentrations ranged from <0.01 to 0.73 ng/g with an average of 0.11 ng/g; the highest concentration was found near industry (Site 8). These values are lower than in sediments along the lower Mississippi River where Laska et al.[27] detected <0.7 to nearly 900 ng HCB per gram sediment in samples from levees and ditches. Although the range is greater, the average HCB in the Corpus Christi region is similar to the average in Portland (Table 4), while both range and average are lower than measured at San Luis Pass (Table 7).

PCP in these samples ranged from 0.05 to 1.1 ng/g with an average of 0.29 ng/g (Table 10). This average is slightly higher than that (0.22 ng/g) for San Luis Pass (Table 7), but lower than that (0.83 ng/g) for Portland (Table 4) or that (3.3 ng/g) reported for sediment from a pond.[24] Sediment from site 3, near the sewage outfall, was highest in PCP.

PCB (as Aroclor 1260) concentrations in the Corpus Christi region varied from 0.11 to 31 ng/g and averaged 4.7 ng/g. This range is comparable to that (0.2 to 35 ng/g) reported

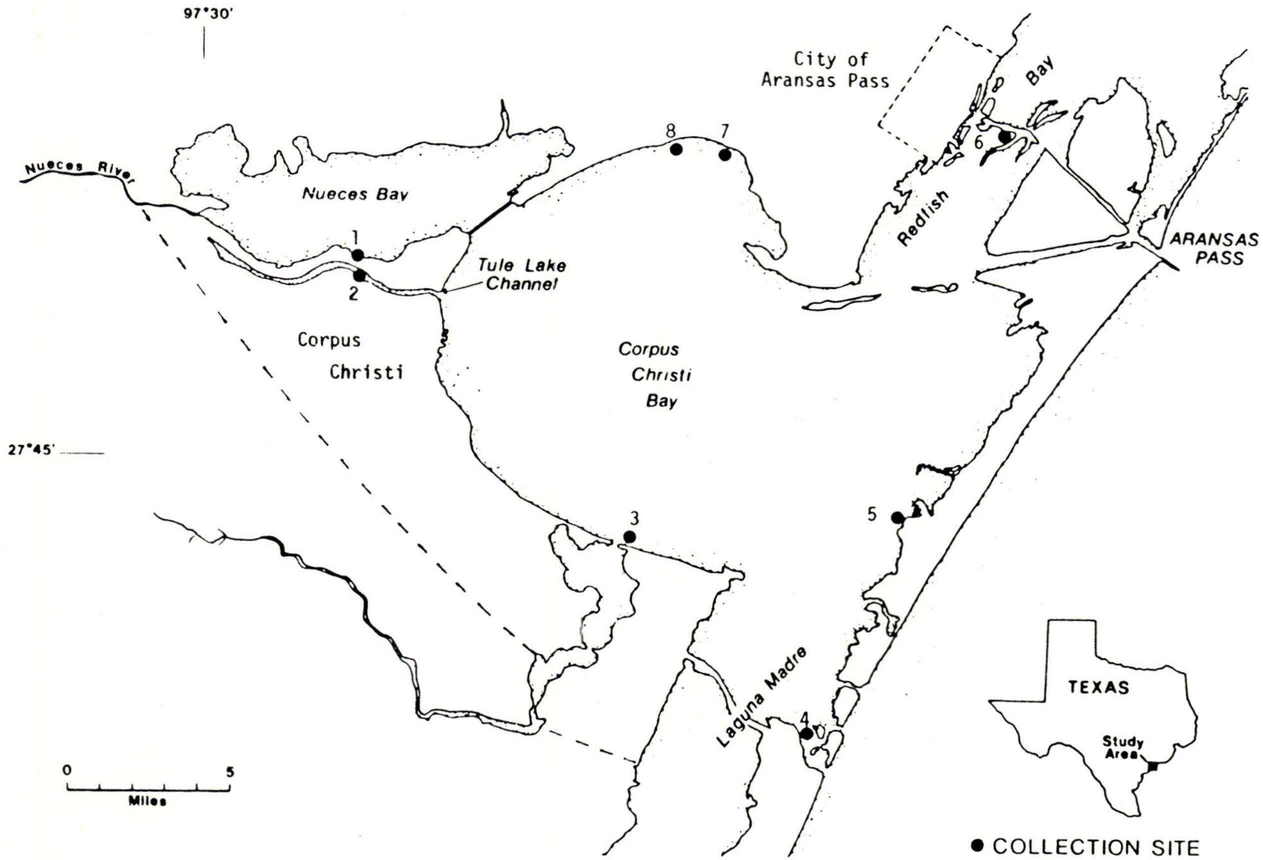

FIGURE 6. The Nueces Estuary/Corpus Christi Bay system. Samples were collected at the sites indicated. (Reprinted with permission from *Chemosphere*, 12, Ray, L. E., Murray, H. E., and Giam, C. S., Analysis of water and sediment from the Nueces Estuary/Corpus Christi Bay (Texas) for selected organic pollutants, Copyright 1983, Pergamon Press, Ltd.)

Table 9
CONCENTRATIONS OF SELECTED POLLUTANTS IN WATER SAMPLES FROM THE NUECES ESTUARY, TEXAS

Site	Description	HCB[a]	PCP	PCB	DBP	DEHP	ΣDDT	αHCH	LIND	CHLO	DIEL
1	Power plant	0.35	4.4	5.1	120	310	1.1	0.49	0.07	0.04	0.05
2	Refinery	<0.01	4.6	3.1	110	770	0.40	0.30	0.13	0.25	0.13
3	Sewage outfall	0.61	68	5.4	150	390	2.5	0.16	0.11	0.93	0.09
4	Residential	0.57	7.8	9.8	140	330	3.1	0.30	<0.01	0.07	0.01
5	Small channel to Gulf	<0.01	1.2	7.2	90	210	1.6	<0.05	<0.01	0.05	0.01
6	Tidal marsh	0.35	1.4	2.8	230	370	0.21	0.06	<0.01	0.03	0.03
7	Industrial	<0.01	43	1.6	70	220	0.45	0.37	0.10	0.04	0.03
8	Industrial	<0.01	42	2.9	90	220	1.2	0.41	0.11	0.03	<0.01
Average		0.24	22	4.7	130	350	1.3	0.26	0.07	0.18	0.03

Note: Concentrations are in ng/ℓ; 2-ℓ samples were analyzed. Values have been corrected for blanks.

[a] Abbreviations: HCB, hexachlorobenzene; PCP, pentachlorophenol; PCB, polychlorinated biphenyls (reported as Aroclor 1260); DBP, di(*n*-butyl)phthalate, DEHP, di(2-ethylhexyl)phthalate; ΣDDT, various isomers cf DDT and DDE; αHCH, α-hexachlorocyclohexane; LIND, lindane; CHLO, chlordane; DIEL, dieldrin.

Table 10
CONCENTRATIONS OF SELECTED ORGANIC POLLUTANTS IN SEDIMENT FROM THE NUECES ESTUARY, TEXAS

Site	HCB[a]	PCP	PCB	DBP	DEHP	ΣDDT	αHCH	LIND	CHLO	DIEL
1	<0.01	0.25	0.52	31	40	0.12	<0.06	<0.01	0.04	<0.01
2	0.01	0.10	1.2	6.7	120	0.13	<0.06	<0.01	0.03	0.03
3	<0.01	1.1	31	28	330	9.2	<0.06	0.16	3.8	0.94
4	0.06	0.34	0.73	36	54	0.66	<0.06	<0.01	2.1	<0.01
5	0.06	0.25	0.50	27	110	0.44	<0.06	<0.01	0.07	<0.01
6	0.02	0.05	0.11	4.2	97	0.03	<0.06	<0.01	<0.01	<0.01
7	0.02	0.12	0.41	13	16,000	0.69	<0.06	<0.01	0.07	0.06
8	0.73	0.12	3.7	12	3,000	0.36	<0.06	0.01	0.01	0.06
Average	0.11	0.29	4.7	20	2,500	1.5	<0.06	0.03	0.77	0.14

Note: Concentrations are in ng/g dry wt and have been corrected for blanks; 50-g aliquots were analyzed.

[a] Abbreviations: HCB, hexachlorobenzene; PCP, pentachlorophenol; PCB, polychlorinated biphenyls (reported as Aroclor 1260); DBP, di(*n*-butyl)phthalate; DEHP, di(2-ethylhexyl)phthalate; ΣDDT, various isomers of DDT and DDE; αHCH, α-hexachlorocyclohexane; LIND, lindane; CHLO, chlordane; DIEL, dieldrin.

Reprinted with permission from *Chemosphere*, 12, Ray, L. E., Murray, H. E., and Giam, C. S., Analysis of water and sediment from the Nueces Estuary/Corpus Christi Bay (Texas) for selected organic pollutants, Copyright 1983, Pergamon Press, Ltd.

for the Mississippi Delta,[25] but less than that (2.1 to 850 ng/g) for the Portland, Maine region (Table 4). Much higher values (up to 100 μg/g) have been measured in sediment from the Lower Fox River.[26] However, the range at Corpus Christi is higher than that (0.2 to 6 ng/g) for the Gulf Coast[25] and that (0.25 to 0.78) for San Luis Pass (Table 7).

Total PAEs in these sediments ranged from 71 to 16,000 ng/g with an average of 2500 ng/g (Table 10). This average is highly skewed due to Site 7, and falls to 540 ng/g if this sample is omitted. Both of these values are substantially higher than reported for the Mississippi Delta (82 ng/g),[25] the Gulf Coast (14 ng/g),[25] or San Luis Pass (159 ng/g; Table 7). Sites 7 and 8 are higher than any tested at Portland, Maine, but the other sites at Corpus Christi are lower than, to about the same as, those at Portland (Table 4). DEHP was the most abundant phthalate at all sites in the Corpus Christi region. This was also true for the water samples (Table 9), although the differences were generally greater for the sediment samples.

Total DDT concentrations (range, 0.03 to 9.2 ng/g; average, 1.5 ng/g) are similar to those reported for sediments along the Mississippi Delta (range, 0.2 to 9.3 ng/g; average, 4.2 ng/g) and the Gulf Coast (range, 0.2 to 4.0 ng/g; average, 1.3 ng/g).[25] The values for Corpus Christi are lower than those (range, <0.03 to 42 ng/g; average 8.0 ng/g) for Portland, Maine (Table 4). Total CHPs ranged from 0.09 to 14 ng/g with an average of 2.4 ng/g (Table 10). The highest concentration (14 ng/g) was detected at site 3 and included total DDT, 9.2 ng/g; lindane, 0.16 ng/g; chlordane, 3.8 ng/g; and dieldrin, 0.94 ng/g.

In comparing the different pollutant concentrations, this study revealed that total PAEs > PCP > PCB > total CHPs > HCB in water and total PAEs ≫ PCB > total CHPs > PCP > HCB in sediment. Except for a shift in PCP, the relative ranking is the same in water and sediment. In water, the ranking of concentrations corresponds closely to the water solubilities of the compounds, although the measured values are well below (one thousandth) the maximum solubility of each compound. It has been reported that the sorption of PAEs is proportional to chain length;[54] our observation that the concentration of DEHP in sediment is greater than that of DBP supports this suggestion. The fact that DEHP exceeds DBP in the water samples, even though DBP is more soluble than DEHP, probably reflects the greater amount of DEHP in the sediment; the presence of suspended material may also contribute to this difference.

No single site in the Nueces/Corpus Christi Bay system was highest or lowest in regard to all the pollutants measured. However, Site 3, the city sewage outfall, was the highest in three of the five pollutants in water and four of the five pollutants in sediment. Site 7, near industry, was the lowest in regard to three of the five pollutants in water, and Site 6, the tidal marsh, was the lowest in regard to four of the five pollutants in sediment.

ACKNOWLEDGMENTS

The authors wish to thank the following individuals who participated in various phases of the work reported here: Drs. S. Kira and H. E. Murray; Y. Hrung, M. L. Diaz-Madero, D. Allen, D. A. Trujillo, and D. Sullivan. Thanks are also extended to Drs. E. Atlas and J. E. Leonard for helpful discussions.

This work was supported in part by grants OCE77-25663, OCE78-02023, and OCE80-19601 from the National Science Foundation, and was completed by the authors at the Department of Chemistry, Texas A & M University, College Station.

REFERENCES

1. **Giam, C. S. and Chan, H. S.,** Control of blanks in the analysis of phthalates in air and ocean biota samples, *Natl. Bur. Stand. Spec. Publ.*, 422, 701, 1976.
2. **Giam, C. S., Chan, H. S., and Neff, G. S.,** Problems in analyses of organic pollutants in open ocean samples, *Proc. Int. Conf. Environ. Sensing Assess*, 2, 21-1-1, 1975.
3. **Giam, C. S. and Wong, M. K.,** Problems of background contamination in the analysis of open ocean biota for chlorinated hydrocarbons, *J. Chromatogr.*, 72, 283, 1972.
4. **Edgerton, T. R., Moseman, R. F., Linder, R. E., and Wright, L. H.,** Multi-residue method for the determination of chlorinated phenol metabolites in urine, *J. Chromatogr.*, 170, 331, 1979.
5. **Giam, C. S., Trujillo, D. A., Kira, S., and Hrung, Y.,** Simplified monitoring procedures for benzo(a)pyrene, hexachlorobenzene and pentachlorophenol in water, *Bull. Environ. Contam. Toxicol.*, 25, 824, 1980.
6. **Guerrero, H., Biehl, E. R., and Kenner, C. T.,** High-pressure liquid chromatography of benzo(a)pyrene and benzo(ghi)perylene in oil-contaminated shellfish, *J. Assoc. Off. Anal. Chem.*, 59, 989, 1976.
7. **Isensee, A. R., Holden, E. R., Woolson, E. A., and Jones, G. E.,** Soil persistence and aquatic bioaccumulation potential of hexachlorobenzene (HCB), *J. Agric. Food Chem.*, 24, 1210, 1976.
8. **Schimmel, S. C., Patrick, J. M., Jr., and Faas, L. F.,** Effects of sodium pentachlorophenate on several estuarine animals: toxicity, uptake, and depuration, in *Pentachlorophenol: Chemistry, Pharmacology and Environmental Toxicology,* Rao, K. R., Ed., Plenum Press, New York, 1978, 147.
9. **Giam, C. S.,** Trace analyses of phthalates and chlorinated hydrocarbons in marine samples, in *Strategies for Marine Pollution Monitoring,* Goldberg, E. D., Ed., Wiley-Interscience, New York, 1976, 61.
10. **Giam, C. S., Chan, H. S., and Neff, G. S.,** Sensitive method for determination of phthalate ester plasticizers in open ocean biota samples, *Anal. Chem.*, 47, 2225, 1975.
11. **Harvey, G. R. and Giam, C. S.,** Polychlorobiphenyls and DDT compounds, in *Strategies for Marine Pollution Monitoring,* Goldberg, E. D., Ed., Wiley-Interscience, New York, 1976, 35.
12. **Murray, H. E., Neff, G. S., Hrung, Y. C., and Giam, C. S.,** Determination of benzo(a)pyrene, hexachlorobenzene and pentachlorophenol in oysters from Galveston Bay, Texas, *Bull. Environ. Contam. Toxicol.*, 25, 663, 1980.
13. **Giam, C. S., Murray, H. E., Ray, L. E., and Kira, S.,** Bioaccumulation of hexachlorobenzene in killifish *(Fundulus similis), Bull. Environ. Contam. Toxicol.*, 25, 891, 1980.
14. **Kanazawa, J.,** Measurement of the bioconcentration factors of pesticides by freshwater fish and their correlation with physicochemical properties or acute toxicities, *Pest. Sci.*, 12, 417, 1981.
15. **Sugiura, K., Washino, T., Hattori, M., Sato, E., and Goto, M.,** Accumulation of organochlorine compounds in fishes. Difference of accumulation factors by fishes, *Chemosphere,* 6, 359, 1979.

16. **Veith, G. D., DeFoe, D. L., and Bergstedt, B. V.,** Measuring and estimating the bioconcentration factor of chemicals in fish, *J. Fish. Res. Board Can.,* 36, 1040, 1979.
17. **Sanborn, J. R., Childers, W. F., and Hansen, L. G.,** Uptake and elimination of (^{14}C)hexachlorobenzene (HCB) by the green sunfish, *Lepomis cyanellus* Raf., after feeding contaminated food, *J. Agric. Food Chem.,* 25, 551, 1977.
18. **Trujillo, D. A., Ray, L. E., Murray, H. E., and Giam, C. S.,** Bioaccumulation of pentachlorophenol by killifish *(Fundulus similis), Chemosphere,* 11, 25, 1982.
19. **Kobayashi, K. and Akitake, H.,** Studies on the metabolism of chlorophenols in fish. I. Absorption and excretion of PCP by goldfish, *Bull. Jpn. Soc. Sci. Fish.,* 41, 87, 1975.
20. **Pruitt, G. W., Grantham, B. J., and Pierce, R. H., Jr.,** Accumulation and elimination of pentachlorophenol by the bluegill, *Lepomis macrochirus, Trans. Am. Fish. Soc.,* 106, 462, 1977.
21. **Hattula, M. L., Wasenius, V. M., Reunanen, H., and Arstila, A. U.,** Acute toxicity of some chlorinated phenols, catechols and cresols to trout, *Bull. Environ. Contam. Toxicol.,* 26, 295, 1981.
22. **Holmberg, B., Jensen, S., Larsson, A., Lewander, K., and Olsson, M.,** Metabolic effects of technical pentachlorophenol (PCP) on the eel, *Anguilla anguilla, Comp. Biochem. Physiol.,* 43(B), 171, 1972.
23. **Glickman, A. H., Statham, C. N., Wu, A., and Lech, J. J.,** Studies on the uptake, metabolism, and disposition of pentachlorophenol and pentachloroanisole in rainbow trout, *Toxicol. Appl. Pharmacol.,* 41, 649, 1977.
24. **Pierce, R. H., Jr. and Victor, D. M.,** The fate of pentachlorophenol in an aquatic ecosystem, in *Pentachlorophenol: Chemistry, Pharmacology and Environmental Toxicology,* Rao, K. R., Ed., Plenum Press, New York, 1978, 41.
25. **Giam, C. S., Chan, H. S., Neff, G. S., and Atlas, E. L.,** Phthalate ester plasticizers: a new class of marine pollutant, *Science,* 199, 419, 1978.
26. **Sullivan, J. R., Delfino, J. J., Buelow, C. R., and Sheffy, T. B.,** Polychlorinated biphenyls in the fish and sediment of the Lower Fox River, Wisconsin, *Bull. Environ. Contam. Toxicol.,* 30, 58, 1983.
27. **Laska, A. L., Bartell, C. K., and Laseter, J. L.,** Distribution of hexachlorobenzene and hexachlorobutadiene in water, soil, and selected aquatic organisms along the lower Mississippi River, Louisiana, *Bull. Environ. Contam. Toxicol.,* 15, 535, 1976.
28. **Giam, C. S., Richardson, P. L., Taylor, D., and Wong, M. K.,** DDT, DDE and PCBs in the tissues of reef dwelling groupers *(Serranidae)* in the Gulf of Mexico and the Grand Bahamas, *Bull. Environ. Contam. Toxicol.,* 11, 189, 1974.
29. **Butler, P. A. and Schutzmann, R. L.,** Fish, wildlife, and estuaries. Residues of pesticides and PCBs in estuarine fish, 1972 to 1976 — National Pesticide Monitoring Program, *Pest Monitor. J.,* 12, 51, 1978.
30. **Rosales, M. T. L., Botello, A. V., Bravo, H., and Mandelli, B. F.,** PCBs and organochlorine insecticides in oysters from coastal lagoons of the Gulf of Mexico, Mexico, *Bull. Environ. Contam. Toxicol.,* 21, 652, 1979.
31. **Rosales, M. T. L. and Escalona, R. L.,** Organochlorine residues in organisms of two different lagoons of northwest Mexico, *Bull. Environ. Contam. Toxicol.,* 30, 456, 1983.
32. **Gossett, R. W., Puffer, H. W., Arthur, R. H., Jr., and Young, D. R.,** DDT, PCB and benzo(a)pyrene levels in white croaker *(Genyonemus lineatus)* from southern California, *Mar. Pollut. Bull.,* 14, 60, 1983.
33. **Ray, L. E., Murray, H. E., and Giam, C. S.,** Organic pollutants in marine samples from Portland, Maine, *Chemosphere,* 12, 1031, 1983.
34. **Baird, R. C., Thompson, N. P., Hopkins, T. L., and Weiss, W. R.,** Chlorinated hydrocarbons in mesopelagic fish of the eastern Gulf of Mexico, *Bull. Mar. Sci.,* 25, 473, 1975.
35. **Roberts, A. E., Hill, D. R., and Tifft, E. C., Jr.,** Evaluation of New York Bight lobsters for PCBs, DDT, petroleum hydrocarbons, mercury, and cadmium, *Bull. Environ. Contam. Toxicol.,* 29, 711, 1982.
36. **Deubert, K. H., Rule, P., and Corte-Real, I.,** PCB residues in *Mercenaria mercenaria* from New Bedford Harbor, 1978, *Bull. Environ. Contam. Toxicol.,* 27, 683, 1981.
37. **Giam, C. S., Chan, H. S., and Neff, G. S.,** Phthalate ester plasticizers, DDT, DDE and polychlorinated biphenyls in biota from the Gulf of Mexico, *Mar. Pollut. Bull.,* 9, 249, 1978.
38. **Kuehl, D. W. and Dougherty, R. C.,** Pentachlorophenol in the environment. Evidence for its origin from commercial pentachlorophenol by negative chemical ionization mass spectrometry, *Environ. Sci. Technol.,* 14, 447, 1980.
39. **Zitko, V., Hutzinger, O., and Choi, P. M. K.,** Determination of pentachlorophenol and chlorobiphenylols in biological samples, *Bull. Environ. Contam. Toxicol.,* 12, 649, 1974.
40. **Niimi, A. J.,** Hexachlorobenzene (HCB) levels in Lake Ontario salmonids, *Bull. Environ. Contam. Toxicol.,* 23, 20, 1979.
41. **Goldberg, E. D., Ed.,** *The Health of the Oceans,* UNESCO Press, Paris, 1976.
42. **Murray, H. E., Ray, L. E., and Giam, C. S.,** Analysis of marine sediment, water and biota for selected organic pollutants, *Chemosphere,* 10, 1327, 1981.

43. **Rubinstein, N. I., Lores, E., and Gregory, N. R.,** Accumulation of PCBs, mercury and cadmium by *Nereis virens, Mercenaria mercenaria* and *Palaemonetes pugio* from contaminated harbor sediments, *Aquat. Toxicol.*, 3, 249, 1983.
44. **Giam, C. S., Atlas, E., Powers, M. A., Jr., and Leonard, J. E.,** Phthalic acid esters, in *The Handbook of Environmental Chemistry, Vol. 3, Part C, Anthropogenic Compounds,* Hutzinger, O., Ed., Springer-Verlag, New York, 1984.
45. **Bieri, R. H., Cueman, M. K., Smith, C. L., and Su, C. W.,** Polynuclear aromatic and polycyclic aliphatic hydrocarbons in sediments from the Atlantic Outer Continental Shelf, *Int. J. Environ. Anal. Chem.*, 5, 293, 1978.
46. **Giger, W. and Schaffner, C.,** Determination of polycyclic aromatic hydrocarbons in the environment by glass capillary gas chromatography, *Anal. Chem.*, 50, 243, 1978.
47. **Larsen, P. F., Gadbois, D. F., Johnson, A. C., and Doggett, L. F.,** Distribution of polycyclic aromatic hydrocarbons in the superficial sediments of Casco Bay, Maine, *Bull. Environ. Contam. Toxicol.*, 30, 530, 1983.
48. **Heit, M., Klusek, C. S., and Miller, K. M.,** Trace element radionuclide and polynuclear aromatic hydrocarbon concentrations in Unionidae mussels from northern Lake George, New York, U.S.A., *Sci. Technol.*, 14, 465, 1979.
49. **Cahnmann, H. J. and Kuratsune, M.,** Determination of polycyclic aromatic hydrocarbons in oysters collected in polluted water, *Anal. Chem.*, 29, 1312, 1957.
50. **Mix, M. C. and Schaffer, R. L.,** Benzo(a)pyrene concentrations in mussels *(Mytilus edulis)* from Yaquina Bay, Oregon, during June 1976 to May 1978, *Bull. Environ. Contam. Toxicol.*, 23, 677, 1979.
51. **Mix, M. C. and Schaffer, R. L.,** Concentrations of unsubstituted polycyclic aromatic hydrocarbons in softshell clams from Coos Bay, Oregon, U.S.A., *Mar. Pollut. Bull.*, 14, 94, 1983.
52. **Dunn, B. P.,** Techniques for determination of benzo(a)pyrene in marine organisms and sediments, *Environ. Sci. Technol.*, 10, 1018, 1976.
53. **Ray, L. E., Murray, H. E., and Giam, C. S.,** Analysis of water and sediment from the Nueces Estuary/Corpus Christi Bay (Texas) for selected organic pollutants, *Chemosphere,* 12, 1039, 1983.
54. **Callahan, M. A., Slimak, M. W., Gabel, N. W., May, I. P., Fowler, C. F., Freed, J. R., Jennings, P., Durfee, R. L., Whitmore, F. C., Maestri, B., Mabey, W. R., Holt, B. R., and Gould, C.,** Water-related environmental fate of 129 priority pollutants, EPA-440/4-79-29b, U.S. Environmental Protection Agency, Washington, D.C., 1979, 94-1.

Chapter 3

CYTOCHROME P-450 AND MIXED-FUNCTION OXYGENASE SYSTEMS IN MARINE INVERTEBRATES

R. Lee and L. Quattrochi

TABLE OF CONTENTS

I. Isolation of Cytochrome P-450 and Mixed-Function Oxygenases ... 52
- A. Collection and Maintenance of Animals ... 52
  - 1. Crabs ... 52
  - 2. Polychaetes ... 52
  - 3. Diet Preparation and Feeding ... 52
- B. Cytochrome P-450 from Crabs ... 52
  - 1. Dissection and Tissue Preparation ... 52
  - 2. Preparation and Solubilization of Crab Microsomes ... 52
  - 3. Purification of Multiple Forms of Cytochrome P-450 ... 53
  - 4. Spectral Studies ... 54
- C. Cytochromes P-450 and Mixed-Function Oxygenases from Polychaetes ... 54
- D. General Methods ... 54

II. Cytochrome P-450 and Mixed-Function Oyxgenase Systems in Polychaetes ... 54

III. Cytochrome P-450 and Mixed-Function Oxygenase Systems in Marine Crabs ... 56

IV. Partial Purification of Cytochromes P-450 from Crabs ... 58

V. Spectral Binding Studies of Crab Cytochrome P-450 ... 60

Acknowledgment ... 63

References ... 64

I. ISOLATION OF CYTOCHROME P-450 AND MIXED-FUNCTION OXYGENASES

A. Collection and Maintenance of Animals

1. Crabs

Blue crabs, *Callinectes sapidus* Rathbun, were collected by trapping or trawling from estuaries of coastal Georgia or were purchased from local crab fisherman. The crabs were maintained in 750-ℓ fiberglass tanks with flowing seawater drawn from the Skidaway River (4 ℓ/min; salinity 20‰) and filtered through a settling tank and sand filter system. The water was delivered through jets to provide aeration.

2. Polychaetes

Nereis virens were collected from mudflats in Maine. They were maintained in trays with an area of 0.1 m^2 that received 200 mℓ/min of filtered, temperature-controlled (20°C) sea water. Each tray was layered to a depth of 3 cm with fine sand (<0.3 mm).

3. Diet Preparation and Feeding

Crabs and worms were fed an artificial diet modified from one used by aquarists. The diet included 2 chicken eggs, 15 mℓ cod liver oil, 4 mℓ PolyVisol vitamin drops, 800 mg each of vitamins B_1, B_{12}, and C, 600 mg paprika, 1 large carrot, 1 g NaCl, 56 g creamed spinach, 400 g fishmeal, 400 g pablum blended in 700 mℓ water, and 36 g gelatin dissolved in 300 mℓ water.

The control diet was prepared by thoroughly blending the first ten ingredients. Up to 700 mℓ water was added as needed when blending in the pablum and fishmeal. The gelatin was dissolved on low heat in 300 mℓ water and then stirred as the food mixture from the blender was added. The mixture was heated slowly with constant stirring until steam was readily apparent. After chilling, the mixture became solid and easy to cut into appropriate feeding portions. Clam juice was occasionally substituted for an equal amount of water to enhance the flavor and mask the presence of pollutant. Pollutants were added and blended into the hot mixture before refrigeration. Worms were fed by placing small cubes of the food on top of their burrows. During cold periods in our area (December to March), blue crabs will not eat. During the remainder of the year, they are aggressive eaters.

B. Cytochrome P-450 from Crabs

1. Dissection and Tissue Preparation

Blue crabs may be dissected with or without chilling (4°C). By holding either the last leg or the pereiopod at its base, the crab can be handled safely and efficiently. The claws or chelipeds are then removed by bone cutters. The carapace sims from the spines forward are cut using bone cutters. The posterial margins are then snipped with surgical scissors. With the carapace sims cut, the halves are split apart by hand.

2. Preparation and Solubilization of Crab Microsomes

Microsomes were prepared from crab hepatopancreas using a stabilization buffer [0.1 *M* potassium phosphate, pH 7.3, containing 1.0 m*M* phenobarbital, 1.0 m*M* EDTA, 1.0 m*M* barbital, 1.0 m*M* DTT, 1% bovine serum albumin (BSA), 0.2 m*M* phenylmethylsulfonyl fluoride (PMSF), and 20% glycerol] to minimize proteolytic degradation of cytochrome P-450. The hepatopancreas was removed from the crab, rinsed twice in about 250 mℓ 0.15% KCl and once in the stabilization buffer, and homogenized with 2 volumes of stabilization buffer in a Potter-Elvehjem homogenizer equipped with a Teflon pestle. The homogenate was centrifuged twice (10 min each) at 10,000 g to sediment the microsomes. The microsomal pellet was resuspended to approximately 50 mg microsomal protein per milliliter in 0.1 *M* potassium phosphate buffer, pH 7.3, containing 1.0 m*M* DTT, 1.0 m*M* EDTA, and 20% glycerol (Buffer I), and stored at liquid nitrogen temperatures for up to 1 month.

Microsomes were solubilized with 0.4% sodium cholate in buffer I at a final concentration of 6 mg/mℓ suspension. The suspension was stirred for 20 min at 4°C, and polyethylene glycol (PEG) 8000 was added dropwise to give a final PEG concentration of 10%. The resulting suspension was centrifuged for 1 hr at 10,000 g, and the pellet was resuspended in a minimal amount of DE53 equilibration buffer (0.01 *M* potassium phosphate, pH 7.4, 0.2% cholate, and 20% glycerol). The 0 to 10% PEG precipitate contained 60 to 65% of the initial microsomal cytochrome P-450 and had a specific content twice that of the initial P-450. All general purification procedures were carried out between 0 to 4°C.

3. Purification of Multiple Forms of Cytochrome P-450

DEAE-cellulose (Whatman DE53), a high capacity anion exchanger, was used to resolve the various forms of cytochrome P-450. The 0 to 10% PEG precipitate, resuspended in DE53 equilibration buffer, was loaded onto a DE53 column (2.25 × 40 cm). The column was washed with 1 bed volume of equilibration buffer, followed by a 500-mℓ linear gradient of 0 to 0.5 *M* KCl in equilibration buffer. The flow rate was 0.5 to 0.6 mℓ/min, and 6-mℓ fractions were collected. The eluate was monitored by measuring absorbance at 418 nm and fractions containing cytochrome P-450 were pooled. The recovery of cytochrome P-450 from the column was approximately 90%.

The major cytochrome P-450 peak from DE53 adsorbed to a phenyl-Sepharose 4B column (1.75 × 12 cm), previously equilibrated with 0.26 *M* potassium phosphate (pH 7.4), containing 0.1 m*M* DTT, 0.1 m*M* EDTA, 0.05% sodium cholate, and 20% glycerol. The column was washed with 1 volume of this same buffer and 3 volumes of 0.005 *M* potassium phosphate (pH 7.4), containing the same components as the wash buffer. Cytochrome P-450 was eluted in a broad peak with 0.2% Triton N101 in the 0.005 *M* phosphate buffer. The recovery of cytochrome P-450 from the column was 50 to 70%. Fractions containing cytochrome P-450 were pooled and applied directly to a hydroxyapatite column (1.75 × 6 cm), previously equilibrated with 0.01 *M* potassium phosphate (pH 7.4), containing 0.1 m*M* DTT, 0.1 m*M* EDTA, 0.05% sodium cholate, and 20% glycerol. The column was washed with 2 bed volumes of 50 m*M* phosphate, pH 7.4, and then eluted with a stepwise potassium phosphate gradient of 0.1, 0.25, and 0.5 *M* potassium phosphate. The recovery of cytochrome P-450 from the column was approximately 50%. Fractions containing cytochrome P-450 were frozen at liquid nitrogen temperatures for further purification or analyses.

A second procedure was used to separate forms of cytochrome P-450 from spider crabs. The microsomes were solubilized with 0.4% sodium cholate. After centrifugation at 100,000 g for 1 hr, the supernatant was applied to a phenyl-Sepharose column. The P-450 was eluted with a phosphate buffer (0.025 *M*, pH 7.4) containing 0.3% sodium cholate and 0.2% Triton N101. Fractions containing P-450 (based on absorbance at 418 nm) were applied to a DEAE-Sephacel column which had previously been equilibrated with 0.010 *M* phosphate buffer (pH 7.4) containing cholate (0.3%), Triton N101 (0.2%), and glycerol (20%). The P-450 forms were eluted with a KCl gradient (0.0 to 0.25 *M*). Two forms were collected and dialyzed overnight in 0.01 *M* phosphate buffer (pH 7.4) containing EDTA (0.1 m*M*), DTT (0.1 m*M*), glycerol (20%), and sodium cholate (0.05%). The P-450 forms were applied to hydroxyapatite columns and P-450 eluted with a stepwise potassium phosphate gradient of 0.05, 0.1, and 0.25 *M* potassium phosphate. All fractions were pooled, dialyzed in the buffer described above, and stored in liquid nitrogen.

4. Spectral Studies

Cytochrome P-450 concentrations were obtained from the CO-reduced difference spectra using an Aminco DW-2a Spectrophotometer[1] and an extinction coefficient of 91 cm^{-1} mM^{-1}.

The actual extinction coefficient for crab cytochrome P-450 may differ from this value. The absolute spectrum of oxidized hemoprotein[2] was determined using buffer I as a reference for dialyzed P-450.

Binding spectra were obtained with oxidized microsomes by adding ligand (dissolved in phosphate/glycerol to acetone, 50:50) to the sample cuvette and an equal volume of buffer to acetone to the reference cuvette. The total concentration of acetone never exceeded 2%. Binding data were analyzed as described by Jefcoate et al.[3] and Comai and Gaylor.[4]

C. Cytochromes P-450 and Mixed-Function Oxygenases from Polychaetes

Tissues of *N. virens* were dissected, weighed, and homogenized in 0.15 *M* KCl in 0.05 *M* Tris buffer (pH 7.5) using a Potter-Elvehjem homogenizer. Cell debris and nuclei were removed by centrifugation at 700 g for 10 min at 4°C. Mitochondria were collected by centrifugation of the supernatant at 8000 g for 10 min. The supernatant from the mitochondrial fraction was centrifuged at 140,000 g for 60 min in an ultracentrafuge (Beckman Model L-40) to sediment microsomes. Cytochrome P-450 and MFO assays were carried out as described earlier.

D. General Methods

Protein concentrations were determined by the method of Lowry et al.,[5] with appropriate corrections for buffer components. Sodium dodecyl (lauryl) sulfate-polyacrylamide gel electrophoresis was performed using the method of Laemmli.[6] The gels (8% acrylamide, 1.5 mm thick) were electrophoresed using about 15 mA per slab until the protein fully migrated into the gel, and subsequently at 40 mA per slab. Gels were fixed for 60 min in isopropanol/glacial acetic acid/water (25:10:55), stained for 60 min in isopropanol/glacial acetic acid/0.05% Brilliant Blue G stain, and destained overnight in isopropanol/glacial acetic acid/water (10:10:80). Protein standards of known molecular weights (BSA, 66,000; egg albumin, 45,000; glyceraldehyde-3-phosphate dehydrogenase, 36,000; carbonic anhydrase, 29,000) were run on the same gel as the samples.

Mixed-function oxygenase (MFO) was assayed using benzo(a)pyrene as substrate by the procedure of Nebert and Gelboin[7] as modified by Singer et al.[8]

II. CYTOCHROME P-450 AND MIXED-FUNCTION OXYGENASE SYSTEMS IN POLYCHAETES

The cytochrome P-450-mediated MFO system was detected in polychaete annelids. Mixed-function oxygenase, i.e., benzo(a)pyrene hydroxylase, from *N. virens* was associated with the lower portions of the intestine with little or no enzyme activity in the pharynx, esophagus, or upper portions of the intestine (Table 1). Subcellular fractionation of intestine extracts indicated that the MFO system was associated with the endoplasmic reticular (i.e., microsomal) fraction. Full enzyme activity required the addition of NADPH, MG^{+2}, and oxygen. Polychaete MFO is similar to that of vertebrates being composed of phospholipid, cytochrome P-450, and NADPH-cytochrome P-450 reductase.[9] A carbon monoxide difference spectrum showing the presence of cytochrome P-450 in microsomes of the intestines of *N. virens* is shown in Figure 1. In *N. virens* exposed to BaP or a mixture of polychlorinated biphenyls (i.e., Aroclor 1254) for 4 weeks, the MFO activity and cytochrome P-450 content were higher in the Aroclor-treated worms relative to control worms, but was not higher in BaP-exposed worms (Table 2). By 8 weeks, both the BaP- and Aroclor-exposed worms had significantly higher ($p < 0.05$) cytochrome P-450 content than control worms.

Table 1
TISSUE DISTRIBUTION OF MFO ACTIVITY IN THE POLYCHAETE *NEREIS VIRENS*[24]

Tissue	Specific activity[a] (units/mg protein)	Total activity[a] (units)
Pharynx	0	0
Esophagus	0	0
Upper portion of intestine	35	150
Lower portion of intestine	230	13,200

[a] One unit of enzyme activity equals the fluorescence equivalent to 1×10^{-2} mol 3-hydroxybenzo(a)pyrene produced in a 60-min incubation at 28°C.

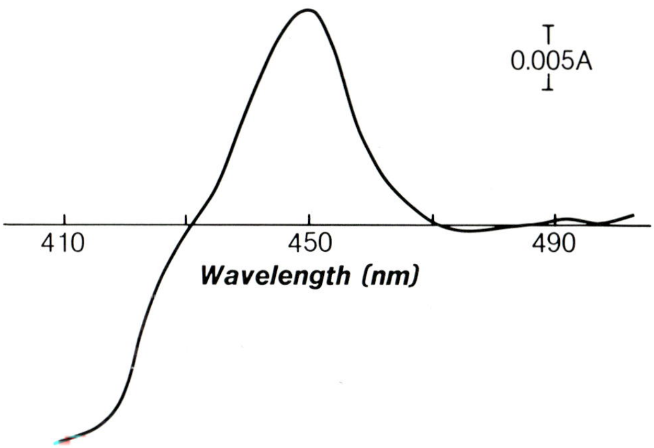

FIGURE 1. Carbon monoxide-difference spectra of dithionite-reduced intestinal microsomes from *Nereis virens*. The concentration of cytochrome P-450 was 0.18 nmol/mℓ.

Table 2
CHANGES IN MFO ACTIVITY AND CYTOCHROME P-450 CONTENT IN *NEREIS VIRENS* FED FOOD CONTAINING BENZO(A)PYRENE AND AROCLOR 1254

Compound[a]	Weeks	MFO (enzyme units/mg protein)	Cytochrome P-450 (nmol/mg protein)
None (control)	4	1.22 ± 0.65	0.10 ± 0.02
	8	1.07 ± 0.27	0.12 ± 0.03
Benzo(a)pyrene (1 mg/g)	4	1.1 ± 0.3	0.09 ± 0.04
	8	1.2 ± 0.4	0.21 ± 0.07
Aroclor 1254 (0.5 mg/g)	4	3.6 ± 0.2	0.22 ± 0.04
	8	3.7 ± 0.7	0.27 ± 0.08

[a] The concentration of the compounds in the food is given in parentheses. The intestinal tissues of 10 to 15 worms were pooled and used for each assay. Each assay was repeated 3 times (n = 3) using a separate pool of worms; the average value and the standard deviation are given in the table. One unit of MFO (benzo(a)pyrene hydroxylase) activity equals the fluorescence equivalent to 1×10^{-7} mol of 3-hydroxybenzo(a)pyrene produced in a 60-min incubation at 28°C.

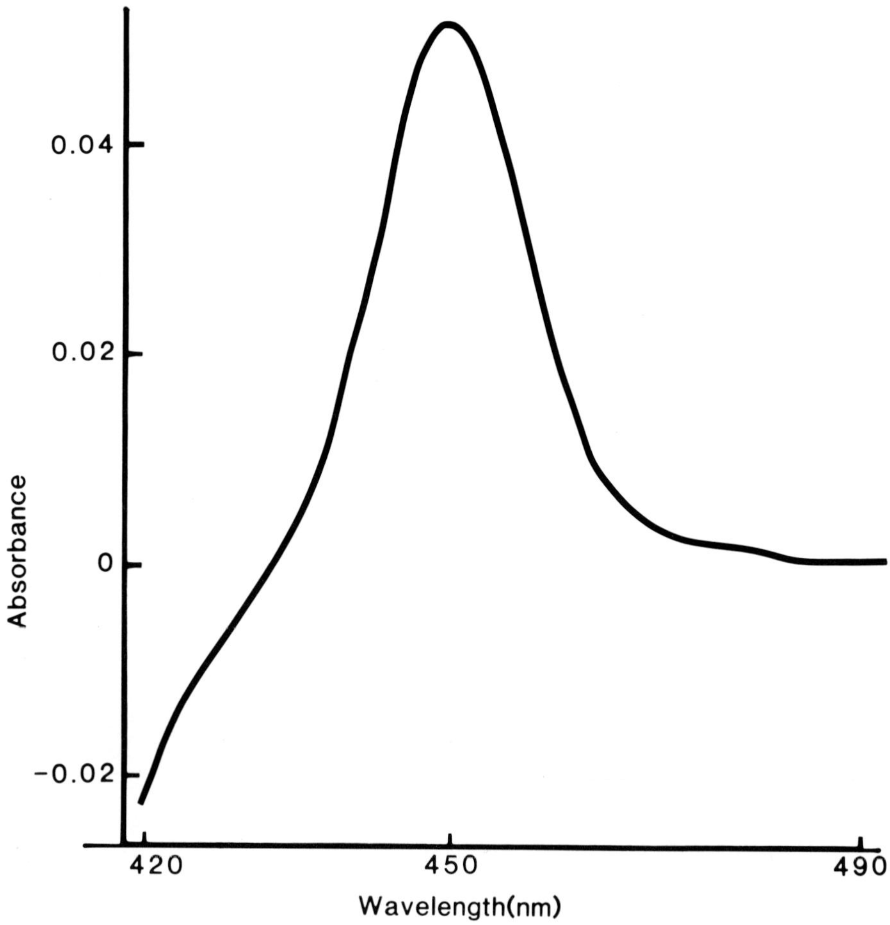

FIGURE 2. Carbon monoxide-difference spectrum of dithionite-reduced hepatopancreas microsomes from female stone crabs, *Menippe mercenaria*. Cytochrome P-450 concentration was 0.82 nmol/mℓ.

III. CYTOCHROME P-450 AND MIXED-FUNCTION OXYGENASE SYSTEMS IN MARINE CRABS

The crab MFO system is similar to that in vertebrate systems since it is associated with the microsomal fraction and requires phospholipid, cytochrome P-450, and NADPH-cytochrome P-450 reductase for activity.[8] Highest MFO, i.e., benzo(a)pyrene hydroxylase, activity in blue crabs is in the stomach and green gland with low activity in blood, gill, reproductive tissues, eyestalk, cardiac muscle, and hepatopancreas.[9] Low MFO activity in the hepatopancreas of many marine crustaceans is due to inhibition by digestive juices released during homogenization of the tissues.[10] Most earlier work on crustaceans showed significant amounts of cytochrome P-240, an inactive form of P-450, in microsomal preparations.[11] An isolation medium consisting of protease inhibitors, albumin, phenobarbital, and various protecting agents allowed us to isolate cytochrome P-450 from crabs with no evidence of cytochrome P-420 (Figure 2).

Various estuarine crab species were collected from the high marsh *(Sesarma cinerum)*, lower marsh *(Una pugnax)*, oxyster reefs *(Menippe mercenaria* and *Libinia emarginata)*, and tidal creeks *(C. sapidus)*. Cytochrome P-450 and MFO activity (BaP hydroxylase) were detected in the hepatopancreas of all crabs assayed (Table 3). Cytochrome P-450 content

Table 3
CYTOCHROME P-450 CONTENT AND MFO (BENZO(A)PYRENE HYDROXYLASE) ACTIVITY IN ESTUARINE CRABS[25]

Species and sex[a]	Month collected	Individual body wt (g)	P-450 (nmol/mg microsomal protein)	P-450 (nmol/g hepato-pancreas)	P-450 (nmol/crab)	MFO activity (nmol/hr/mg microsomal protein)
U. pugnax, male	Aug.	3—4	0.23	0.07	—	—
	Oct.	3—4	0.14 ± 0.07 (3)	0.18	0.015	8.0 ± 1.1 (3)
U. pugnax, female	Oct.	2—3	0.17 ± 0.11 (3)	0.25	0.22	31.0 ± 2.8 (3)
S. cinerum, male	Apr.	2—3	0.31	—	—	—
	Apr.	2—3	0.34	—	—	—
S. cinerum, female	Oct.	2—3	0.51 ± 0.17 (3)	2.5	0.16	0.20
M. mercenaria, male	Aug.	240	1.0	2.1	24	0.45
	Aug.	130	0.7	2.4	17	0.50
M. mercenaria, female	Oct.	300	0.20	1.2	4	0.50
C. sapidus, male	July	—	0.04	0.14	1.1	0.05
	Aug.	200—350	0.06 ± 0.04 (3)	—	4.9 ± 1.6 (3)	0.06
C. sapidus, female	Sept.	200	0.18	0.55	4.5	0.05
	Oct.	100—200	0.19 ± 0.11 (4)	1.3	7.6 ± 3.1 (4)	0.05
C. sapidus, juvenile	Sept.	40	0.17	0.56	1.7	0.02
Libinia sp., male	Nov.	40	0.36	2.7	9.4	0.66
Libinia sp., female	Nov.	30	0.56	5.1	10.7	0.14

[a] Crabs were collected from an estuary along the Wilmington River, Georgia. The MFO activity is given in nanomoles of 3-hydroxybenzo(a)pyrene produced in a 60-min incubation at 28°C. Each *Uca* or *Sesarma* assay was the result of 30 pooled crabs. In case of more than one assay, the values are the mean ± SD with the number of separate groups of 30 crabs assayed in parentheses. For *Menippe*, *Callinectes*, and *Libinia*, each individual was assayed separately.

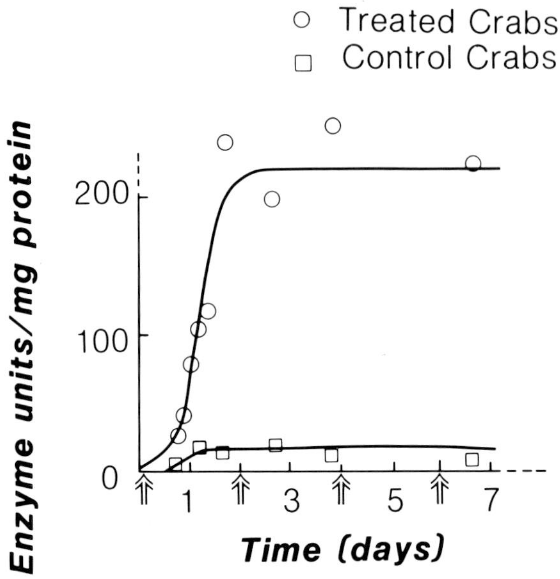

FIGURE 3. Induction of benzo(a)pyrene hydroxylase in blue crabs. Adult male crabs were injected with 1-mg doses of benz(a)anthracene dissolved in corn oil or corn oil alone. Arrows (↑) mark times of injection. The hepatopancreas of each crab was homogenized and assayed for enzyme activity by the method of Nebert and Gelboin.[7] One unit of BaP hydroxylase activity is defined as that amount catalyzing the formation in a 60-min incubation period at 28°C of hydroxylated product causing fluorescence equivalent to that of 1×10^{-12} mol of 3-hydroxybenz(a)pyrene.

varied from 0.05 nmol/mg microsomal protein in the blue crab, *C. sapidus,* to 1.08 nmol/mg in the stone crab *M. mercenaria.* The turnover number for hepatopancreas BaP hydroxylase activity varied from 182 units/mol P-450 in the fiddler crab, *U. pugnax,* to 0.12 units/mol in juvenile *C. sapidus.* On an individual basis, cytochrome P-450 was highest in *M. mercenaria,* which contained 17 to 24 nmol P-450 per crab or 2.1 to 2.4 nmol P-450 per gram hepatopancreas.

Several experiments were conducted to determine the responses of the MFO system to organic pollutants in the food. The marsh crab *S. cinerum* was fed food containing no. 5 fuel oil at a concentration of 1 mg/g. The MFO activities (BaP hydroxylase) of the hepatopancreas after 23 weeks were 81 and 8 enzyme units/mg protein in crabs fed oiled and control diets, respectively. In a second experiment, male blue crabs, *C. sapidus,* were fed food containing Aroclor 1254 at a concentration of 70 mg/g. After 1 week, the cytochrome P-450 was 0.12 ± 0.03 ($n = 5$) and 0.06 ± 0.04 ($n = 3$) in Aroclor-fed and control crabs, respectively. Aroclor 1254 concentrations were 85 and 0.1 mg/g hepatopancreas for Aroclor-fed and control crabs, respectively. As a result of injecting benz(a)anthracene into *C. sapidus,* the extracts of hepatopancreas had higher MFO activity (BaP) after 2 days (Figure 3). These results, i.e., higher MFO activity and P-450 content after exposure to inducers, were not consistently found when experiments were repeated.

IV. PARTIAL PURIFICATION OF CYTOCHROMES P-450 FROM CRABS

Our initial purification sequence for P-450 involved cholate digestion of the microsomes followed by chromatography on n-octylamine Sepharose 4B.[12,13] The cytochromes P-450 were eluted from the column with 0.1% Triton N101 and the fraction with cytochrome P-450 was chromatographed on a hydroxyapatite column. From *Libinia* and *Callinectes* two

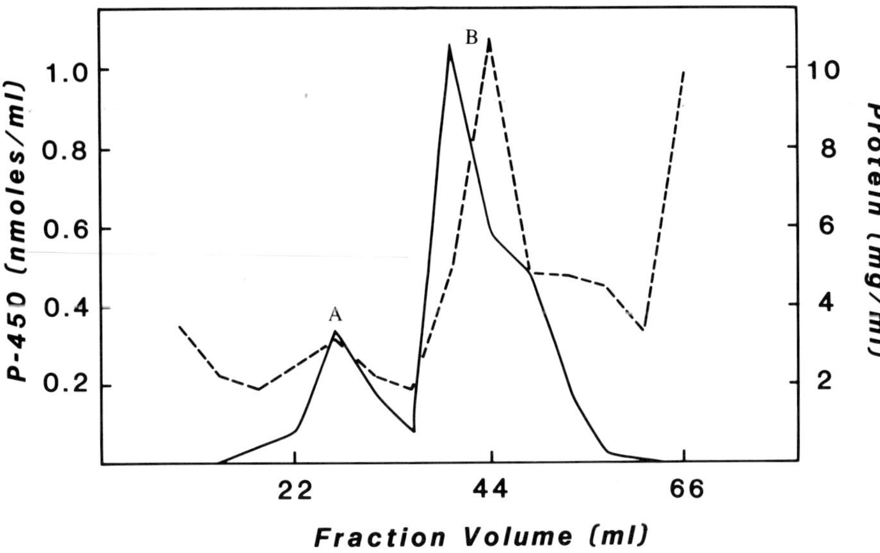

FIGURE 4. Hydroxyapatite column chromatography of blue crab cytochrome P-450. The solid line represents cytochrome P-450; the dotted line represents protein. Proteins were eluted by a stepwise elution going from 0.012 to 0.5 *M* potassium phosphate buffer. The smaller peak is referred to as peak A and the larger peak as peak B.[12]

cytochromes P-450 eluted from the hydroxyapatite column. Figure 4 shows the profile obtained from *C. sapidus* with the two peaks labeled A and B. For untreated crabs, peak A was always a minor P-450 accounting for 10 to 20% of the total P-450 eluted from hydroxyapatite. The CO-difference spectra of peaks A and B showed that both had absorbance maximum at 451 nm. Peaks A and B showed different electrophoretic mobility on polyacylamide gel electrophoresis. Peaks A and B were only partially purified and, in addition to a major protein band at approximately 50,000 daltons, faint bands at 24,000 daltons were observed. The highest specific content of cytochrome P-450 obtained was 1.8 nmol/mg. For blue crabs, only one peak was observed when the void volume fraction of the affinity column was chromatographed on hydroxyapatite. This peak eluted slightly before peak B. Using DEAE-cellulose, James and Little[14] found two cytochromes P-450 in hepatopancreas microsomes of the spiny lobster, *Panulirus argus*. The major P-450 had a molecular weight of 53,000, while the minor peak was 48,000 daltons. We are not aware of other studies showing multiple forms of cytochrome P-450 in other marine invertebrates.

Insect microsomes have a number of cytochromes P-450 and new forms have been observed after exposure to various inducers.[15,16] However, in our studies, there was no consistent appearance of new forms of P-450 after exposure to Aroclor 1254.

For the spider crab, *L. emarginata*, two forms of cytochrome P-450 were eluted from a DEAE-Sephacel column and are referred to as P-450_{I} and P-450_{II} (Figure 5). The major form in male crabs was P-450_{II} ($\approx$54% of total), while P-450_{I} ($\approx$20%) was a minor form. Cytochrome P-450_{II} from male and female spider crabs was partially purified to a specific content of 5 nmol P-450 per milligram protein (Table 4).

The absolute spectra of partially purified cytochrome P-450_{II} in the oxidized, reduced, and CO-reduced state from male and female spider crabs are shown in Figures 6 and 7. Differences between male and female crabs were observed with a peak in the oxidized spectra of 417 and 418 nm for the male and female crabs, respectively. After reduction, the peaks were shifted to lower wavelengths, 416 nm, in both male and female. Reduction in the

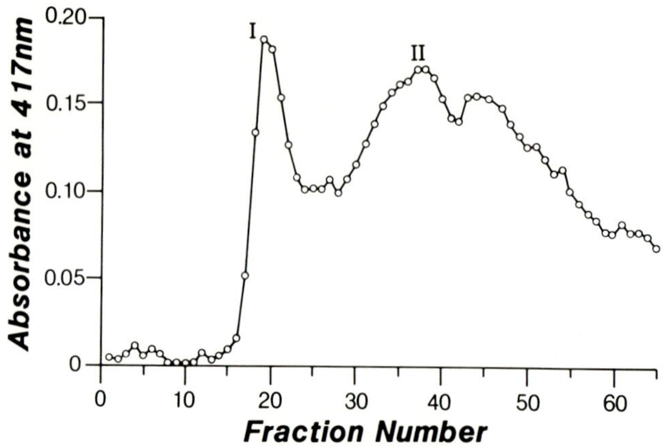

FIGURE 5. Cytochrome P-450 in *Libinia*. Cytochromes P-450 from the spider crab, *L. emarginata,* were separated by passage down a DEAE-Sephacel column. A potassium chloride gradient (0.0 to 0.25 *M*) was used to elute cytochromes P-450.

Table 4
PARTIAL PURIFICATION OF CYTOCHROME $P\text{-}450_{II}$ FROM SPIDER CRAB HEPATOPANCREAS MICROSOMES

Purification step	Cytochrome P-450[a] (nmol/mg protein)
Microsomes	0.1
Cholate solubilization, 0—10% Peg 6000	0.2
DEAE-cellulose	0.5
Phenyl-Sepharose 4B	1.0
Hydroxyapatite	2.0
Matrix Gel Red A	5.0

[a] The concentration of cytochrome P-450 was determined from the reduced CO-difference spectrum using an extinction coefficient of 91 cm^{-1} mM^{-1}.

presence of carbon monoxide produced a CO-complex with absorption maxima of 449 and 450 nm for male and female crabs, respectively. The wavelength maxima at around 417 nm indicated the oxidized $P\text{-}450_{II}$ was in the low-spin state. Lack of an absorbance shoulder between 390 and 400 nm indicated absence of high-spin ferric hemoprotein.

Molecular weight determinations by SDS-polyacrylamide gel electrophoresis indicated a major band at 54,000 daltons for both male and female cytochrome $P\text{-}450_{I}$. The gel electrophoresis patterns of partially purified cytochromes $P\text{-}450_{II}$ from spider crabs are shown in Plate 1.

V. SPECTRAL BINDING STUDIES OF CRAB CYTOCHROME P-450

The binding of numerous organic compounds to oxidized cytochrome P-450 causes spectral changes in the Soret region for the heme prosthetic group. Such organic compounds include

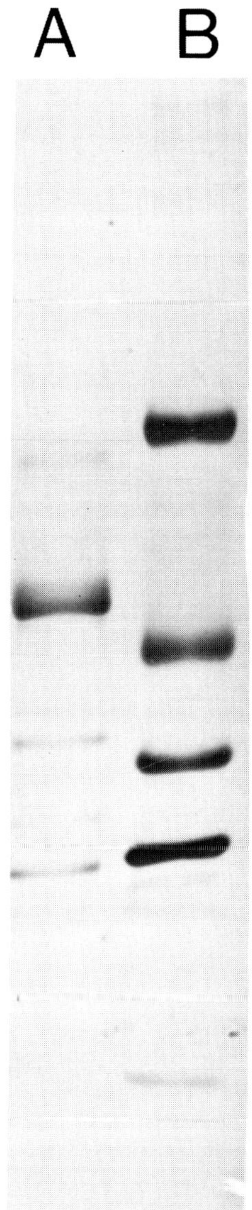

PLATE 1. SDS-polyacrylamide gel pattern of cytochromes P-450_{II} from *Libinia*. Lane A contained standard proteins (6 μg each); lane B contained 15 μg P-450_{II}. The cytochrome P-450 was obtained from hydroxyapatite chromatography (5 nmol/mg protein).

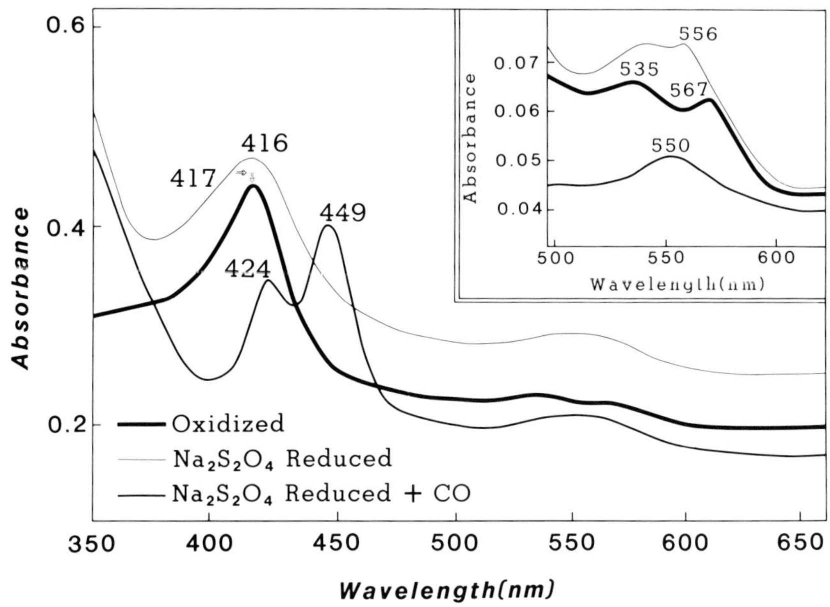

FIGURE 6. Absolute spectra of cytochrome P-450 from male spider crab. The sample cuvette contained cytochrome P-450 (1.3 nmol/mℓ) diluted in 0.1 *M* potassium phosphate buffer (pH 7.4). Reference cuvette contained only phosphate buffer.

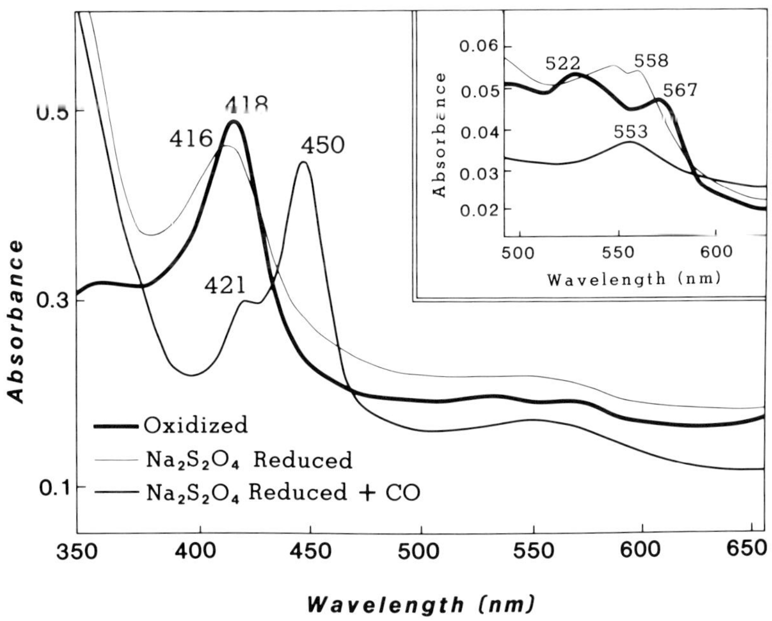

FIGURE 7. Absolute spectra of cytochrome P-450 from female spider crab. The sample cuvette contained cytochrome P-450 (1.8 nmol/mℓ) diluted in 0.1 *M* potassium phosphate buffer (pH 7.4). Reference cuvette contained only phosphate buffer.

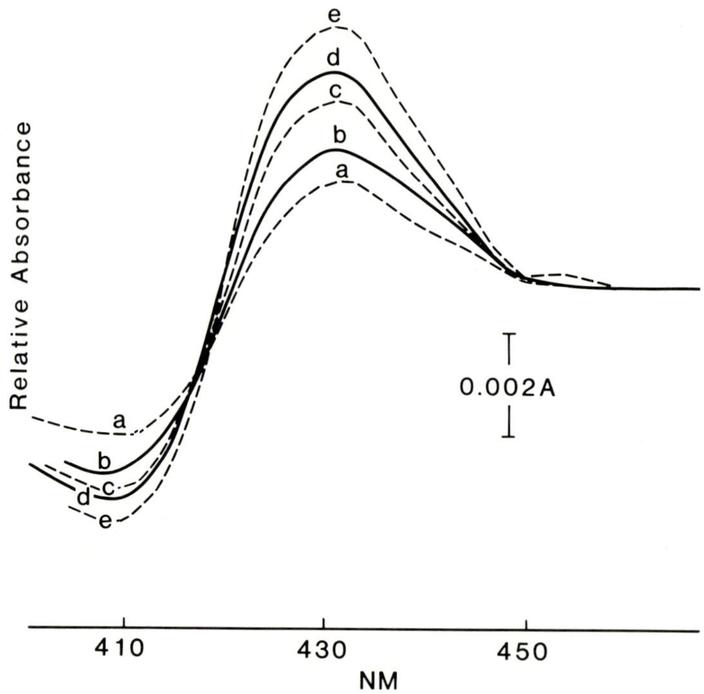

FIGURE 8. The effect of serial additions of *n*-octylamine on the magnitude of the difference spectrum of cytochrome P-450 from female crab microsomes prepared by ultracentrifugation. The cuvette contained 0.26 nmol of cytochrome P-450 and (a) 0.012, (b) 0.036, (c) 0.048, (d) 0.096, and (e) 0.215 m*M* *n*-octylamine, respectively.

pesticides, anesthetics, aliphatic and aromatic amines, certain drugs, and steroids.[17-22] Differences spectrophotometry has been used to classify the spectral changes which occur when a ligand is added to microsomes or partially purified cytochrome P-450. Type I changes have maxima in the 385 to 398-nm region and minima around 420 nm. Ligands which elicit type I spectral changes are generally molecules with extensive hydrophobic regions in their interactions between the ligand and a binding locus in the cytochrome P-450 structure. Electron paramagnetic resonance spectroscopy has shown type I binding to result from shifting equilibrium of the spin states of ferric iron in cytochrome P-450 from low to high spin.[23] Type II spectral changes are characterized by minima in the 390- to 410-nm region and maxima in the 430-nm region, properly termed ''heme ligand binding'' since type II binding involves heteroatom ligand binding directly to the heme iron at the same site as carbon monoxide and oxygen.

The binding spectra for *n*-octylamine to blue crab microsomes are shown in Figure 8. As ligand concentrations approach saturation, the absorption maxima became less well defined. Analysis by an Eadie-Hofstee plot (Figure 9) of A/C (change in absorbance (maximum-minimum/concentration)) vs. A showed two linear regions with a transition between them. The negative reciprocal of the slope of these lines gives the apparent ligand-cytochrome P-450 dissociation constants, K_s^{app}. The two K_s^{app} values for *n*-octylamine were 0.011 and 1.1 m*M*. The two binding processes involve a high-affinity form of cytochrome P-450 that saturates at lower ligand concentration and a low-affinity cytochrome P-450 that saturates at high-ligand concentration.

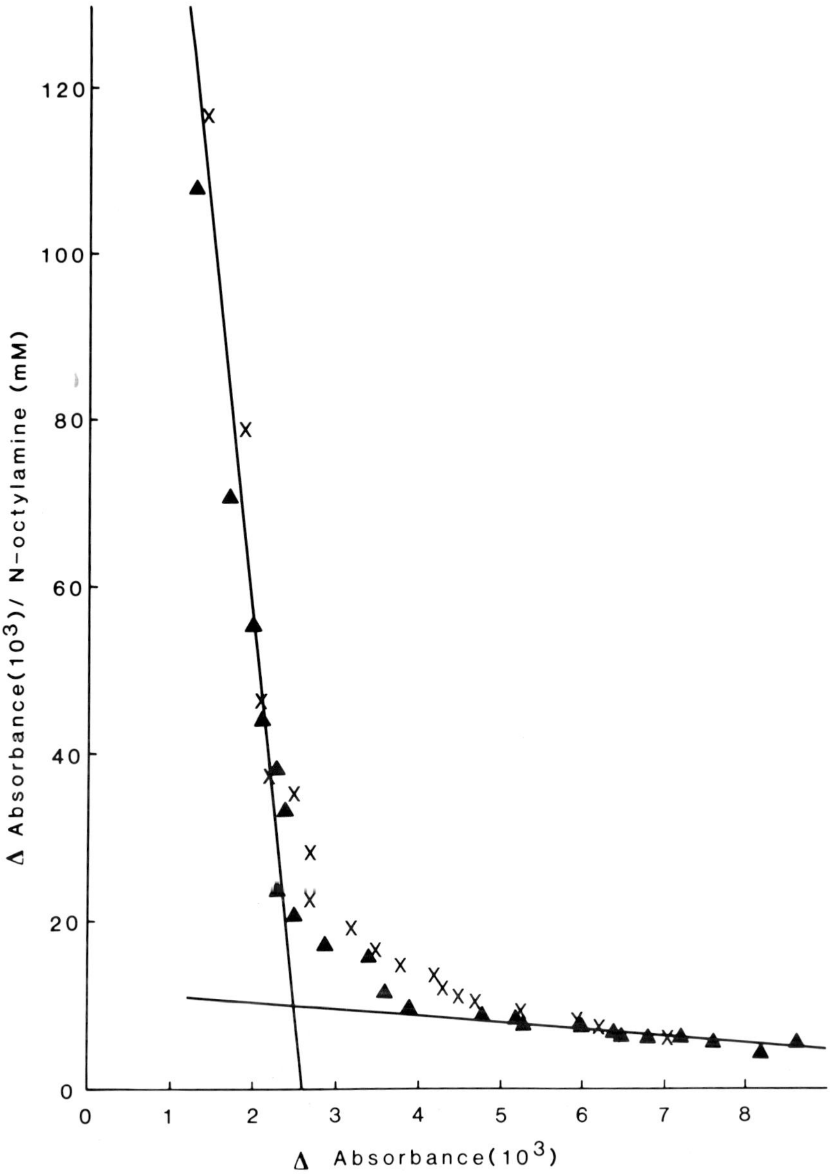

FIGURE 9. Eadie-Hofstee plot of *n*-octylamine binding to cytochrome P-450 from crab hepatic microsomes. Hepatic microsomes from male crabs were prepared by ultracentrifugation (x) and polyethylene glycol precipitation (▲). *N*-octylamine concentration was varied from 0.012 to 1.91 m*M*.

ACKNOWLEDGMENT

This work was supported in part by grant OCE-8310507 from the National Science Foundation.

REFERENCES

1. **Omura, T. and Sato, R.,** The carbon monoxide-binding pigment of liver microsomes, *J. Biol Chem.*, 239, 2379, 1964.
2. **Agosin, M., Norello, A., White, R., Repetto, Y., and Pedemonte, J.,** Multiple forms of noninduced rat liver cytochrome P-450: metabolism of 1-(4′-ethylphenoxy)3,7-dimethyl-6,7-epoxy-trans-2-octene by reconstituted preparations, *J. Biol. Chem.*, 254, 9915, 1979.
3. **Jefcoate, C. R. E., Gaylor, J. L., and Calabrese, R. L.,** Ligand interactions with cytochrome P-450. I. Binding or primary amines, *Biochemistry*, 8, 3455, 1969.
4. **Comai, K. and Gaylor, J. L.,** Existence and separation of three forms of cytochrome P-450 from rat liver microsomes, *J. Biol. Chem.*, 248, 4947, 1979.
5. **Lowry, O. H., Rosebrough, N. J., Farr, A. L., and Randall, R. J.,** Protein measurement with the Folin phenol reagent, *J. Biol. Chem.*, 193, 265, 1951.
6. **Laemmli, U. K.,** Cleavage of structural proteins during the assembly of the head of bacteriophage T4, *Nature (London)*, 227, 680, 1970.
7. **Nebert, D. W. and Gelboin, H. V.,** Substrate-inducible microsomal arylhydroxylase in mammalian cell culture. I. Assay and properties of inducible enzyme, *J. Biol. Chem.*, 243, 6242, 1968.
8. **Singer, S. C., March, P. E., Gonsoulin, F., and Lee, R. F.,** Mixed function oxygenase activity in the blue crab, *Callinectes sapidus:* characterization of enzyme activity from stomach tissue, *Comp. Biochem. Physiol.*, 65C, 129, 1980.
9. **Lee, R. F.,** Mixed function oxygenases (MFO) in marine invertebrates, *Mar. Biol. Lett.*, 2, 98, 1981.
10. **Pohl, R. J., Bend, J. R., Guarino, A. M., and Fouts, J. R.,** Hepatic microsomal mixed-function oxidase activity of several marine species from coastal Maine, *Drug. Metab. Dispos.*, 2, 454, 1974.
11. **Singer, S. C., March, P. E., Jr., Gonsoulin, F., and Lee, R. F.,** Mixed function oxygenase activity in the blue crab, *Callinectes sapidus:* characterization of enzyme activity from stomach tissue, *Comp. Biochem. Physiol.*, 65C, 129, 1979.
12. **Conner, J. W. and Singer, S. C.,** Purification scheme for cytochrome P-450 of blue crab, *Callinectes sapidus* Rathbun., *Aquat. Toxicol.*, 1, 271, 1981.
13. **Conner, J. W. and Lee, R. F.,** Multiple forms of cytochrome P-450 in marine crabs, in *Cytochrome P-450, Biochemistry, Biophysics and Environmental Implications,* Hietanen, E., Laitinen, M., and Hanninen, O., Eds., Elsevier Biomedical Press, New York, 1982, 245.
14. **James, M. O. and Little, P. J.,** Characterization of cytochrome P-450 dependent mixed-function oxidation in the spiny lobster, *Panulirus argus,* in *Biochemistry, Biophysics and Regualtion of Cytochrome P-450,* Gustafsson, J. A., Carlstedt-Duke, J., Mode, A., and Rafter, J., Eds., Elsevier, Amsterdam, 1980, 113.
15. **Hallstrom, I., Black, A., Grafstrom, R., Rannug, U., and Sundvall, A.,** Cytochrome P-450 and xenobiotic metabolism in *Drosophila melanogaster,* in *Biochemistry, Biophysics and Regulation of Cytochrome P-450,* Gustafsson, J. A., Carlstedt-Duke, J., Mode, A., and Rafter, J., Eds., Elsevier, Amsterdam, 1980, 109.
16. **Naquira, C., White, R. A., and Agosin, M.,** Multiple forms of *Drosophila* cytochrome P-450, in *Biochemistry, Biophysics and Regulation of Cytochrome P-450,* Gustafsson, J. A., Carlstedt-Duke, J., Mode, A., and Rafter, J., Eds., Elsevier, Amsterdam, 1980, 105.
17. **James, M. O., Khan, M. A. Q., and Bend, J. R.,** Hepatic microsomal mixed-function oxidase activities in several marine species common to coastal Florida, *Comp. Biochem. Physiol.*, 62C, 155, 1979.
18. **Hlavica, P., Mietaschk, J., and Baden, I.,** Interactions of ligands with cytochrome P-450, *Biochem. J.*, 204, 425, 1982.
19. **Schenkman, J. B., Remmer, H., and Estabrook, R. W.,** Spectral studies of drug interaction with hepatic microsomal cytochrome, *Mol. Pharmacol.*, 3, 113, 1967.
20. **Burke, M. D.,** Cytochrome P-450: a pharmacological necessity or a biochemical curiosity, *Biochem. Pharmacol.*, 30, 181, 1981.
21. **Ebel, R. E.,** *In vitro* effects of chlordecon (Kepone) on hepatic microsomal cytochrome P-450, *Pest. Biochem. Physiol.*, 14, 221, 1980.
22. **Ebel, R. E., O'Keefe, D. H., and Peterson, J. A.,** Substrate binding to hepatic microsomal cytochrome P-450, *J. Biol. Chem.*, 253, 3888, 1978.
23. **Kumaki, K., Sato, M., Kon, H., and Nebert, D. W.,** Correlation of type I, type II and reverse type I difference spectra with absolute changes in spin state of hepatic microsomal cytochrome P-450 iron from five mammalian species, *J. Biol. Chem.*, 253, 1048, 1978.
24. **Lee, R. F. and Singer, S. C.,** Detoxifying enzymes system in marine polychaetes: increases in activity after exposure to aromatic hydrocarbons, *Rapp. P. -v. Reun. Cons. Int. Explor. Mer.*, 179, 29, 1980.
25. **Lee, R. F., Conner, J. W., Page, D., Ray, L. E., and Giam, C. S.,** Cytochrome P-450 dependent mixed-function oxygenase system in marsh crabs, in *Physiological Mechanisms of Marine Pollutant Toxicity,* Vernberg, W. B., Calabrese, A., Thurberg, F. P., and Vernberg, F. J., Eds., Academic Press, New York, 1982, 145.

Chapter 4

MONOOXYGENASE SYSTEMS IN MARINE FISH

J. J. Stegeman

TABLE OF CONTENTS

I. Objectives 66

II. Isolation of Monooxygenases 67
- A. Animal Data 67
- B. Sample Preparation 68
  1. Tissue Homogenization 69
  2. Isolation of Enzymes and Subcellular Fractions 69
  3. Embryo Fractionation 71
  4. Livers from Eleutheroembryos, Larvae, and Juveniles 72

III. Enzyme Purification 72
- A. Purification of NADPH-Cytochrome P-450 Reductase 73
- B. Purification of Cytochromes P-450 74
- C. Purification of Cytochrome P-450E 75
- D. Nonionic Detergent Removal 75

IV. Evaluation of Monooxygenase Systems 76
- A. Spectroscopic Analysis 76
- B. Catalytic Functions 76
  1. Chemicals 77
  2. Benzo(a)pyrene Monooxygenase 77
  3. Ethoxyresorufin O-Deethylase 79
  4. Aminopyrine N-Demethylase 79
  5. Ethoxycoumarin O-Deethylase 80
  6. NADPH-Cytochrome c Reductase 81
  7. NADH-Cytochrome c Reductase 81
  8. Glucose-6-Phosphatase 81
  9. Succinate Cytochrome c Reductase 81
  10. Limits of Detection 81
- C. High Pressure Liquid Chromatography 82
  1. Benzo(a)pyrene 82
  2. Testosterone 82
  3. 7,8-Benzoflavone and 5,6-Benzoflavone 82
- D. Reconstitutions 83
- E. Electrophoresis 83
- F. Immunochemical Studies 84

V. Results 84
- A. Environmental Induction 84
- B. Significance of Induction 87

Acknowledgments 89

References.. 89

I. OBJECTIVES

Cytochrome P-450-dependent monooxygenases play a major role in the biotransformation or metabolism of myriad organic xenobiotics, such as pesticides, hydrocarbons, chlorinated hydrocarbons, and numerous biological molecules, including steroid hormones. The metabolism of both xenobiotics and endogenous substrates can result either in activation or inactivation of these substrates, with quite different consequences for an organism. Thus, the metabolism of the environmental procarcinogen benzo(a)pyrene, well-known from mammalian studies, can result in more rapid elimination of the compound, but at the same time may form the active carcinogenic derivatives of this compound (Figure 1). The biological roles of xenobiotic monooxygenases indicate an important involvement in the effects of environmental chemicals in fish, including the likelihood that the neoplasms occurring in fish in various regions manifest an activity of cytochrome P-450 in activating carcinogens. A broad objective, then, of studying monooxygenase systems in marine animals is to understand the mechanisms of biotransformation and their role in detoxication, toxication, and effects of pollutants. Among more specific objectives are the following:

1. To estimate the relative capacity for foreign compound metabolism, both activation and detoxication, in species under different environmental conditions, in different organs, in different developmental stages, etc.
2. To evaluate whether metabolism of specific compound is adaptive or maladaptive.
3. To assess the feasibility of using characteristics of monooxygenase systems to signal the exposure of animals to biochemically significant levels of certain xenobiotics.
4. To determine how xenobiotic exposure can affect monooxygenase functions, including those involved in metabolism of both environmental chemicals and numerous types of endogenous regulatory molecules, e.g., steroids. The role of specific forms of cytochrome P-450 and their regulation are important to meeting these objectives.

The methods of studying the nature of monooxygenase systems have been developed principally with mammalian preparations over the past 10 to 15 years. Consequently, many of the methods employed in studies on these systems in aquatic species have been adapted from the mammalian literature. The methods described here are among the principal ones that we have used for the studies in this program, and include methods we have developed and others modified from the literature. Other investigators, of course, are also actively studying monooxygenase systems in aquatic (both marine and freshwater) species. Many of these investigators have described in the primary literature methods for studying microsomal electron transport components and monooxygenase activities, including some of the same aspects described here and additional aspects that we have not considered. In addition, there are major compilations of methods for analyzing features of monooxygenase systems, such as those appearing in *Methods in Enzymology*.[1] It should be noted that the present compilation is not meant to be all inclusive, or to supplant the primary sources. Rather, it is our intention

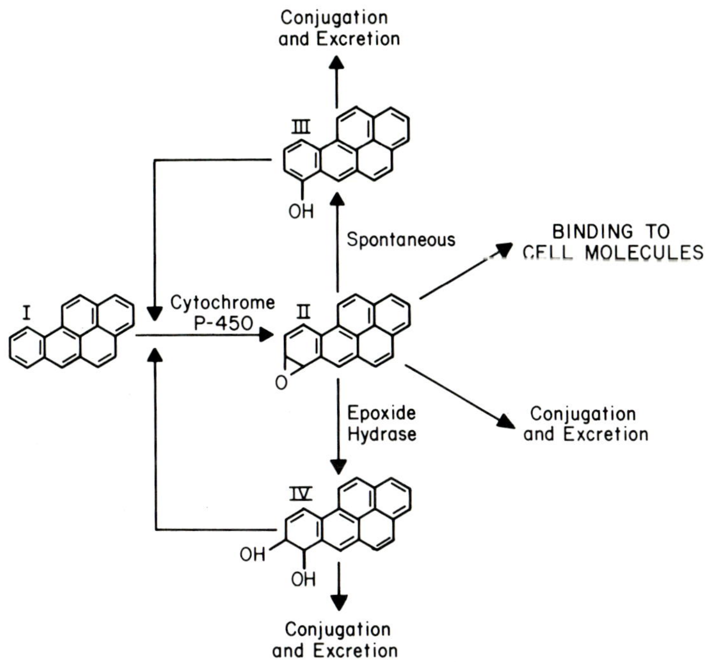

FIGURE 1. Simplified scheme for metabolism of benzo(a)pyrene by fish. Epoxides (II) can rearrange to phenols (III) or be hydrated to dihydrodiols (IV). Epoxidation can occur at several sites. Products III and IV can again serve as substrates, in the latter case leading to diolepoxides known to include active carcinogenic derivatives.

that this chapter be considered a guide to methods and approaches, including some precautions, which form a basic paradigm for those who may wish to undertake similar lines of research with marine species. Further, our aim is to provide a selection of data obtained employing some of these methods and to indicate some research questions. Described here are methods applied to studies with fish, but they can be applied to invertebrate systems, as well, as we have done with crustaceans and molluscs.

II. ISOLATION OF MONOOXYGENASES

A. Animal Data

One objective of studying monooxygenases is to assess these functions in field-sampled animals. However, valid interpretation of such data generally requires comparison with experimental results.

Because numerous factors can influence monooxygenase systems, certain precautions should be observed in experimental systems. Thus, certain characteristics of both experimental and field-sampled animals, and their environments, must be noted in order to allow any comparison of results with different species, or from different laboratories. Sex, season, temperature, age, and diet are among the factors known to influence monooxygenase functions in many animal species, both vertebrate and invertebrate. The influence of some of these has been established in several teleost species as part of this program (e.g., see References 2 to 4). An example of how strongly such factors can influence teleost monooxygenases is provided in Table 1, which describes sex differences seen in one species.

As a general rule, one should not only observe the condition of these variables, but report

Table 1
HEPATIC MICROSOMAL CYTOCHROME P-450 CONTENT AND MONOOXYGENASE ACTIVITIES IN GONADALLY MATURE TELEOST FISH[3]

	Winter flounder (*P. americanus*)	
Character	**Male**	**Female**
Cytochrome P-450[a]	1.28 ± 0.49[b]	0.15 ± 0.02
ER O-deethylase[c]	1.76 ± 0.68	0.06 ± 0.03
BaP hydroxylase[c]	0.21	0.18 ± 0.09

[a] nmol cytochrome P-450/mg microsomal protein.
[b] Values are means ± SD, with N greater than 3 in each case. Single values represent means of replicate assays on pooled samples.
[c] nmol product/min/nmol cytochrome P-450.

them. Not only should sex, season, and age be noted, but also the extent of gonadal development and the habitat temperature. Many of these factors will influence both the liver weight/body weight ratio and the content of microsomal protein, affecting the overall capacity of an animal to metabolize xenobiotics. Diet, which has been poorly studied, could also weigh heavily in both the expression and interpretation of data. Dietary insufficiency can contribute to a rapid decline in liver weight/body weight ratio, with effects on the yield of microsomal protein (milligrams protein per gram liver), the specific content or activity of cytochrome P-450, and the content normalized to a unit of organ weight. We have found that in scup *(Stenotomus chrysops),* a 30% decline in liver weight/body weight ratio can occur within 3 days after capture. It is possible that the quality of diet could influence pharmacokinetics and/or the effect of experimentally administered compounds in teleosts. In general, measurements of liver weight/body weight ratios of nonspawning animals should reveal whether or not there are questions about the sufficiency of the diet.

B. Sample Preparation

The dissection of organs and preparation of subcellular fractions are important to obtaining valid results when measuring monooxygenase activity. Organs should be removed from animals immediately after they are killed, placed in buffer on ice, and homogenized as soon as possible thereafter. Care must be taken to exclude samples contaminated by bile, which contains surfactants that might disrupt membranes, or contaminated by gut contents, which can also contain surfactants or digestive enzymes.

It is possible to freeze small pieces of tissue in liquid N_2 for later homogenization, but we have observed such thawed tissue to have slightly altered sedimentation characteristics. However, in our hands the characteristics of monooxygenase activity in membrane fractions prepared from liver frozen in liquid N_2 are in general not distinguishable from those obtained with preparations of fresh tissue. Although it is advisable to establish this empirically for each new species, it is also possible that characteristics of microsomal enzymes themselves prepared from quick frozen organs can indicate whether degradation has likely occurred. Thus, the appearance of cytochrome P-420 may be such an indication. Freezing on dry ice or at about $-20°C$ in an ordinary freezer should always be avoided because of extensive degradation that can accompany slow freezing. Data obtained with such samples must be considered highly suspect, unless it is proved that such treatment does not alter the enzymes in the species under study. Conditions for freezing of any samples should always be reported.

1. Tissue Homogenization

Homogenization of teleost liver is easily accomplished by standard procedures, but these must sometimes be modified for extrahepatic tissues to avoid including components that might introduce artifacts into data normalized either to protein content, organ weight, or body weight. In our studies[5] with extrahepatic organs, animals were killed by decapitation and tissues were immediately excised and placed in ice-cold 0.1 *M* phosphate, pH 7.3. With gills, only the filaments were taken; gut samples were immediately opened and the lumen extensively rinsed with buffer. Tissues were scissors-minced and homogenized in 4 or more volumes of appropriate buffer per gram of tissue weight using a Potter-Elvehjem tissue grinder with a motor-driven pestle. Including protease inhibitors in homogenization is also important with some extrahepatic organs.

Organs with more connective tissue often required more vigorous homogenization, but whatever the protocol, it was consistent for each organ or tissue type. Portions of any organ not effectively homogenized, e.g., connective tissue, were removed from whole homogenates and weighed, so that data expressed per gram of organ would reflect only that portion contributing to the sample. Homogenization of tissues rich in mitochondria, e.g., heart, was less vigorous, and carried out in larger volumes of buffer to avoid undue rupture of mitochondria.

2. Isolation of Enzymes and Subcellular Fractions

Isolation of appropriate membrane fragments of tissues can be accomplished by differential centrifugation[6] using either an ultracentrifuge or a high-speed centrifuge capable of sufficient speeds and force. Microsomes, fragments of the endoplasmic reticulum, can also be prepared by density gradient centrifugation or by calcium precipitation.[7] It has been our experience that calcium precipitation, while theoretically more convenient, is less straightforward than differential centrifugation, and is not recommended, although at least one investigator found it useful in preparing microsomes from cod liver.[8]

To establish the subcellular distribution of monooxygenase components, we generally employed the following procedure: tissue was homogenized on ice either in 10 volumes of 0.25 *M* sucrose, 0.1 *M* TRIS-HCl, pH 7.4, or in 5 volumes of 0.1 *M* phosphate, pH 7.4 with 1.15% KCl. The pH of TRIS buffers is temperature sensitive, and all buffers should be prepared with this in mind. Two, four, and two strokes of the motor-driven pestle were used to homogenize the tissue at 800, 1350, and 1850 rpm, respectively. Subcellular fractions were prepared by centrifuging (Sorvall RC2-B centrifuge SS-34 rotor,) at 600 *g* for 10 min (nuclear), 5,000 *g* for 10 min (heavy mitochondrial), 13,000 *g* for 10 min (light mitochondrial), and 40,000 *g* for 90 or 120 min (microsomal). The pellets were resuspended in the homogenizing buffer, recentrifuged, and the supernatants were combined at each step. The washed pellets were resuspended in 0.05 *M* TRIS-HCl, pH 7.8, 1.0 m*M* EDTA, 10 m*M* mercaptoethanol, 20% glycerol, or in 0.1 *M* phosphate, pH 7.3, to a concentration of 1 to 13 mg protein per milliliter. Characteristics of microsomes prepared by high-speed centrifugation like that above and ultracentrifugation employing 100,000 *g* are compared in Table 2. The slight difference clearly indicates that suitable microsomal fractions can be obtained without an ultracentrifuge.

Although most interest may be in microsomal function, it is important to evaluate the quality of other subcellular fractions presumably enriched in one membrane type or another. This is usually accomplished by the analysis of selected marker enzymes and by electron microscopic analysis of the isolated fractions (e.g., see References 19 and 34). The methods of de Duve et al.[6] include calculation and plotting of the "relative specific activity", which is used to determine the subcellular localization of an enzyme (see Figure 2).

As an alternative to preparation of microsomes, monooxygenase activities can be deter-

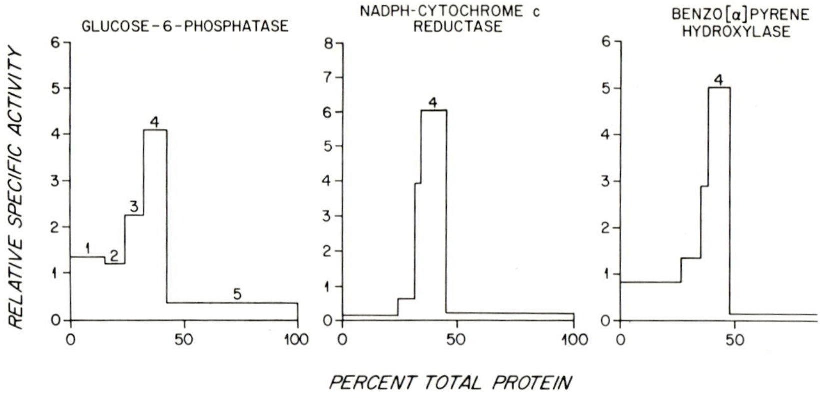

FIGURE 2. Subcellular distribution of enzyme activities. The data were converted to relative specific activity (RSA) according to de Duve et al.[6] RSA values of 4 or greater demonstrate primary localization of the activity in that fraction. Activity in other fractions could reflect cross-contamination and incomplete recovery or minor activity inherent in those fractions. Fraction 1, nucleii and debris; 2, mitochondria; 3, light mitochondria; 4, microsomal vesicles; 5, cytosol.

Table 2
COMPARISON OF MITOCHONDRIAL SUPERNATANT AND MICROSOMES PREPARED BY DIFFERENT METHODS

Component or activity	10 k × *g* supernatant	40 k × *g* microsomes	105 k × *g* microsomes
Protein (mg/g liver)	89 ± 2	16.2 ± 1.2	16.5 ± 0.6
NADH-cytochrome c reductase (units/mg)	39 ± 4	160 ± 8	181 ± 10
NADPH-cytochrome c reductase (units/mg)	31 ± 2	109 ± 9	115 ± 3
Cytochrome P-450 (nmol/mg)	0.08 ± 0.01	0.41 ± 0.01	0.43 ± 0.01
BaP hydroxylase (units/mg)	0.097 ± 0.016	0.343 ± 0.068	0.302 ± 0.070
Aminopyrine N-demethylase (units/mg)	0.35 ± 0.13	3.1 ± 0.68	2.8 ± 0.85

Note: Livers from 7 fish were pooled, minced, and homogenized in 5 volumes of buffer, and postmitochondrial supernatant (PMS) was prepared by centrifugation at 700 × *g* for 10 min and then 9000 × *g* for 15 min. Portions of the PMS were centrifuged at either 40,000 × *g* for 90 min to obtain 40 × *g* microsome or at 105,000 × *g* for 90 min to obtain 105 k × *g* microsomes. Both microsomal preparations were resuspended in 3 mℓ of buffer/g of liver. Units = nmol product formed/min.

mined either in whole homogenates, or in postmitochondrial supernatant fluids (PMS; also referred to as "S-9") generally obtained after centrifugation at 9000 *g* for 10 min. Preparations analogous to PMS can also be obtained by filtration with Millipore filters, which we have done with samples of midwater fish obtained at sea. Microsomes can be harvested at a later date from such preparations frozen in liquid N_2 or PMS (S-9) can be assayed as is. However, enzymatic activity measured in such fractions may be more usefully expressed per gram tissue or per nanomole P-450. First, these fractions will contain the bulk of cellular protein in addition to that found in the endoplasmic reticulum. Normalizing data to protein content would therefore yield low activities (Table 2), tending to minimize or even completely mask real differences in activity of microsomal enzymes between samples. Any comparison with microsomal data would also be precluded. Second, cytosolic factors can contribute to a higher activity than is seen in microsomes alone (Table 3). These factors might include Krebs cycle intermediates in the presence of mitochondrial fragments[9] or even glutathione.

Table 3
INFLUENCE OF CYTOSOLIC MATERIAL ON MICROSOMAL AHH ACTIVITY IN *FUNDULUS* LIVER

Reaction addition	Aryl hydrocarbon hydroxylase (nmol/min/g liver)
Postmitochondrial supernatant	1.80 ± 0.06
Microsomes	1.24 ± 0.09
Microsomes plus postmicrosomal supernatant[a]	1.74 ± 0.13

[a] Microsomal and soluble (microsomal supernatant) fractions were remixed in proportions to approximate the content of the postmitochondrial supernatant from which they were prepared. Soluble fractions typically possess negligible activity. The soluble fraction here contained 0.25 units/g liver.

Table 4
CYTOCHROME P-450 AND MICROSOMAL ENZYME ACTIVITIES IN SAMPLES FROZEN IN LIQUID N_2

	Scup liver microsomes analyzed	
Activity	September 1979 (fresh)	September 1982 (stored)
Cytochrome P-450 (nmol/mg)	0.81	0.82
AHH (nmol/min/mg)	1.18 ± 0.20	1.34 ± 0.22

Although the mechanism of such possible stimulation is not yet certain, it also can alter the perception of activity compared to that in microsomes.

Microsomal fractions can be resuspended in varied volumes of buffer per gram of liver, but care should be taken to accurately measure and record such volumes, accounting for losses, to allow reasonably accurate normalization to tissue or organ weight. Preparations of some organs from well-fed animals often have a very clear, less colored, sticky pellet beneath the microsomal pellet. This is glycogen, and care should be taken to avoid inclusion of the glycogen pellet.

It is possible to freeze subcellular preparations in small aliquots in liquid N_2 for future analysis. Addition of glycerol to the buffer is recommended as it seems to enhance stability. Early studies have considered factors in storage of microsomes.[10] Once frozen, such samples can be held for extended periods in liquid N_2. Quick frozen samples can also be archived at $-80°C$, but storage at warmer temperatures is not advised. Appropriately archived samples exhibit little change over time. We have observed that a sample of hepatic microsomes prepared in September 1979 from freshly caught scup and archived in liquid N_2 possessed characteristics of optical properties and catalytic functions several years later that were indistinguishable from those in the freshly archived microsomes (e.g., see Table 4).

3. Embryo Fractionation

In these studies, monooxygenase systems were examined in fish embryos and larvae, an area where much additional work is needed, given the prospects for environmental carcinogenesis or teratogenesis resulting from embryo exposure. Several methods had to be modified for use with the small samples. Examples of the methods used with *Fundulus* embryos[11] are discussed here. In preparation for homogenization, embryos were dechorionated with microsurgical scissors, washed in 0.05 *M* TRIS-HCl, pH 7.2, 1 m*M* EDTA, and 0.25 *M* sucrose (TES), and then gently pulsed in and out of a Pasteur pipette tip to rupture yolk sacs and release a substantial portion of the fluid yolk. After this, embryos were washed twice by swirling in TES, and then homogenized in TES (15 to 25 μℓ per embryo) in a Potter-Elvehjem tissue grinder with a Teflon pestle, 8 strokes at about 2500 rpm. Eleutheroembryos were simply washed with TES and then similarly homogenized. All operations were carried out on ice. The number of embryos and eleutheroembryos used varied, but generally was greater than 50. Embryos were weighed using a microanalytical balance.

Centrifugal fractions of embryos were prepared in either 7- or 10-mℓ tubes (Sorvall SS-34 rotor). The standard conditions were equivalent to 5 min at a maximum of 200 *g* (low speed), 15 min at a maximum of 10,000 *g* (10 k), and 180 min at a maximum of 40,000 *g* (40 k or microsomal) for full tubes. However, in all cases, tubes were spun with substantially less than full volumes and sedimentation conditions were adjusted using the "K" factor method (DuPont Instrument, Centrifuge Applications Bulletin No. 1). The period required to sediment microsomes did not exceed 90 min. Pellets were resuspended directly in a monooxygenase assay buffer, 0.5 *M* TRIS-HCl, pH 7.2 (0.2 to 3.0 μℓ per embryo) when only the microsomal fraction was to be assayed. When the low speed and 10K fractions were also examined, each pellet was washed by resuspension in buffer by hand homogenization and then resedimented. The wash and the original supernatant fluid were combined at each step in the fractionation.

4. Livers from Eleutheroembryos, Larvae, and Juveniles

Livers were dissected from *Fundulus* eleutheroembryos in ice-cold 0.5 *M* TRIS-HCl, pH 7.2, and 0.25 *M* sucrose (TS) using microsurgical instruments. Livers were carefully separated from gall bladders, transferred with a drawn-out Pasteur pipette to a 1-mℓ Potter-Elvehjem tissue grinder, washed once with TS, and then homogenized in TS (1 to 3 μℓ per liver) with a Teflon pestle, 8 strokes at about 2500 rpm. The resulting dilute whole homogenates (about 0.5 μg protein per microliter) were assayed immediately without further fractionation. Homogenates were prepared from pools of livers dissected from 30 to 60 individuals. Eleutheroembryo carcasses remaining after dissection were fractionated like whole embryos. Larvae and juveniles were immobilized by placing in ice-cold seawater, killed by pithing, and then washed with and dissected in ice-cold TS buffer. Livers carefully separated from gall bladders were homogenized like those from eleutheroembryos. The concentration of protein in the liver homogenates ranged from 0.5 to 4.0 μg/μℓ.

In some cases, whole homogenates were immediately assayed for monooxygenase activity. However, the following procedure was established for the preparation of microsomes from larval and juvenile livers. Homogenates were centrifuged in 0.4-mℓ conical-bottom polypropylene tubes (Sorvall SS-34 rotor with Sorvall No. 364 rubber adaptors). A postmitochondrial supernatant was prepared by centrifuging 300 μℓ of homogenate for 10 min at 5000 rpm; microsomes were then sedimented at 18,000 rpm for 60 min. These speeds are approximately equivalent to a maximum of 2,100 and 27,500 g, respectively. Centrifugation times and speeds were corrected when volumes other than 300 μℓ were used. Microsomes were resuspended in 0.5 *M* TRIS-HCl, pH 7.2, to a protein concentration of 0.4 to 0.6 μg/μℓ. A comparison of adult liver microsomes prepared by this microprocedure with those prepared by a more standard fractionation procedure indicated that the methods provide equivalent fractions. Thus, these microprocedures are appropriate to obtain fractions from very small samples, whatever the origin.

III. ENZYME PURIFICATION

A full understanding of cytochrome P-450 monooxygenase systems and their role in pollutant effects will depend on understanding their regulation in fish or other groups of animals, which in turn requires knowledge of specific forms or isozymes of cytochrome P-450. Our efforts to address detailed questions of specific forms have centered on isozymes of cytochrome P-450 purified from scup liver. The methods given here for solubilization of scup liver microsomes were reached after extensive testing of various detergent conditions,[12] and in our hands resulted in the least inactivation or loss of cytochrome P-450 during purification. Alternate methods employing zwitterionic detergents have been used by other investigators with rainbow trout.[13] Purification of cytochrome b_5 is not detailed here, although we have found it important in evaluating the function of purified cytochromes P-450.[14]

Scup hepatic microsomes were prepared from untreated fish with the inclusion of a wash step utilizing 0.1 *M* Na-pyrophosphate, pH 7.4, containing 10 m*M* EDTA. The wash step was effective in removing extraneous proteins as judged by a ~50% increase in cytochrome P-450 specific content. Microsomes were frozen in liquid N_2 and stored in the same or at − 80°C until use. The hepatic monooxygenase components were purified at 0 to 4°C, whereas buffers were routinely prepared at temperature and titrated to the indicated pH values (reported at room temperature) before adding the detergent component from a stock solution. The concentrations of glycerol and Emulgen in the buffers are given as v/v percentages, while the PEG 6000, cholate, and deoxycholate (DOC) concentrations are w/v percentages. Phenylmethanesulfonylfluoride (PMSF) was added as a solid just prior to the solubilizing buffer. All columns were poured, equilibrated, and run in accord with standard chromatographic procedures by gravity feed at rates of 10 to 50 mℓ/hr depending on the nature of the resin and column dimensions. A_{295} (protein) and A_{416} (heme) measurements for column fractions were corrected by subtracting the A_{650}. The protein absorbance was measured at A_{295} to avoid the UV component of the Emulgen (γ_{max} = 276 nm) present in most of the column buffers.

NADPH-cytochrome P-450 reductase was assayed spectrophotometrically based on its ability to reduce the nonphysiological electron acceptor cytochrome c. The purity of this enzyme was evaluated by specific activity (micromole substrate converted to product per minute per milligram protein), A_{277}/A_{456} (protein to flavin) ratio, and appearance on SDS-PAGE. Cytochrome b_5 was assayed by its characteristic reduced minus oxidized difference spectrum and appearance on SDS-PAGE. Purity was evaluated by A_{416}/A_{295} (heme to protein) ratio and specific (heme) content. Cytochromes P-450 were detected by their characteristic CO-ligated, reduced difference spectrum and purified on the basis of their differing apparent molecular weights judged by SDS-PAGE and the γ_{max} of their CO, reduced difference spectrum. Purification was routinely evaluated by A_{416}/A_{295} (heme to protein) ratio, specific content, and appearance on SDS-PAGE.

A. Purification of NADPH-Cytochrome P-450 Reductase

Scup hepatic microsomal NADPH-cytochrome P-450 reductase was purified by modifications of the standard chromatographic procedure on DEAE-Sephadex A25 and 2′,5′-ADP-Sepharose.[15,16] Microsomes were pelleted at 100,000 *g* for 60 min and resuspended to 10 mg protein per milliliter in 0.2 *M* TRIS-HCl, pH 7.7, containing 30% glycerol, 1 m*M* dithiothreitol (DTT), 1 m*M* EDTA, and 0.1 m*M* PMSF. The microsomal proteins were solubilized while stirring on ice by dropwise addition of aqueous 20% Emulgen 913 (Kao-Atlas, Japan) to give 1.5% Emulgen. After 20 min, the solution was treated with 0.03% protamine sulfate, added dropwise from a 1.5% (w/v) aqueous stock solution, and stirring was continued for 10 min. The soluble fraction was obtained by centrifugation at 100,000 *g* for 90 min.

The solubilized enzyme activity was chromatographed on a DEAE-Sephadex A25 column which had been equilibrated in several bed volumes of 50 m*M* HEPES, pH 7.7, containing 20% glycerol, 1 m*M* DTT, 1 m*M* EDTA, and 0.15% Emulgen 913 ("column buffer"). The column was washed with column buffer to elute several red cloudy fractions containing cytochrome P-450 and was then developed with a linear gradient to 0 to 0.4 *M* KCl in 4 bed volumes of column buffer. The NADPH-dependent cytochrome c reducing activity was eluted as a broad peak near the end of the gradient. The fractions containing the highest activity were pooled, supplemented with 2 μ*M* FMN, concentrated severalfold by ultrafiltration in an Amicon stirred cell (XM-50 membrane), and dialyzed overnight against 15 volumes of column buffer without the detergent.

The retained fraction was chromatographed on an 8-mℓ bed volume column of 2′,5′-ADP-Sepharose equilibrated in the dialysis buffer. The column was washed with 5 volumes of 20 m*M* K-phosphate, pH 7.7, containing 20% glycerol, 1 m*M* DTT, and 0.1% DOC (20 m*M* buffer); with 5 column volumes of 100 m*M* K-phosphate, pH 7.7, containing 20% glycerol, 1 m*M* DTT, and 0.1% DOC; and, briefly, with 20 m*M* buffer again. Finally, the enzyme activity was eluted with a linear gradient of 0 to 1.0 m*M* NADP in 5 bed volumes of 20 m*M* buffer. The peak activity was estimated to elute in a fraction corresponding to about 0.13 m*M* NADP. The fractions containing the highest activity were pooled, concentrated by ultrafiltration, and dialyzed against 250 volumes of 20 m*M* K-phosphate, pH 7.5, containing 20% glycerol, 0.5 m*M* EDTA, and 0.5% DOC to remove NADP and DTT. The purified scup NADPH-cytochrome P-450 reductase was stored in the dialysis buffer at −70°C, and the catalytic activity was stable for at least 6 months.

B. Purification of Cytochromes P-450

The overall purification strategy for scup cytochrome P-450 included detergent solubilization of microsomes, fractionation with polyethylene glycol (PEG) 6000, and resolution of the various cytochrome P-450 fractions by ion exchange chromatography on DEAE-cellulose. The resolved cytorochrome P-450 forms were further purified by a combination of rechromatography on DEAE-cellulose and column chromatography on hydroxylapatite and CM-cellulose. Partially purified cytochrome P-450 samples were stable for at least 6 months when frozen in liquid N_2 and stored at −70°C.

Hepatic microsomes (typically 1 g microsomal protein) were suspended at 10 mg protein per milliliter in 0.1 *M* TRIS-HCl, pH 7.4, containing 30% glycerol, 1 m*M* DTT, 1 m*M* EDTA, and 0.1 m*M* PMSF (I_c = 0.25, adjusted with KCl), and solubilized while stirring by the dropwise addition of potassium cholate and Emulgen 911 to final concentrations of 0.5 and 0.2%, respectively. After 30 min, the solution was centrifuged at 100,000 *g* for 75 min and the pellet was discarded. The soluble fraction was then treated with PEG 6000 by the dropwise addition of a 50% aqueous stock solution to a final concentration of 6% while stirring continuously. After 15 min, the cloudy solution was centrifuged at 35,000 *g* for 15 min and the large grey pellet was discarded. The supernatant fraction was adjusted to a final concentration of 20% PEG 6000 by further addition while stirring and, after 15 min, was centrifuged at 100,000 *g* for 90 min. The bright, clear-orange supernatant fraction was discarded and the deep-red pellet was dissolved in a minimum volume of 10 m*M* HEPES, pH 7.5, containing 20% glycerol, 1 m*M* DTT, 0.1 m*M* EDTA, 0.5% cholate, and 0.2 Emulgen 911 (Buffer I).

The concentrated soluble cytochromes P-450 were chromatographed on a DEAE-cellulose column which had been equilibrated with no more than 1 bed volume of Buffer I. The column was washed with 20 m*M* K-phosphate, pH 7.5, containing 20% glycerol, 1 m*M* DTT, 0.1 m*M* EDTA, 0.5% cholate, and 0.2% Emulgen 911 (Buffer II) until no more visible color eluted approximately 1 bed volume), and then the column was developed with a linear gradient of 0 to 0.5 *M* KCl in 4 to 5 bed volumes Buffer II. The KCl gradient was monitored with a YSI Model 31 conductivity bridge. Five fractions containing cytochrome P-450 were eluted from the DEAE-cellulose column and were labeled Fraction A through E in order of their elution (Figure 3). NADPH-dependent cytochrome c reductase activity was found to elute between the last cytochrome P-450 fraction and cytochrome b_5. Complete

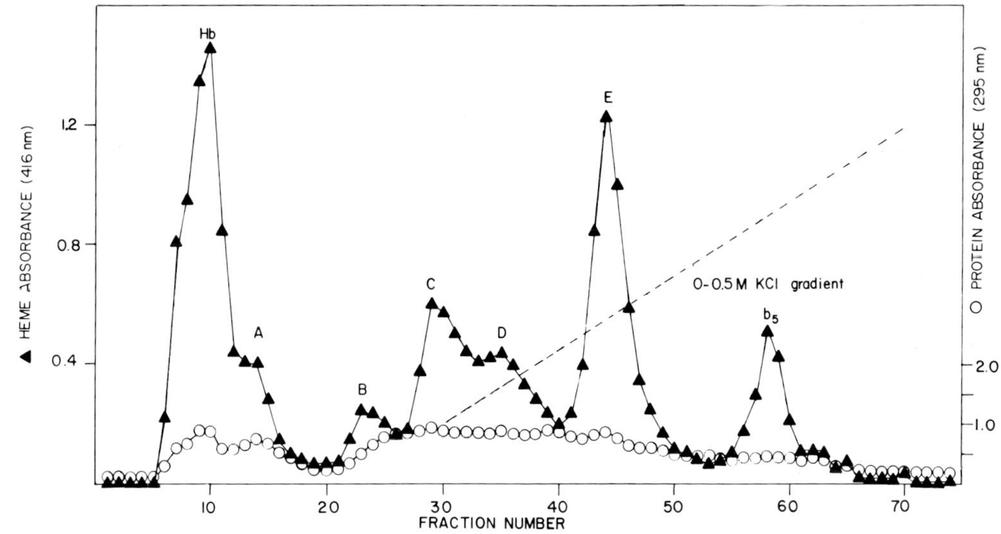

FIGURE 3. DEAE-cellulose ion exchange chromatography of hepatic microsomal cytochromes P-450 from untreated scup. 400 nmol of detergent-solubilized, PEG-fractionated cytochrome P-450 (see methods) were chromatographed at 4°C on a DEAE-cellulose column (80 mℓ bed volume) equalibrated in buffer I. The column was washed with buffer II to elute the hemoglobin and was developed with a 350-mℓ linear gradient of 0 to 0.5 *M* KCl in buffer II at fraction No. 12. Salt concentration was monitored by conductivity bridge. Cytochrome P-450-containing fractions were combined as follows: (A) fraction Nos. 13 to 15; (B) fraction Nos. 22 to 24; (C) fraction Nos. 28 to 32; (D) fraction Nos. 34 to 37; (E) fraction Nos. 42 to 47. Cytochrome b_5 eluted in fraction Nos. 56 to 60. Fractions of approximately 5.5 mℓ were collected throughout the chromatography. The total cytochrome P-450 recovery for the chromatography was greater than 90%. (From Klotz, A. V., Stegeman, J. J., and Walsh, C., *Arch. Biochem. Biophys.*, 226, 578, 1983. With permission.)

resolution of the first cytochrome P-450 from hemoglobin was obtained in later experiments by an initial wash with Buffer I after loading as hemoglobin does not bind to the column in 10 m*M* HEPES.

C. Purification of Cytochrome P-450E

The apparent major hydrocarbon-induced P-450, cytochrome P-450E, was further purified by dialyzing the pooled DEAE-cellulose fractions against 20 volumes of Buffer I (without detergent) overnight and chromatographing the retained fraction on a hydroxylapatite column which had been equilibrated in Buffer I including 0.2% Emulgen 911 but no cholate. The column was developed with 5 bed volumes of 20 m*M* K-phosphate, pH 7.5, containing 20% glycerol, 1 m*M* DTT, 0.1 m*M* EDTA, and 0.2% Emulgen 911 (20 m*M* Buffer III); 5 bed volumes of 50 m*M* Buffer III; and the cytochrome P-450 was eluted as a sharp peak with 100 m*M* Buffer III.

D. Nonionic Detergent Removal

Emulgen 911 was removed from the purified cytochrome P-450 preparations by repeating the hydroxylapatite chromatography, but replacing the Emulgen 911 in the column buffers with 0.4% cholate or 0.1% DOC. Briefly, the cytochrome P-450 sample was dialyzed overnight against 20 volumes of Buffer I without detergent and chromatographed on a hydroxylapatite column which had been equilibrated with Buffer I containing 0.2% Emulgen 911. The column was washed with 5 bed volumes of 20 m*M* Buffer III containing 0.4% cholate or 0.1% DOC until the column effluent fell below an A_{276} of 0.05 (typically, approximately 3 bed volumes). The column could be further washed with 50 m*M* Buffer

III containing 0.4% DOC, and the cytochrome P-450 was finally eluted as a concentrated band with 200 m*M* Buffer III containing 0.1% DOC. The peak cytochrome P-450 fractions were combined, frozen in liquid N_2 as small aliquots, and stored at $-70°C$. Typical recoveries on this column chromatography step were 70% with little change in the specific content of the resulting preparation.

IV. EVALUATION OF MONOOXYGENASE SYSTEMS

A. Spectroscopic Analysis

Various spectral analyses are required to verify and quantitate cytochromes P-450 and cytochrome b_5. The methods reported by Omura and Sato[17] are most often cited, but several modifications exist.[18] Analysis of cytochrome P-450 is accomplished by obtaining either the dithionite difference spectrum of CO-treated preparations, or the CO-difference of dithionite-treated preparations. In general, we have employed the former method to reduce interference by hemoglobin. Samples in the cuvettes have usually contained about 1 mg microsomal protein per milliliter, with a limit of detection of about 20 pmol of cytochrome P-450, which was established by titration of cytochrome P-450 content. Assays for cytochrome b_5 are usually carried out prior to cytochrome P-450 analysis using dilutions or concentrations like those above. After obtaining an NADH difference spectrum, NADH and CO are balanced in the sample and reference cuvettes, then the sample reduced by $Na_2S_2O_4$ to determine the cytochrome P-450. This procedure not only permits quantitation of cytochrome b_5, but effectively eliminates interference by b_5 or hemoglobin in the analysis, and permits observation of any cytochrome P-420. This degradation product of cytochrome P-450, if present, could limit validity of many interpretations. It is also possible to reduce CO-bound microsomes enzymatically with NADPH, but unless anaerobic conditions are obtained, only the fraction of the P-450 temporarily escaping reoxidation will be observed spectrally.

Generally, the above procedures are adequate for hepatic microsomes, but extrahepatic preparations may require modifications to account for additional contaminating chromophores. An example is provided by heart,[19] which generally yields microsomes contaminated by mitochondrial fragments. First, spectral analyses of scup heart mitochondrial preparations were carried out with resuspensions containing about 1 mg mitochondrial protein per milliliter. Heart mitochondria and liver microsomes mixed in resuspensions that contained about 0.5 mg protein from each were assayed as above to determine the limits of cytochrome P-450 detection in the presence of mitochondrial chromophores. In some assays on each type of preparation, succinate and KCN were added to achieve final concentrations of 5 and 1 m*M*, respectively, in both sample and reference cuvettes. This treatment to eliminate interference by mitochondrial cytochromes, particularly cytochromes a-a_3, present as contaminants in heart microsomes, was necessary to obtain reproducible spectra that could be analyzed. The identity and quantity of cytochrome P-450 determined optically were subsequently confirmed by electron paramagnetic resonance spectroscopy.[19]

B. Catalytic Functions

Evaluating the biological role or regulation of cytochromes P-450 in either microsomes or resolved preparations requires characterization of catalytic functions. A suitable spectrum of substrate and inhibitor probes for cytochromes P-450 can be used to indicate the regulation and function of different isozymes. The substrates that might be considered for use should include some that are known to reflect the activity of different isozymes or subsets of cytochrome P-450. An appropriate suite of substrates could help establish the identity of a catalyst and its relationship to other P-450s. Admittedly, this is complicated by the fact that most substrates appear to be catalyzed by more than one form of cytochrome P-450. However, there are some substrates that are more or less selective. Thus, aryl hydrocarbon (benzo(a)pyrene) hydroxylase activity and ethoxyresorufin O-deethylase activity are catalyzed preferentially by methylcholanthrene or aromatic hydrocarbon-type inducible isozymes. By contrast, some N-demethylase reactions are catalyzed preferentially by forms not inducible by these compounds. The use of these and selected other substrates, such as ethoxycoumarin, testosterone, α-naphthoflavone (ANF), and inhibitors such as ANF, SKF-525A, and metyropone, are considered in detail below. The reader should note that there are other useful probes for catalytic function including, e.g., the warfarins, biphenyl, the phenylimidazoles, various suicide substrates, and the use of bacteria mutation assays. While these may also reveal some specificity in analysis of monooxygenases, the assays provided here are considered to be more useful in providing an initial comparison with widely studied systems.

Assay methods for enzymes involved in microsomal electron transport are also given. NADPH-cytochrome P-450 (cytochrome c) reductase (and possibly NADH-cytochrome b_5 reductase and cytochrome b_5) is involved in reduction of cytochrome P-450, essential to catalysis. Insufficient function of such could be a limiting factor in monooxygenase activity. Methods for a few relevant marker enzymes to characterize subcellular fractions are also provided.

1. Chemicals

The quality of chemicals employed in monooxygenase studies can be very important. Small amounts of contaminants or oxidized products can seriously interfere with assays meant to detect similar products that are formed biologically. In general, the best quality compounds should be obtained, and further purification is often required. Procedures we have employed for benzo(a)pyrene, α-naphthoflavone, testosterone, and aminopyrine are provided below. Some substrates require synthesis, one of which is ethoxyresorufin, which gains importance by virtue of its close association with hydrocarbon-induced cytochrome(s) P-450. Methods for synthesis of ethoxyresorufin were originally provided by Prough et al.[20] Our modification[21] of the procedure is described here. Briefly, the sodium salt of resorufin was prepared by dissolving 1.2 g crude resorufin (5.6 mmol) in 15 mℓ anhydrous dimethylformamide containing 2.0 g anhydrous Na_2CO_3 (19 mmol). After dissolution occurred, 4 mℓ ethyl iodide (50 mmol) was passed slowly through a 1-mℓ alumina column and introduced directly into the solution. The reaction was refluxed at 45°C for 24 hr and stopped by the addition of 50 mℓ water. The crude product was extracted with methylene chloride, washed extensively with 0.1 *M* $NaHCO_3$, and evaporated to dryness under reduced pressure. The washed extract was redissolved in a small volume of $CHCl_3/CH_3CN$ (9/1) and chromatographed in the same solvent system on a 2 × 30 cm column of silica gel 60. The orange fractions which eluted with an R_f of approximately 0.3 were individually spotted for thin layer chromatography on silica G in solvent system A. The fractions judged free of other chromophores were combined, evaporated to dryness, and recrystallized in acetone. The 7-ethoxyresorufin obtained (161 mg, 12%) was judged greater than 95% pure by a combination of TLC, HPLC, 90 MHz ^1H-NMR, and elemental analysis.

2. Benzo(a)pyrene Monooxygenase

Benzo(a)pyrene monooxygenase (aryl hydrocarbon hydroxylase, [AHH]) is closely, but not exclusively, associated with hydrocarbon-induced cytochrome(s) P-450. This feature, plus the widespread distribution of this compound in the environment and the sensitivity of assays, makes it a desirable substrate to study. Precautions appropriate to working with a carcinogen, available from the National Institutes of Health, must be employed.

AHH activity can be assessed by determining the production of fluorescent derivatives or measuring the total production of water-soluble derivatives of ^3H-BaP. In our hands, AHH activity was usually assayed by a modification of the method of Van Cantfort et al.,[22] but with several changes in the procedure. The overall procedure is more fully described here.

The substrate, benzo(a)pyrene, was prepared for use in the following way. Generally labeled ^{3}H-BaP (5 mCi) was mixed with unlabeled BaP to a specific activity of about 300 μCi/μmol and dissolved in 6 mℓ of hexane. This solution was then extracted 5 times with 3 mℓ of aqueous 1 *M* KOH/DMSO (15:85; v/v), washed twice with 3 mℓ of deionized water, dried with anhydrous $MgSO_4$, filtered through a hexane-washed 0.5-μm Fluoropore filter, and dried under a stream of nitrogen. The purified ^{3}H-BaP solid was stored at −20°C under nitrogen. All glassware used in the purification procedure was washed with chromic acid cleaning solution and hexane, and the entire procedure was done under red light. The actual specific activity of each preparation was determined by counting aliquots of dilutions of three or four individual stock solutions of the ^{3}H-BaP.

The reaction mixture consisted of 0.5 *M* TRIS-HCl with 1.4 mg/mℓ bovine serum albumin, 0.40 m*M* NADPH or NADH, 70 μM ^{3}H-BaP (approximately 150 μCi/μmol), and about 60 μg of sample protein in a volume of 50 μℓ. This small assay volume reflects our efforts to establish a microassay (see below). The reaction was initiated with 2 μℓ ^{3}H-BaP in methanol, after which the samples were incubated 15 min at appropriate temperatures. Reactions were stopped with 100 μℓ 0.15 *M* KOH in 85% DMSO and were extracted 3 times with 0.5 mℓ hexane. Added to 3 mℓ of scintillation cocktail was 50 μℓ of the aqueous layer containing the metabolites, acidified with 30 μℓ 0.6 *N* HCl and counted in a liquid scintillation counter (Beckman LS-100C). Counting efficiency was determined with internal standards. The influence of α-naphthoflavone on BaP metabolism was determined by adding 2 μℓ ANF in methanol to the BaP monooxygenase reaction mixture just prior to addition of ^{3}H-BaP. BaP metabolism in various preparations was generally linear with time to 20 min and with protein to 0.15 mg/mℓ of reaction mixture; assays were carried out in the linear range. Blanks consisted of reaction mixtures minus cofactor. Assays were carried out in triplicate under red light.

For small volumes of enzyme (e.g., those available from embryos), the reaction mixture was further altered.[11] The AHH assay was carried out in 10 × 75 mm disposable glass tubes with a reaction volume of 25 μℓ. Reaction mixtures consisted of 0.05 *M* TRIS-HCl, pH 7.2 at 30°, 0.40 m*M* NADPH, 0.070 m*M* ^{3}H-BaP, 0.8 mg/mℓ bovine serum albumin, and varying amounts of embryonic protein depending on the fraction assayed. Between 15 to 100 μg of whole embryo microsomal protein, 40 to 10 μg of whole liver protein, or 1 to 5 μg of liver microsomal protein was added per reaction mixture. Blanks consisted of the complete reaction mixture without NADPH. The reaction was initiated on ice by adding BaP in 1 μℓ of acetone, then incubating at reaction temperature. Activity measured with adult *Fundulus* hepatic microsomes preincubated with BaP at the reaction temperature and then initiated by the addition of cofactor was like that with initiation on ice by the addition of BaP. When the influence of ANF or SKF-525A on the reaction was examined, these inhibitors were added with the BaP in 1 μℓ of acetone. Reactions were stopped by placing on ice and adding 50 μℓ of 0.15 *M* KOH in 85% DMSO. The stopped reaction mixtures were extracted 3 times by vortexing for 15 sec with 0.25 mℓ hexane, centrifuging for 5 min at 5000 rpm, and then aspirating off the organic phase. Polar metabolites were quantitated by counting 30 to 50 μℓ of the remaining aqueous phase in 3 mℓ of Aquasol acidified with 10 μℓ of 0.6 *N* HCl. Samples were counted in Beckman LS-100C liquid scintillation counter, and efficiency was determined by internal standardization with ^{3}H-toluene. All micropipetting devices used in the procedure were calibrated gravimetrically using water with correction for evaporation.

The assay was carried out under red light from the addition of ^{3}H-BaP through the first extraction with hexane. Although Van Cantfort et al.[22] indicate that the activity measured in the assay is little affected by exposure to light, the sensitivity of the assay is decreased when the procedure is carried out under white light. The limit of detection of the assay is determined by the sample to blank ratio. Exposure of BaP solutions and reaction mixtures to white light results in elevated blanks, presumably due to photooxidation. The blank was minimized in the procedure described here by reducing the assay volume and using great care to protect the ^{3}H-BaP from light. The assay as described can detect about 1 pmol of metabolites in the final stopped reaction mixture.

AHH activity can also be assayed in a procedure modified from Nebert and Gelboin,[23] which detects phenolic derivatives that fluoresce strongly under basic conditions.[24] Reaction mixtures employed here contained either 0.5 mg NADPH or a generating system consisting of 5 m*M* NADP, 15 m*M* glucose-6-phosphate, 3 m*M* $MgCl_2$, and 2 units of glucose-6-phosphate dehydrogenase in either 0.1 *M* phosphate or 0.1 *M* Tris. After 15 min preincubation, microsomal protein was added to a final concentration ranging from 0.15 to 1.5 mg/mℓ depending on the tissue, with liver usually at 0.37 to 0.45 mg/mℓ. Reactions were initiated by addition of 7.8 μg of BaP in 20 μℓ methanol, and incubation carried out at 29°C for 20 min. Final volume of reaction mixtures was 0.5 mℓ and final pH was 7.0. For determining pH optima, 0.5 mg NADPH replaced the generating system. When studying ANF inhibition, both ANF and BaP were added in a total of 20 μℓ methanol. Reactions were stopped with 1 mℓ acetone and 3 mℓ hexane. Hydroxylated product was extracted from 0.1 to 2.0 mℓ of the hexane, depending on activity, with 0.1 *N* NaOH (3 mℓ), and assayed fluorometrically using a suitable fluorometer (excitation at 396 nm and emission at 520 nm) with authentic 3-OH-BaP as a reference standard. All phases of incubation and product extraction were carried out in the dark or under red light.

3. Ethoxyresorufin O-Deethylase

Ethoxyresorufin O-deethylase (EROD) activity, also linked to induced cytochrome(s) P-450, can be measured by direct assays either fluorometrically or spectrophotometrically.[20,21] The latter method, detailed here, may be more widely useful because of the greater availability of spectrophotometers. It has been compared to the fluorometric assay in detail.[21] The assay employs a 1.0-mℓ reaction mixture containing 2 μ*M* 7-ethoxyresorufin, 0.1 *M* TRIS, pH 8.0, with 0.1 *M* NaCl, and 22 μℓ microsomes (3 to 8 mg/mℓ protein). A saturated stock solution (approximately 400 μ*M*) of 7-ethoxyresorufin in methanol should be prepared fresh daily by dissolving a crystal of 7-ethoxyresorufin in hot methanol, allowing the solution to cool to room temperature, and filtering. The concentration can be verified by dilution into buffer and measuring the A_{482} (ϵ_{482} = 22.5 mM^{-1} cm^{-1}). A cocktail can be prepared, and enzyme added some minutes prior to assay. The reaction is initiated with NADPH at a final concentration of 0.5 m*M*. The appearance of resorufin is monitored at 572 nm using a suitable recording spectrophotometer. The extinction coefficient of resorufin, experimentally determined to be 73 mM^{-1} cm^{-1}, provides a rapid, sensitive assay of this reaction.

4. Aminopyrine N-Demethylase

Aminopyrine N-demethylase (APD) has, to date, not been found to be catalyzed effectively by aromatic hydrocarbon-induced cytochrome P-450 in a number of studies with fish.[25] It can therefore be one of those activities employed to evaluate other cytochrome P-450 forms. Other activities in this category include benzphetamine N-demethylase or ethylmorphine N-demethylase, assays for which are readily reproduced.[26,27] Although APD appears to be principally regulated differently from AHH or EROD activities, APD induction by hydrocarbon inducers is occasionally noted in fish. Studies of APD may therefore be interesting in an effort to identify the nature and latitude of this catalyst.

Aminopyrine N-demethylase has been assayed by using NADPH or a generating system as described above, in 0.166 *M* HEPES, pH 7.6, at a final volume of 1.5 mℓ. Microsomal protein concentrations were 0.1 to 1.0 mg/mℓ of reaction mixture, with liver microsomes added at 0.25 to 0.35 mg/mℓ. Reactions were initiated by adding 15 μmol of aminopyrine that had been twice recrystallized from methanol, incubated at 15 min at 29°C, and stopped by adding 0.5 mℓ 25% saturated $ZnSO_4$ and 0.5 mℓ saturated $Ba(OH)_2$. Formaldehyde was determined using the Nash reagent.[28] Formaldehyde for constructing a standard curve should be prepared from paraformaldehyde (1 or 2 drops of 0.1 *N* NaOH and moderate heating will aid dissolution). A range of standards should be incubated and treated as samples. It should be noted that various substances, including TRIS, can contribute a positive reaction with the Nash reagent, and that blanks must be carefully designed. It has been our experience that HEPES buffer provides the least interference.

A radiometric procedure for analysis of APD[29] was modified for use on a microscale with embryonic material. The substrate used was prepared in the following way. Unlabeled aminopyrine (AP), recrystallized twice from methanol, was mixed with an aqueous solution of ^{14}C-AP to a specific activity of 1.5 μCi/μmol. The AP was extracted from this solution into chloroform, washed 3 times with deionized water to remove any soluble impurities, dried with anhydrous $MgSO_4$, filtered through a chloroform-washed 0.2-μm Fluoropore filter, and dried under a stream of nitrogen, and then under vacuum. The labeled AP was stored as a dry powder, desiccated at −20°C. The specific activity of each preparation was checked by counting aliquots of dilutions of three or four individual stock solutions.

The assay was carried out in 6 × 50 mm tubes with the total reaction volume reduced to 15 μℓ. The reaction mixture consisted of 7.5 m*M* ^{14}C-AP, 7.5 m*M* NADPH, 10 m*M* semicarbazide-HCl, 1 m*M* $MgCl_2$, 0.15 *M* TRIS-HCl, pH 7.2 at 30°C, and varying amounts of protein depending on the sample assayed. About 100 μg of whole eleutheroembryo microsomal protein, 10 μg of juvenile liver microsomal protein, or 20 μg of adult liver microsomal protein was added per reaction mixture. Blanks consisted of the complete reaction mixture without NADPH. Reactions were initiated on ice by the addition of ^{14}C-AP in 2 μℓ of H_2O, incubated at 30°C for 35 min, and stopped by placing on ice and adding 50 μℓ of ice-cold 0.1 *N* NaOH and 0.38 mℓ of chloroform. After addition of chloroform, tubes were vortexed for 10 sec to extract the unmetabolized parent compound, and then the phases were separated by centrifugation for 5 min at 5000 rmp. A 60-μℓ aliquot of the aqueous phase of each tube was transferred to a new tube containing 0.38 mℓ of fresh chloroform, then vortexed and centrifuged again. To quantitate the production of formaldehyde, 40 μℓ of the final aqueous phase was counted in 3 mℓ of Aquasol acidified with 10 μℓ of 0.6 *N* HCl. The scintillation counting efficiency was determined using ^{14}C-toluene as an internal standard. The efficiency of recovery of formaldehyde was measured by adding 200 pmol $^{14}CH_2O$ (10 μCi/μmol) in place of ^{14}C-AP; typically, the recovery was around 80 to 85%. The recovery of formaldehyde was considered when activity was calculated.

5. *Ethoxycoumarin O-Deethylase*

Ethoxycoumarin metabolism is induced in mammals by various types of inducers. In fish it is induced by hydrocarbons, but is also catalyzed to a significant extent by isozymes other than the MC- or BNF-induced forms. Ethoxycoumarin O-deethylase assays were modified from the literature.[30] Reaction mixtures (0.4 mℓ final volume) contained 0.1 *M* phosphate, pH 7.8, 50 to 100 pmol of microsomal cytochrome P-450, and 0.2 m*M* 7-ethoxycoumarin added in 4 μℓ of CH_3OH. Reactions were initiated with NADPH at a final concentration of 1 m*M*. Incubations were performed for the desired time and halted by the addition of 60 μℓ 2 *N* HCl. Reaction mixtures can be reduced in volume to 0.1 mℓ, in which case around 10 pmol of microsomal cytochrome P-450 can be employed. Reduced volume reactions should be stopped with a mixture of 60 μℓ 1 *N* HCl and 340 μℓ of buffer. Reactions were extracted with 0.9 mℓ $CHCl_3$ and the aqueous phase was aspirated away. Finally, 0.6 mℓ of the $CHCl_3$ phase was transferred to a fresh tube containing 2 mℓ of 30 m*M* $Na_2B_4O_7$, pH 9.2, and the fluorescence of the aqueous phase was measured by excitation at 370 nm and monitoring the emission at 455 nm. Quantitation was based on the fluorescence of a 7-hydroxycoumarin standard, which was linear between 20 and 500 pmol under identical conditions.

6. NADPH-Cytochrome c Reductase

Microsomal or purified NADPH-cytochrome P-450 reductase (EC 1.6.2.4) was routinely assayed by a modification of Phillips and Langdon,[31] employing cytochrome c as an electron acceptor. A suitable reaction mixture contains 0.175 m*M* NADPH, 80 μ*M* horse heart cytochrome c, 0.2 *M* phosphate, pH 7.7, ionic strength 0.58 in a total volume of 1.65 mℓ. The reaction was initiated by addition of 0.02 to 0.5 mg microsomal protein depending on the tissue, and reduction of cytochrome c followed at 550 nm using a recording spectrophotometer. Cytochrome c reduced was determined using an extinction coefficient of 21.1 cm^{-1} mM^{-1}.

7. NADH-Cytochrome c Reductase

NADH-cytochrome b_5 reductase activity was assayed in a manner identical to that for NADPH cytochrome c reductase with 0.25 m*M* NADH replacing NADPH. Reference cuvettes in both cases contained reaction mixtures with no enzyme. In some circumstances, such as the possible presence of mitochondrial material, it is advisable to add 1 m*M* KCN to inactive cytochrome oxidase (cytochromes a-a_3) and prevent reoxidation of the reduced cytochrome c in any of the cytochrome c reductase assays.

8. Glucose-6-Phosphatase

Glucose-6-phosphatase (G-6-Ptase; EC 3.1.3.9), generally localized in the endoplasmic reticulum, was assayed by measuring the production of inorganic phosphate in a reaction mixture containing 50 mol cacodylate buffer, pH 6.5, and 30 μmol glucose-6-phosphate in 1.4 mℓ.[5] This mixture was preincubated for 5 min at 30°C, the reaction initiated by addition of 0.1 mℓ of enzyme (0.1 to 1.0 mg protein), and incubated for 10 min at 30°C, after which the reaction was terminated by adding 1 mℓ 10% TCA. A series of "zero-time" control reaction mixtures in which enzyme was added after addition of TCA was also run. The denatured protein was sedimented by centrifuging at 1800 rpm for 9 min or by refrigeration overnight, and inorganic phosphate in the supernatant was determined by the method of Flynn et al.[32]

9. Succinate Cytochrome c Reductase

Succinate cytochrome c reductase (succinate dehydrogenase; EC 1.3.99.1), an indicator of mitochondrial presence, was assayed according to Green et al.[33] Reaction mixtures were 1.65 mℓ and contained 10 mg/mℓ bovine serum albumin, 8 μ*M* cytochrome c, 1 m*M* KCN, 0.02 to 0.1 mg microsomal protein per milliliter, and 0.02 *M* phosphate, pH 7.4. Reactions were initiated by addition of sodium succinate to 5 m*M*, and the reduction of cytochrome c followed at 25°C as above. This activity and that of G-6-Ptase may serve to indicate both localization of important components and contamination of microsomes by mitochondria. However, a complete study of subcellular fractionation requires analysis of a larger suite of presumed marker enzymes, as carried out by Balk et al.[34]

10. Limits of Detection

Practical limits of detection are important to establish the sensitivity of assays and the validity of results, particularly with low activity samples. The limits determined empirically for the assays considered here were established under the conditions used for these assays. These limits are provided only to give examples and would not necessarily be relevant to these assays done elsewhere or with different samples: cytochrome P-450, 0.01 nmol/mg microsomal protein; cytochrome b_5, 0.005 nmol/mg; cytochrome c reductases, 0.2 nmol cytochrome c reduced per minute per milligram; aminopyrine demethylase, 0.13 nmol formaldehyde produced per minute per milligram (spectrophotometrically) or 0.1 nmol/min/mg (radiometrically); EROD, 0.015 nmol resorufin produced per minute per milligram; and

AHH activity, 1.0 pmol/min/mg (radiometrically) or 0.5 pmol 3-OH-BaP equivalents produced per minute per milligram (fluorometrically). Note that data should always be corrected for extraction efficiency of the material being assayed in all assays that are not direct measurements.

C. High Pressure Liquid Chromatography

Analysis of multiple products of some substrates can establish a regioselectivity of monooxygenases, and with some can indicate whether activity might be more likely to have functions in toxication or in detoxication. Assays for four such substrates are provided here.

1. Benzo(a)pyrene

Methods for analysis of BaP metabolites have flourished in the past 10 years.[35] Metabolites of ^{3}H-BaP have been obtained in our studies by in vitro incubation of microsomes prepared from various organs of fish. Reaction mixtures of 1-mℓ containing 60 μ*M* BaP, 0.375 mg NADPH per milliliter, and 2.0 to 5.9 mg microsomal protein per milliliter were incubated for 20 min. The reactions were stopped with 1 volume of ice-cold acetone at the end of the incubation, either anthraquinone (approximately 0.5 μg) or 6β-hydroxytestosterone added as an internal standard, and the reaction mixtures were extracted twice with 2 volumes of ethyl acetate. The combined organic extracts were dried under a stream of N_2 and redissolved in a small volume of acetonitrile or methanol. Zero-time and boiled enzyme blanks were treated in a similar fashion. Separation of BaP metabolites was achieved using a DuPont LC 859 chromatograph fitted with a 50-μℓ injection loop, a 25-cm DuPont Zorbax ODS column, and a filter photometer UV detector operating at 254 nm. Metabolites were eluted with a gradient of 40 to 85% acetonitrile in water. Metabolites can be identified based on coelution with authentic standards and quantitated by counting the labeled fractions eluting from the column. All procedures were carried out under red light. Recovery was determined for each metabolite by isolating labeled metabolites previously produced and carrying these individually through the complete procedure used for zero-time blanks. Recoveries ranged between 61 and 83%, with a mean at about 70%.

2. Testosterone

A method for analysis of steroid metabolites by HPLC was also modified[36] from the literature.[37] Reaction mixtures contained 0.5 m*M* NADPH and 25 μ*M* 4-^{14}C-testosterone (52 μCi/μmol) or 100 μ*M* 1,2,6,7,16,17-^{3}H-testosterone (25 μCi/μmol) in a final volume of 0.4 mℓ 0.1 *M* K-phosphate, pH 7.8. The reactions were quenched with 1 volume of ice-cold acetone, extracted twice with 2 mℓ of ethyl acetate, and the combined organic extracts were evaporated to dryness. Prior to extraction for HPLC analysis, each sample was spiked with unlabeled 17-epitestosterone as an internal standard. The extracts were resuspended in a small volume of methanol and analyzed on a C-18 reverse phase column. Testosterone metabolites were separated with a gradient of 0 to 30% solvent D (20 min), 30 to 80% solvent D (30 min), and 100% solvent D (30 min) at a flow of 1 mℓ/min. The weak solvent (C) was composed of CH_3OH/H_2O (45:55, v/v) and the strong solvent (D) was composed of $CH_3OH/CH_3CN/H_2O$ (55:10:35, v/v/v). The order of elution from the HPLC column was 6α-OH-, 15α-OH-, 7α-OH-, 6β-OH-, 14α-OH-, 16α-OH-, 11α-OH-, 16β-OH-, 11β-OH-, and 2β-OH-testosterone; androstendione; testosterone; and 17-epitestosterone under these conditions. All standards were resolved to baseline except the 15α-/7α-OH- and 2β-/11β-OH-testosterone paris. Products were identified relative to standards and quantitated by liquid scintillation counting, corrected for recovery of the unlabeled internal standard.

3. 7,8-Benzoflavone and 5,6-Benzoflavone

Metabolism of 7,8-benzoflavone (7,8-BF; α-naphthoflavone, ANF) or of 5,6-benzoflavone (5,6-BF; β-naphthoflavone, BNF) provides further information on the active site(s) of various cytochromes P-450. Oxidation of ANF or BNF was detected by the HPLC method of Stegeman and Woodin.[38] Alternate methods for HPLC analysis of flavone metabolites have been published.[39] Reaction mixtures in 0.1 *M* TRIS, pH 7.5, contained microsomes and 100 μ*M* ANF or BNF. The reactions were initiated with NADPH, stopped with one volume of ice-cold acetone, and extracted twice with two volumes of ethyl acetate. The combined organic extracts were dried under a stream of N_2 or under reduced pressure and resuspended in a small volume of acetonitrile. The products were separated by HPLC on a reverse-phase C-18 column utilizing a 30 min linear gradient of 40 to 70% acetonitrile in water at 2 mℓ/min. Products were quantitated by peak areas (Hewlett-Packard 3390A integrator) and activities were expressed in arbitrary integration units/min/nmol cytochrome P-450 because no product standards were available for quantitation. Analysis of the products is establishing their identity as a suite of dihydrodiols and phenols.[42]

D. Reconstitutions

Characterization of the catalytic functions of purified isozymes of cytochrome P-450 is required in order to make subsequent interpretations regarding the significance of patterns of regulation of that isozyme. Reconstitutions of purified cytochrome P-450 with reductase and lipid can be accomplished in any of several ways; each has technical or interpretative advantages or disadvantages. The technically more practical is reconstitution of cytochrome P-450 with the reductase in a nonmembranous aggregate with dilaurolphosphatidylcholine. This method requires an unphysiologically high ratio of reductase to cytochrome P-450, typically 1:1, compared to 1:20 or so which usually exists in microsomes. An alternative method is to form liposomes with a molar excess of cytochrome P-450 to reductase. We have employed the former method in our studies to date.[13] Reconstitutions were typically accomplished in as small a volume as practical, and analysis of products was accomplished using the methods above. It is particularly important in effecting reconstitutions to consider the limits of detection and anticipated turnover of a substrate based on microsomal studies to properly evaluate the conditions for reconstitution. It is further important to know that small amounts of detergent can strongly affect the activity in reconstitutions.

E. Electrophoresis

Electrophoretic analysis should accompany studies of monooxygenases, and is required to monitor the quality of fractions during purification. SDS-polyacrylamide disc gel electrophoresis (SDS-PAGE) to resolve microsomal proteins is most often carried out according to the methods of Laemmli et al.[41] using 8% polyacrylamide gels. In our hands, 50 μg of microsomal protein per well and a current of 2 mA per gel maintained for the duration of the run generally provided good results. Gels can be stained for protein for 30 min with 0.2% Coomassie Blue in 10% acetic acid per 50% methanol, and destained by diffusion, although lower concentration and longer staining may reveal more minor bands. Before staining for protein, gels can also be stained for peroxidase activity, which indicates the presence of heme, with tetramethylbenzidine according to the method of Thomas et al.[36] Suitable standards for molecular weight determination (all from Sigma Chemical Co.) are bovine serum albumin (66,000), glutamate dehydrogenase (53,000), ovalbumin (45,000), aldolase (39,000), and trypsinogen (24,000). Many other standards are also available. Molecular weights can be estimated by comparison with standards run both externally and internally with unknowns. A few standards of higher and lower molecular weight can be mixed with a sample before it is applied to a gel, which can improve accuracy in estimating molecular weight. Densitometric analysis of destained gels can be accomplished by scanning with a suitable densitometer.

F. Immunochemical Studies

Immunochemical analysis of cytochromes P-450 structure and function is a very fruitful area of research. Specific antibodies can be used to analyze structure, function, and regulation of cytochrome P-450 isozymes. We have developed both poly- and monoclonal antibodies to the form of cytochrome P-450 from scup identified as the major hydrocarbon-inducible isozyme. Our development of these has been subsequent to PRIMA, and descriptions will appear in the primary literature.

V. RESULTS

Research on cytochrome P-450 monooxygenases in the PRIMA program followed several lines, including detailed characterization of the systems, interorgan and interspecific differences, regulation by environmental and biological variables, characterization of chemically induced states, and the evaluation of monooxygenase characteristics in field-sampled organisms. These and other lines of research are continuing. Briefly considered here are some results and conclusions of our studies and those of others which emphasize environmental induction and suggest areas where research is needed.

As indicated in several recent reviews,[43,44] monooxygenases in different fish species are quite similar with respect to the range of substrates metabolized. Review of the literature also reveals that pH (near 7.4) and in some cases ionic strength optima for specific reactions in vitro are rather similar in different species. Temperature optima for monooxygenase reactions generally vary with habitat temperature, ranging from 25 to 30°C for fish from northern temperature waters[45] to near 40°C for fish from subtropical waters.[46] The distinctions in temperature optima are like those seen in many other enzymes,[47] but studies to identify the most temperature-sensitive component of monooxygenase systems are in some disagreement. Williams et al.[48] suggest that membrane lipids may dictate the temperature response, while Philpot et al.[49] indicate that the NADPH-cytochrome P-450 reductase per se is the sensitive moiety of the system. Studies to establish the quantitative relationships between in vitro and in vivo rates and the mechanism of temperature effect require further research.

Regulation of monooxygenase function by biological variables was addressed in several marine species. As indicated above, sex differences in hepatic monooxygenases have been identified in several teleosts (e.g., see References 2 and 3). The roles of circulating steroids, principally estradiol, have also been described in teleosts,[36] but the mechanism is not yet established. In addition to steroids and reproductive influences, the regulation during development has been established. The patterns of AHH activity during development from embryo to eluetheroembryo, first established in *F. heteroclitus,*[11] are important in considerations of developmental effects of xenobiotics and the use of embryos in carcinogenesis studies. Induction of AHH activity at very early stages of development was also indicated.[50,51] As with temperature, the significance of xenobiotics to in vivo disposition is not fully understood.

A. Environmental Induction

One objective of studies of the forms, functions, and regulation of cytochromes P-450 in PRIMA was to define the characteristics that would enable clear identification of animals induced in the environment. The search for specific substrates and inhibitors of activity in microsomes from experimentally induced fish in several laboratories has provided a pattern that thus far is largely consistent. First, 3-methylcholanthene-type inducers (aromatic hydrocarbons, β-naphthoflavone-BNF, certain congeners of chloro- and bromobiphenyls) are very active in fish, whereas phenobarbital and phenobarbital-type inducers have no clearly demonstrable inductive effect (for reviews, see References 43 and 44). An example of

Table 5
HEPATIC MICROSOMAL ELECTRON TRANSPORT COMPONENTS IN *MICROPOGON UNDULATUS* TREATED WITH BENZO(A)PYRENE

Character	Control (4)	Benzo(a)pyrene (5)
Cytochrome P-450 (nmol/mg)	0.12 ± 0.02	0.35 ± 0.02[a]
Fe^{2+}-CO λ_{max} (nm)	450	449
Cytochrome b_5 (nmol/mg)	0.003	0.009[a]
NADPH-cytochrome c reductase (units/mg)[b]	79 ± 7	109 ± 12
NADH-cytochrome c reductase (units/mg)[b]	128 ± 3	147 ± 5
BaP hydroxylase (AHH) (units/mg)	0.026 ± 0.002	0.420 ± 0.021[a]
Percent AHH activity with 10^{-4} *M* 7,8-BF	128%	8%[a]

Note: Experimental animals were injected intraperitoneally with 10 mg BP/kg and sacrificed 24 hr later.

[a] Significantly different from controls at $P \leq 0.001$.
[b] Units equal nmol cytochrome c reduced/min, or nmol ^{3}H-BP metabolized/min. Reactions were carried out at optimal conditions and in the linear range of protein and time. BaP monooxygenase blank reactions lacked NADPH. Data are mean ± SD.

Table 6
EFFECTS OF α-NAPHTHOFLAVONE, METYRAPONE, AND SKF 525-A ON CROAKER LIVER MONOOXYGENASE ACTIVITY

		AHH activity in:	
Inhibitor	Concentration (μ*M*)	Control	(BaP) treated
None	—	0.024 ± 0.003[a]	0.420 ± 0.021[a]
ANF	0.125	124 ± 12[b]	106 ± 10
	5.0	106 ± 5	36 ± 15
	100.0	115 ± 10	8 ± 2
MT	1.0	71 ± 5	49 ± 25
	10.0	83 ± 5	116 ± 5
	100.0	158 ± 13	92 ± 32
SKF 525A	1.0	121 ± 20	117 ± 5
	10.0	71 ± 10	135 ± 7
	100.0	133 ± 16	69 ± 31

[a] Activities are nmol/min/mg microsomal protein, and are derived from samples referred to in Table 5.
[b] Values are percent of activity obtained without additions. Means of replicate assays, ± SD of the assay.

induction is provided in Table 5. As stated above, induction by 3-methylcolanthene-type inducers has been identified in embryos as well as adults. Second, induced systems characteristically have elevated levels of benzo(a)pyrene hydroxylase activity and ethoxyresorufin O-deethylase activities, and are more sensitive to inhibition by α-naphthoflavone. However, other common inhibitors, e.g., SKF-525A and metyrapone, are generally not selective in fish. They can have effects in some species, but do not have any apparent effect in others (Table 6). The relationship between levels of catalytic activity and ANF inhibition seen in

Table 7
RELATIONSHIP BETWEEN AHH ACTIVITY, EROD ACTIVITY, AND INHIBITION BY α-NAPHTHOFLAVONE IN FLOUNDER *(PLATICHTHYS FLESUS)*

AHH activity group	(N)	Percent activity with 100 μ*M* ANF	AHH activity (nmol/min/nmol P-450)	EROD activity (nmol/min/mg)
<0.025	(5)	370 ± 140	0.061 ± 0.018	-ND-
0.05—0.075	(6)	100 ± 16	0.228 ± 0.037	0.046 ± 0.008
>0.100	(3)	54 ± 12	0.513 ± 0.068	0.081 ± 0.026

Note: Samples were arbitrarily grouped according to AHH activity in nmol per mg microsomal protein. Data are means ± SD.

Data obtained by collaboration among Stegeman, J. J., Tarlebo, J., and Solbakken, J. E.

some samples (e.g., Table 7) suggests that the crossover from stimulation by ANF to inhibition by ANF might occur at very low AHH activities, suggesting that even these low rates might reflect the activity of an induced isozyme. This is consistent with our findings that antibodies against induced cytochrome P-450 in scup can inhibit more than 95% of the AHH activity in microsomes (see below).

A combination of these catalytic and other characteristics can be taken as strong evidence for the presence of induced cytochrome P-450 in fish. Within the limits imposed by possible influence of natural chemical or biological factors on these features, it has been concluded that induction by chemicals in the environment does occur, and that such induction may in fact be widespread geographically among teleost species.[52-54] These conclusions are based on varied catalytic evidence, on spectrophotometric and electrophoretic evidence, and on comparisons with experimentally induced samples. At present, such a rigorous and full assessment is generally required for substantive interpretations. Nevertheless, it would appear that high levels of ethoxyresorufin O-deethylase activity alone might be used as a first indication of induction. However, the apparent dissociation of EROD activity from AHH activity seen in some cases[52] needs to be explained. It could reflect multiple induced forms.

Resolution and purification of cytochromes P-450 from scup[12] and trout[13] have provided a much clearer picture of induced cytochromes P-450 in fish. The resolution of cytochromes P-450 by ion exchange chromatography (e.g., Figure 1) has revealed that five or more isozymes are present in scup liver. The highly purified forms differ in physicochemical and catalytic properties[14] (Table 8). However, reconstitution studies and studies with polyclonal antibodies[55] show that the form termed cytochrome P-450E is the major catalyst of benzo(a)pyrene metabolism in scup and the apparent major AH-induced form. This induced isozyme accounts for as much as 50% of the microsomal cytochrome P-450 in some fish sampled directly from the environment.[12] Scup cytochrome P-450E shares many properties with major BNF-induced forms from other fish and rats (Table 9), but there is strong immunochemical relatedness only among fish (Table 10). Similar results are seen with antibodies to trout BNF-P-450.[56] Monoclonal antibodies presently available against cytochrome P-450E[60] should be useful in establishing further the relationships between scup P-450E and other forms,[61] and the regulation of this form in the environment.

Although the occurrence of extensive environmental induction seems to be established, and the major induced catalyst in some cases identified, there remain important questions concerning the origin of this induction. Microsomal studies indicate that different inducers, such as BNF, 3-methylcolanthene, or 3,4,5,3′4′5′-hexachlorobiphenyl, elicit a very similar

Table 8
PROPERTIES OF PURIFIED EMULGEN-FREE SCUP CYTOCHROMES P-450[71]

Form	MW[a] (kd)	UV-visible absorbance peaks (nm) Fe^{2+}-CO	Oxidized direct			
E	54.3	447	278.5	417.5	534	567
A	52.7	447.5	278	417	532	565
B	45.9	449.5	278	416	532	567
Fraction D	—[b]	451	279	419	536	566

[a] MW determined by SDS-PAGE.
[b] This preparation was heterogeneous, but was enriched in a band near 50 kd.

Table 9
SOME CHARACTERISTICS OF MAJOR MC- OR BNF-INDUCIBLE CYTOCHROME P-450 FROM VARIOUS SPECIES

Species	Isozyme designation	MW (kd)	Fe^{2+}-CO λ_{max}(nm)	Prominent activities	Ref.
Fish					
Scup	P-450E	54	447	AHH, EROD	12
Rainbow trout	P-450 LM$_{1b}$	58	447	AHH, EROD	57
Cod	P-450c	58	447	Phe,[a] EROD	58
Mammals					
Rat	P-450c	56	447	AHH, EROD	59

[a] Phenanthrene; the activities for AHH were not given.

response in fish, but definitive work to establish that these compounds induce proteins that are identical has not been done. However, immunological studies now underway in our laboratory, including analysis by western blot, indicate identity between the enzymes induced by different MC-type compounds. Given that the major proteins induced by some different 3-methylcolanthene-type inducers are identical, the origin of such induction in fish in the environment might be linked to the chemistry in a given area. Correlation with PCB residues in some areas suggests that these compounds could be responsible. Further studies are important to establish probable causes of widespread induction and whether some active compound(s) is suitably bioavailable and of sufficient concentration to elicit induction.

B. Significance of Induction

The significance of induction to the health of individuals or populations also remains to be established. It has been demonstrated in one study that cytochrome P-450 activity might determine the extent of elimination of aromatic hydrocarbons.[62] However, the functions of oxidative metabolism in both toxication and detoxication suggest that induction could override an animal's capacity to detoxify reactive products by conjugation reactions, or may result in a depletion of cellular glutathione, with toxic consequences. A compilation of data on these latter enzymes in marine species appears in James et al.,[63] and a discussion of the role of these enzymes in xenobiotic effects in aquatic species is also available.[64] However, studies are necessary to determine how regulation of these various enzymes, including

Table 10
EFFECT OF POLYCLONAL ANTIBODIES AGAINST SCUP P-450E ON HEPATIC MICROSOMAL EROD ACTIVITY IN VARIOUS SPECIES

Microsome source	mg IgG/nmol P-450	% activity remaining[a]
BNF		
Scup	40[b]	8 ± 1
Grunt	60	16 ± 1
Trout	40	15 ± 1
Rat	60	105 ± 3

[a] 100% activity in each case is that obtained upon incubation with partially purified Ig from preimmune serum. EROD (ethoxyresorufin O-deethylase) activities were scup, 6.4; grunt, 2.4; trout, 3.05; and rat, 4.3 nmol/min/nmol P-450. Values are means of triplicate assays ± SD.

[b] Points of maximal inhibition on a titration curve, except for grunt, which was analyzed at 2 points.

From Stegeman, J. J., Woodin, B. R., Park, S. S., Kloepper-Sams, P. K., and Gelboin, H. V., *Mar. Environ. Res.*, in press. With permission.

induction of cytochrome P-450 isozymes and activities, can be linked to any such imbalance in fish or other marine animals.

Higher rates of metabolism of benzo(a)pyrene have not been associated with any differences in the profile of benzo(a)pyrene metabolites in studies to date (e.g., Table 11; Reference 65). In our studies, the patterns of BaP metabolites were remarkably conserved in virtually all preparations from teleost fish, with a strong preference for formation of benzo-ring metabolites. These are usually seen as trans-dihydrodiols, although in some individual fish, the pattern is of benzo-ring phenols correlated with low epoxide hydrolase activity.[66] Efficient activation to mutagens[67] and binding to DNA[68] suggest that stereoselectivity of induced cytochrome P-450 and of epoxide hydrolase in fish may be like that in mammals.[69]

The induction we have demonstrated in the embryos of fish could play a role in the activation of mutagens or procarcinogens in developmental stages. Altered metabolism of endogenous substrates is also a possible pathway for effects of environmental induction; e.g., in rainbow trout, the metabolism of steroids can be altered by induction.[70] The metabolism of steroids by marine teleost liver cytochrome P-450 has been established in the course of PRIMA studies. Results obtained with *Fundulus* and winter flounder microsomes indicate that the principal catalyst(s) for testosterone hydroxylase activities and those responsible for some xenobiotic metabolism are differentially regulated.[3] This is consistent with the identity of the apparent major testosterone hydroxylase (cytochrome P-450A) distinct from that of the major benzo(a)pyrene hydroxylase (cytochrome P-450E) purified from scup, another marine teleost species.[14] However, metabolism of testosterone is also catalyzed by the induced cytochrome P-450E. The hepatic metabolism of estrogens has yet to be carefully addressed in marine species, but alteration of both estrogens and androgens could affect reproductive processes.

The studies briefly considered here reflect substantial gains in knowledge of cytochrome P-450 in marine species. Outstanding questions are concerned with the nature of cytochrome

Table 11
BENZO(a)PYRENE METABOLITES FORMED BY SCUP LIVER, KIDNEY, AND GILL MICROSOMES

	Metabolites formed by:		
Metabolite	Liver	Kidney	Gill
9,10-Dihydrodiol	341 (29)	71 (31)	9 (28)
4,5-Dihydrodiol	ND	ND	ND
7,8-Dihydrodiol	331 (28)	49 (22)	8 (25)
1,6-Quinone	77 (6)	7 (3)	2 (6)
3,6-Quinone	108 (9)	6 (3)	1 (3)
9-Hydroxy	32 (3)	15 (6)	0.5 (1)
3-Hydroxy	307 (26)	80 (35)	11 (34)
Total activity (nmol BaP transformed/ min/mg protein)	1.19	0.228	0.032

Note: Values are pmol product/min/mg protein with percent of total in parentheses. Data for liver were obtained with 2 fish pooled. Values for kidney and gill represent data obtained with tissues pooled from more than 20 fish. Volume of reaction mixtures was 0.25 mℓ with protein concentrations of 0.5 mg/mℓ (liver), 1.9 mg/mℓ (kidney), and 3.1 mg/mℓ (gill). The percent of total BaP metabolized during incubations was 8.5% (liver), 2.0% (kidney), and 0.5% (gill). ND, not detected.

From Stegeman, J. J., Woodin, B. R., and Binder, R. L., *Natl. Cancer Inst. Monogr.*, 65, 371, 1984. With permission.

P-450 isozymes other than PAH-inducible cytochrome P-450 in marine species, their relationship to mammalian forms, and their regulation. Development of probes for cytochrome P-450 genes from fish should accelerate the study of such questions. Further research on function and regulation of specific forms of cytochrome P-450 and other enzymes, including flavoprotein monooxygenases, should establish their significance in the biological effects of marine pollutants.

ACKNOWLEDGMENTS

Support of the National Science Foundation is gratefully acknowledged. Work supported by NSF and summarized or reported here represents efforts by a number of students and co-workers, particularly Robert Binder, Alan Klotz, and Bruce Woodin, to whom great appreciation is extended. Contribution No. 4899 from the Woods Hole Oceanographic Institution.

REFERENCES

1. **Fleischer, S. and Packer, L., Eds.,** *Methods in Enzymology,* Vol. 52 (Part C), Academic Press, New York, 1978.
2. **Stegeman, J. J. and Chevion, M.,** Sex differences in cytochrome P-450 and mixed-function oxygenase activity in gonadally mature trout, *Biochem. Pharmacol.*, 28, 553, 1980.
3. **Stegeman, J. J. and Woodin, B. R.,** Differential regulation of hepatic xenobiotic and steroid metabolism in marine teleost species, *Mar. Environ. Res.*, 14, 422, 1984.

4. **Stegeman, J. J.,** Temperature influence on basal activity and induction of mixed function oxygenase activity in *Fundulus heteroclitus, J. Fish. Res. Board Can.*, 36, 1400, 1979.
5. **Stegeman, J. J., Binder, R. L., and Orren, A.,** Hepatic and extrahepatic miscrosomal electron transport components and mixed-function oxygenases in the marine fish *Stenotomus versicolor, Biochem. Pharmacol.*, 28, 3431, 1979.
6. **De Duve, C., Wattiaux, R., and Baudhin, P.,** Distribution of enzymes between subcellular fractions in animal tissues, *Adv. Enzymol.*, 24, 291, 1962.
7. **Kamath, S. A. and Ananth Narayan, K.,** Interaction of Ca^{++} with endoplasmatic reticulum of rat liver: a standard procedure for the isolation of rat liver microsomes, *Anal. Biochem.*, 48, 53, 1972.
8. **Goksøyr, A.,** The Cytochrome P-450 System of the Cod, *(Gadus morhua)*, Cand. Scient. dissertation, University of Bergen, Norway, 1984.
9. **Cinti, D. L., Richie, A., and Schenkman, J. B.,** Hepatic organelle interaction. II. Effect of tricarboxylic acid cycle intermediates on N-demethylation and hydroxylation reactions in rat liver, *Mol. Pharmacol.*, 8, 339, 1972.
10. **Borton, P., Carson, R., and Reed, D. J.,** Stability of suspensions of rat liver microsomes quick-frozen with liquid nitrogen and stored at −85°, *Biochem. Pharmacol.*, 23, 2332, 1974.
11. **Binder, R. L. and Stegeman, J. J.,** Microsomal electron transport and xenobiotic monooxygenase activities during the embryonic period of development in the killifish, *Fundulus heteroclitus, Toxicol. Appl. Pharmacol.*, 73, 432, 1984.
12. **Klotz, A. V., Stegeman, J. J., and Walsh, C.,** An aryl hydrocarbon hydroxylating hepatic cytochrome P-450 from the marine fish *Stenotomus chrysops, Arch. Biochem. Biophys.*, 226, 578, 1983.
13. **Williams, D. E. and Buhler, D. R.,** Purification of cytochromes P-448 from β-naphthoflavone-treated rainbow trout, *Biochim. Biophys. Acta*, 717, 398, 1982.
14. **Klotz, A. V., Stegeman, J. J., and Walsh, C.,** Multiple isozymes of hepatic cytochrome P-450 from the marine teleost fish scup *(Stenotomus chrysops), Mar. Environ. Res.*, 14, 402, 1984.
15. **Strobel, H. W. and Dignam, J. D.,** Purification and properties of NADPH-cytochrome P-450 reductase, *Methods Enzymol.*, 52, 89, 1978.
16. **Yasukochi, Y. and Masters, B. S. S.,** Some properties of a detergent-solubilized NADPH-cytochrome *c* (cytochrome P-450) reductase purified by biospecific affinity chromatography, *J. Biol. Chem.*, 251, 5337, 1976.
17. **Omura, T. and Sato, R.,** The carbon monoxide-binding pigment of liver microsomes, *J. Biol. Chem.*, 239, 2370, 1965.
18. **Johannesen, K. A. M. and DePierre, J. W.,** Measurement of cytochrome P-450 in the presence of large amounts of contaminating hemoglobin and methemoglobin, *Anal. Biochem.*, 86, 725, 1978.
19. **Stegeman, J. J., Woodin, B. R., Klotz, A. V., Wolke, R. W., and Orme-Johnson, N. R.,** Cytochrome P-450 and monooxygenase activity in cardiac microsomes from the fish *Stenotomus chrysops, Mol. Pharmacol.*, 21, 517, 1982.
20. **Prough, R. A., Burke, M. D., and Mayer, R. T.,** Direct fluorometric methods for measuring mixed-function oxidase activity, *Methods Enzymol.*, 52, 372, 1978.
21. **Klotz, A. V., Stegeman, J. J., and Walsh, C.,** An alternative 7-ethoxyresorufin O-deethylase activity assay: a continuous visibile spectrophotometric method for measurements of cytochrome P-450 monooxygenase activity, *Anal. Biochem.*, 140, 138, 1984.
22. **Van Cantfort, J., DeGraeve, J., and Gielen, J. E.,** Radioactive assay for aryl hydrocarbon hydroxylase, improved method and biological importance, *Biochem. Biophys. Res. Commun.*, 79, 505, 1977.
23. **Nebert, D. W. and Gelboin, H. V.,** Substrate inducible microsomal aryl hydrocarbon hydroxylase in mammalian cell culture. I. Assay and properties of induced enzyme, *J. Biol. Chem.*, 234, 6242, 1968.
24. **Holder, G., Yagi, H., Levin, W., Lu, A. Y. H., and Jerina, D. M.,** Metabolism of benzo[a]pyrene III. An evaluation of the fluorescence assay, *Biochem. Biophys. Res. Commun.*, 65, 1975.
25. **Schwen, R. J. and Mannering, G. J.,** Hepatic cytochrome P-450-dependent monooxygenase systems of the trout, frog and snake. III. Induction, *Comp. Biochem. Physiol.*, 71B(3), 445, 1982.
26. **Elcombe, C. R. and Lech, J. J.,** Induction of monooxygenation in rainbow trout by polybrominated biphenyls: a comparative study, *Environ. Health Perspect.*, 23, 309, 1978.
27. **James, M. O. and Bend, J. R.,** Effect of polynuclear aromatic hydrocarbons and polyhalogenated biphenyls on hepatic mixed-function oxidase activity in marine fish, in *Carcinogenic Polynuclear Aromatic Hydrocarbons in the Marine Environment,* Richards, N., Ed., Proc. Conf. Pensacola, Fla., Aug. 1978, U.S. Environmental Protection Agency, Washington, D.C., 1981.
28. **Cochin, J. and Axelrod, J.,** Biochemical and pharmacological changes in the rat following chronic administration of morphine, nalophine and normorphine, *J. Pharmacol. Exp. Ther.*, 125, 105, 1959.
29. **Poland, A. P. and Nebert, D. W.,** A sensitive radiometric assay of aminopyrine N-demethylation, *J. Pharmacol. Exp. Ther.*, 184, 269, 1973.

30. **Greenlee, W. F. and Poland, A.,** An improved assay of 7-ethoxycoumarin O-deethylase activity: induction of hepatic enzyme activity in C57BL/6J and DBA/2J mice by phenobarbital, 3-methylcholanthrene and 2,3,7,8-tetrachlorodibenzo-*p*-dioxin, *J. Pharmacol. Exp. Ther.*, 205, 596, 1978.
31. **Phillips, A. H. and Langdon, R. G.,** Hepatic triphosphopyridine nucleotide-cyt-c reductase isolation, characterization, and kinetic studies, *J. Biol. Chem.*, 237, 2652, 1962.
32. **Flynn, R. M., Jones, M. E., and Lipmann, F.,** A colorimetric determination of inorganic phosphate, *J. Biol. Chem.*, 211, 791, 1954.
33. **Green, D. E., Mu, G., and Kohout, P. M.,** Studies on the terminal electron transport system. I. Succinate dehydrogenase, *J. Biol. Chem.*, 217, 551, 1955.
34. **Balk, L. Maner, S., Bergstrand, A., and DePierre, J.,** Preparation and characterization of subcellular fractions suitable for studies of drug metabolism from the trunk kidney of the northern pike, (Esox lucius) and assay of certain enzymes of xenobiotic metabolism in these subfractions, *Biochem. Pharmacol.*, 33(15), 2447, 1984.
35. **Selkirk, J. K., Croy, R. G., Roller, P. P., and Gelboin, H. V.,** High-pressure liquid chromatographic analysis of benzo[a]pyrene metabolism and covalent binding and the mechanism of action of 7,8-benzoflavone and 1,2-epoxy-3,3,3-trichloropropane, *Cancer Res.*, 34, 3474, 1974.
36. **Stegeman, J. J., Pajor, A. M., and Thomas, P.,** Influence of estradiol and testosterone on cytochrome P-450 and monooxygenase activity in immature brook trout, *Salvelinus fontinalis, Biochem. Pharmacol.*, 31, 3979, 1982.
37. **Van der Hoeven, T. A.,** Separation of metabolites of testosterone hydroxylation by high performance liquid chromatography, *J. Chromatogr.*, 196, 494, 1980.
38. **Stegeman, J. J. and Woodin, B. R.,** The metabolism of α-naphthoflavone (7,8-benzoflavone) by hepatic microsomes from the marine fish *Stenotomus versicolor, Biochem. Biophys. Res. Commun.*, 95, 328, 1980.
39. **Stegeman, J. J., Melancon, M. J., and Woodin, B. R.,** In vitro and in vivo metabolism of α-naphthoflavone (ANF) and β-naphthoflavone (BNF) in fish, *Pharmacologist*, 26, 1984.
40. **Vyas, K. P., Shibata, T., Highet, R. J., Yeh, H. J., Thomas, P. E., Ryan, D. E., Levin, W., and Jerina, D. M.,** Metabolism of α-naphthoflavone and β-naphthoflavone by rat liver microsomes and highly purified reconstituted cytochrome P-450 systems, *Biol. Chem.*, 258(9), 5649, 1983.
41. **Laemmli, U. K.,** Cleavage of structural proteins during the assembly of the head of bacteriophage T4, *Nature (London)*, 227, 680, 1970.
42. **Thomas, P. E., Ryan, D., and Levin, W.,** An improved staining procedure for the detection of the peroxidase activity of cytochrome P-450 on sodium dodecyl sulfate polyacrylamide gels, *Anal. Biochem.*, 75, 168, 1976.
43. **Stegeman, J. J.,** Polynuclear aromatic hydrocarbons and their metabolism in the marine environment, in *Polycyclic Hydrocarbons and Cancer*, Vol. 3, Gelboin, H. V. and Ts'o, P. O. P., Eds., Academic Press, New York, 1981, 1.
44. **Lech, J. J., Vodicnik, M. J., and Elcombe, C. R.,** Induction of monooxygenase activity in fish, in *Aquatic Toxicology*, Weber, L. J., Ed., Vol. 1, Raven Press, New York, 1982, 107.
45. **Pohl, R. J., Bend, J. R., Guarino, A. M., and Fouts, J. R.,** Hepatic microsomal mixed-function oxidase activity of several marine species from coastal Maine, *Drug. Metab. Dispos.*, 2, 545, 1974.
46. **James, M. O., Kahn, M. A. Q., and Bend, J. R.,** Hepatic microsomal mixed-function oxidase activities in several marine species common to coastal Florida, *Comp. Biochem. Physiol.*, 62C, 155, 1979.
47. **Hazel, J. R. and Prosser, C. L.,** Molecular mechanisms of temperature compensation in poikilotherms, *Physiol. Rev.*, 54, 620, 1974.
48. **Williams, D. E., Becker, R. R., Potter, D. W., Guengerich, F. P., and Buhler, D. R.,** Purification and comparative properties of NADPH-cytochrome P-450 reductase from rat and rainbow trout: differences in temperature optima between reconstituted and microsomal trout enzymes, *Arch. Biochem. Biophys.*, 225, 55, 1983.
49. **Pohl, R. J., Serabjit-Singh, C. J., Slaughter, S. R., Albro, P. W., Fouts, J. R., and Philpot, R. M.,** Hepatic microsomal NADPH-cytochrome P-450 reductase from little skate, *Raja erinacea*. Comparison of thermolability and other molecular properties with a mammalian enzyme, *Chem.-Biol. Interact.*, 45, 283, 1983.
50. **Binder, R. L. and Stegeman, J. J.,** Induction of aryl hydrocarbon hydroxylase activity in embryos of an estuarine fish, *Biochem. Pharmacol.*, 29, 949, 1980.
51. **Binder, R. L., Stegeman, J. J., and Lech, J. J.,** Induction of cytochrome P-450-dependent monooxygenase systems in embryoes and eleutheroembryoes of the killifish, *Fundulus heteroclitus, Chem.-Biol. Interact.*, 55, 185, 1985.
52. **Stegeman, J. J., Klotz, A. V., Woodin, B. R., and Pajor, A. M.,** Induction of hepatic cytochrome P-450 in fish and the indication of environmental induction in scup *(Stenotomus chrysops), Aquat. Toxicol.*, 1, 197, 1981.
53. **Spies, R. B., Felton, J. S., and Dillard, J.,** Hepatic mixed-function oxidases in California flatfishes are increased in contaminated environments and by oil and PCB ingestion, *Mar. Biol.*, 70, 117, 1982.

54. **Foureman, G. L., White, N. B., Jr., and Bend, J. R.,** Biochemical evidence that winter flounder *(Pseudopleuronectes americanus)* have induced hepatic cytochrome P-450-dependent monooxygenase activities, *Can. J. Fish. Aquat. Sci.*, 40, 854, 1983.
55. **Stegeman, J. J., Woodin, B. R., Park, S. S., Kloepper-Sams, P. K., and Gelboin, H. V.,** Microsomal cytochrome P-450 function in fish evaluated with polyclonal and monoclonal antibodies to cytochrome P-450 from scup *(Stenotomus chrysops), Mar. Environ. Res.*, 17, 83, 1985.
56. **Williams, D. E., Buhler, D. J., Klotz, A. V., and Stegeman, J. J.,** in preparation.
57. **Williams, D. E., Bender, R. C., Morrissey, M. T., Selivonchick, D. P., and Buhler, D. R.,** Cytochrome P-450 isozymes in salmonids determined with antibodies to purified forms of P-450 from rainbow trout, *Mar. Environ. Res.*, 14, 13, 1984.
58. **Goksøyr, A.,** Purification of hepatic microsomal cytochromes P-450 from B-napthoflavone treated cod (*Gadus morhua*), a marine teleost fish, *Biochim. Biophys. Acta,* 840, 409, 1985.
59. **Ryan, D. E., Thomas, P. E., Korzeniowski, D., and Levin, W.,** Separation and characterization of highly purified forms of liver microsomal cytochrome P-450 from rat treated with polyclorinated biphenyls, phenobarbital, and 3-methylcholanthrene, *J. Biol. Chem.*, 254, 1365, 1979.
60. **Park, S. S., Klotz, A. V., Miller, H., Gelboin, H. V., and Stegeman, J. J.,** Monoclonal antibodies to a hydrocarbon hydroxylating liver microsomalcytochrome P-450 of marine fish scup *(Stenotomus chrysops), Pharmacologist,* 26, 124, 1984.
61. **Cheng, K.-C., Krutzsch, H. C., Park, S. S., Grantham, P. H., Gelboin, H. V., and Friedman, F. K.,** Amino-terminal sequence analysis of six cytochrome P-450 isozymes purified by monoclonal antibody directed immunopurification, *Biochem. Biophys. Res. Commun.*, 123, 1201, 1984.
62. **Lu, P., Metcalf, R. L., Plummer, N., and Mandel, D.,** The environmental fate of three carcinogens: benzo[a]pyrene, benzidine and vinyl chloride evaluated in laboratory model ecosystems, *Arch. Environ. Contam. Toxicol.*, 6, 129, 1977.
63. **James, M. O., Bowen, E. R., Dansette, P. M., and Bend, J. R.,** Epoxide hydrase and glutathione S-transferase activities with selected alkene and arene oxides in several marine species, *Chem. Biol. Interact.*, 25, 321, 1979.
64. **Lech, J. J. and Bend, J. R.,** The relationship between biotransformation and the toxicity and fate of xenobiotic chemicals in fish, *Environ. Health Perspect.*, 34, 115, 1980.
65. **Stegeman, J. J., Woodin, B. R., and Binder, R. L.,** Patterns of benzo[a]pyrene metabolism by varied species, organs, and developmental stages of fish, *Natl. Cancer Inst. Monogr.*, 65, 371, 1984.
66. **Stegeman, J. J. and James, M. O.,** Individual variation in patterns of benzo(a)pyrene metabolism in the marine fish scup *(Stenotomus chrysops), Mar. Environ. Res.*, 17, 122, 1985.
67. **Stegeman, J. J., Skopek, T. R., and Thilly, W. G.,** Bioactivation of polynuclear aromatic hydrocarbons to cytotoxic and mutagenic products by marine fish, in *Carcinogenic Polynuclear Aromatic Hydrocarbons in the Marine Environment,* Richards, N., Ed., Proc. Conf. Pensacola, Fla., Aug. 1978, U.S. Environmental Protection Agency, Washington, D.C., 1981.
68. **Varanasi, U. and Gmur, D. J.,** Metabolic activation and covalent binding of benzo(a)pyrene to deoxyribonucleic acid catalyzed by liver enzymes of marine fish, *Biochem. Pharmacol.*, 29, 753, 1980.
69. **Jerina, D. M., Michaud, D. P., Feldmann, R. J., Armstrong, R. N., Vyas, K. P., Thakker, D. R., Yagi, H., Thomas, P. E., Ryan, D. E., and Levin, W.,** Stereochemical modeling of the catalytic site of cytochrome P-450c, in *Microsomes, Drug Oxidations, and Drug Toxicity,* Sato, R. and Kato, R., Eds., Japan Scientific Societies Press, Tokyo, 1978, 195.
70. **Hansson, T., Rafter, J., and Gustafsson, J.-A.,** Effects of some common inducers on the hepatic microsomal metabolism of androstenedione in rainbow trout with special references to cytochrome P-450-dependent enzymes, *Biochem. Pharmacol.*, 29, 583, 1980.
71. **Klotz, A. V.,** Purification and Charting of the Hepatic Microsomal Monooxygenase System from the Coastal Marine Fish *Stenotomus chrysops,* Ph.D. thesis, Woods Hole Oceanographic Institute/Massachusetts Institute of Technology, Woods Hole, 1983.

Chapter 5

IMMUNOCOMPETENCE IN INVERTEBRATES

R. S. Anderson

TABLE OF CONTENTS

I. Methods for Assessment of Immunocompetence in Invertebrates 94
 A. Blood Cells and Their Functions 94
  1. Total and Differential Blood Cell Counts 94
  2. Phagocytosis 94
  3. Bacterial Clearance 95
  4. Coelomocyte-RBC Rosette Formation 95
 B. Humoral Factors 96
  1. Hemagglutinin 96
  2. Hemolysin 96
  3. Bacterial Agglutinin 96
  4. Lysozyme 96
  5. *Glycera* Bactericidal Activity 97
 C. Dosing Procedures 97

II. Immunological Responses of Invertebrates 98
 A. *Mercenaria mercenaria* (Hard Clam) 98
  1. Hemolytic Activity 98
  2. Hemagglutination and Bacterial Agglutination 100
  3. Lysozyme 101
  4. Bacterial Clearance 102
 B. *Glycera dibranchiata* (Polychaete) 103
  1. Hemagglutinins and Bacterial Agglutinins 103
  2. Hemolysin 104
  3. Cytotoxic Factor 104
  4. Bactericidin 104
  5. Lysozyme 106
  6. Bacterial Clearance 106
  7. In Vitro Phagocytosis 107
  8. Total and Differential Coelomocyte Counts 107
  9. Amebocyte-Erythrocyte Rosetting 107
 C. Summary 107

Acknowledgments 108

References 108

I. METHODS FOR ASSESSMENT OF IMMUNOCOMPETENCE IN INVERTEBRATES

A. Blood Cells and Their Functions

1. Total and Differential Blood Cell Counts

Mercenaria mercenaria were held on ice until hemolymph samples were taken. The anterior adductor muscle was made accessible by filing the shell, and 0.15 cc of hemolymph was withdrawn using a syringe fitted with an 18-gauge needle. Then 0.15 cc of cold, sterile 0.5% neutral red solution in sea saline was also drawn up into the syringe. The blood and stain were mixed by ejecting the contents into a small dish on ice and drawing it back into the syringe. An aliquot of the mixture was placed in a hemocytometer, and the cells were allowed to adhere in a moist chamber for 15 min at room temperature. Granulocytes (those cells containing red-staining granules) and hyalinocytes (cells not containing granules) were quantified by microscopic observation.

In the case of *Glycera dibranchiata,* the worm was placed in a petri dish and the posterior end held with forceps. A small incision, approximately 10 mm in length, was made in the dermis of the anterior end with a scalpel. The worm was held over the dish and another pair of forceps was run gently down the length of the body, from posterior to anterior, to expel the coelomic fluid into the dish. The coelomic fluid was transferred from the petri dish to a small plastic tube with a Pasteur pipette, and diluted 1:100 with sterile sea saline. The total white and red blood cell counts were determined using a hemocytometer.

2. Phagocytosis

In addition to studying the uptake of untreated test particles, bacteria or mammalian RBCs were frequently treated with formaldehyde to enhance their phagocytosis.

Erythrocyte solutions were centrifuged at 800 g for 3 min at 4°C. After discarding the supernatant, the packed erythrocytes were resuspended in 10 volumes of Hanks' balanced salt solution (HBSS) and centrifuged at 800 g for 3 min at 4°C. The packed cells were resuspended in 10 volumes of a 6% formalin solution. The red cells were incubated at 4°C for 24 hr (for sheep erythrocytes) or 48 hr (for rabbit erythrocytes), with occasional gentle mixing. The cells were then washed 5 times with 5 volumes of HBSS (800 g, 3 min, 4°C). Then packed cells were resuspended in 9 volumes of 0.15 *M* potassium phosphate buffer, pH 7.2. The solution was stored at 4°C and diluted further when needed.

The formalin treatment for bacteria varied from that for erythrocytes in that 3% formalin was used, the incubation was 24 hr, and the final suspension was in HBSS adjusted to an absorbance of 0.700 at 650 nm.

In vivo phagocytosis was determined in *Mercenaria* after injection into the anterior adductor muscle of 0.1 mℓ of a 1% suspension of formalin-treated rabbit erythrocytes in sterile sea saline. The clams were returned to the aquarium and removed only for the period of time necessary to withdraw hemolymph. At timed intervals, a small amount of hemolymph was removed from the muscle with a syringe. A drop was placed on a coverslip and incubated in a moist chamber at room temperature for 15 min.

The coverslip was gently washed with sea saline, fixed in methanol, stained with hemotoxylin and eosin, cleared with xylene, and mounted on a slide with Permount. When dry, the slides were viewed under a microscope (magnification × 400), and phagocytic index, average number of erythrocytes per cell, degree of adhesion of erythrocytes, and presence of extracellular erythrocytes were determined.

To study in vitro phagocytosis, the anterior end of the clam shell was filed and hemolymph was extracted from the anterior adductor muscle using a syringe with an 18-gauge needle. A drop of hemolymph was placed on a coverslip and incubated in a moist chamber at room temperature for 15 min. The coverslip was very gently washed with sterile sea saline, and a few drops of a 0.1% suspension of formalin-treated rabbit erythrocytes in sea saline were added. It was incubated in a moist chamber for 30 min at room temperature. Excess erythrocytes were very gently washed off with sea saline, the coverslips were fixed in methanol, stained with hemotoxylin and eosin, cleared with xylene, and mounted on a slide with Permount. The slides were viewed under a microscope (magnification × 400) and scored as described above.

To obtain *Glycera* coelomocytes for in vitro phagocytosis, whole coelomic fluid was allowed to stand at room temperature for 45 min, after which time the clear fluid and white coelomocyte layers could be gently pipetted off and 1 to 2 drops placed on a coverslip. This coverslip was placed in a moist chamber and incubated for 15 min at room temperature to allow the cells to adhere. A few drops of sea saline were very gently added to the coverslip and then all excess fluid was removed by tipping the coverslip onto blotting paper. A few drops of 0.05% suspension of formalin-treated rabbit erythrocytes in sea saline were added to the coverslip, and it was incubated in a moist chamber for 90 min. The excess fluid was gently blotted and the coverslip was processed as described in the protocol for quantifying phagocytosis in *M. mercenaria*.

3. Bacterial Clearance

The ability of an organism to eliminate viable microorganisms from its blood gives some measure of its general ability to cope with infection. An assay for this function in *Mercenaria* is given in this section. *Mercenaria* hemolymph is usually quite free of bacteria, hence the fate of experimentally introduced bacteria is relatively easy to follow. *Glycera* coelomic fluid, on the other hand, apparently contains a variety of naturally occurring microorganisms. This "background" makes it technically difficult to study bacterial clearance in *Glycera*.

Flavobacterium 807098 originally isolated from an oyster holding tank was grown overnight in marine broth at 37°C. After being centrifuged at 1100 g for 10 min at 40°C, the bacteria were resuspended in sterile sea saline and adjusted to an absorbance of 0.7 at 650 nm. After filing the shell, 0.05 mℓ of the bacterial suspension was injected into the clam's anterior adductor muscle. At 30 min postinjection, 0.1-mℓ hemolymph was withdrawn from the anterior adductor muscle, placed in a tube containing 9.9 mℓ sterile water, and mixed to osmotically lyse the hemocytes. Of this mixture, 1 mℓ was transferred to a tube containing 9.0 mℓ sterile water and mixed. Then, 0.1 mℓ of this mixture was placed onto each of 2 petri dishes. Molten agar (50°C) was added, the dishes were covered, and the bacteria present were incorporated into the agar by gently rotating the plates. Another 0.1-mℓ hemolymph sample was withdrawn 4 hr after injection of bacteria. It was subjected to the same dilution and plating as the 30-min sample. After all plates had hardened, they were inverted and transferred to a 37°c incubator for 48 hr. The plates were read on a colony counter and the amount of bacterial clearance was determined. Several dilutions of the injection suspension were plated and counted to determine the number of bacteria originally injected.

4. Coelomocyte-RBC Rosette Formation

After allowing the coelomic fluid to stand at room temperature for 45 min, the serum and white blood cells were removed and diluted 1:4 with Alsevers solution which had been previously adjusted to sea salinity. It was found that of many media tested, only Alsevers was effective in preventing clumping of the coelomocytes. The cells were centrifuged at 500 g for 3 min at 4°C and resuspended to the original volume with sea salinity Alsevers. This wash proved necessary to prevent occasional agglutination of the erythrocytes due to the presence of *Glycera* hemagglutinin. An aliquot of cells was mixed with an equal volume of 1% fresh rabbit erythrocytes in sea salinity Alsevers. After 1 hr, a small drop of the coelomocyte-erythrocyte suspension was placed on a hemocytometer and the number of rosettes per 100 white blood cells was determined. A coelomocyte with >5 RBCs attached to its surface was considered to be a rosette-forming cell.

B. Humoral Factors

1. Hemagglutinin

One milliliter of hemolymph was withdrawn from the exposed adductor muscle of *Mercenaria* and centrifuged at 500 g for 10 min at 4°C. The supernatant (cell-free hemolymph) was placed in a 50-μℓ-per-drop bottle. Using U-bottom microtiter plates, 1 drop of HBSS (50 μℓ) was placed into wells #2 to 12, and 2 drops of cell-free hemolymph were placed in the first well. The hemolymph was serially diluted through well #11 using a serial dilution mixer; well #12 was a hemolymph-free control. One 50-μℓ drop of a 20% formalin-treated rabbit erythrocyte solution in 0.15 *M* saline was added to each chamber. After gently vibrating the plate in order to mix the chamber contents, the plate was covered and left at room temperature for 60 min. The plate was then placed on a microtiter magnifier apparatus. The titer for a given sample was expressed as the reciprocal of the greatest dilution producing definite hemagglutination.

Glycera cells were removed from the coelomic fluid by centrifugation at 1100 *g* (5 min, 4°C). The hemagglutinin titer was determined as outlined above, except that 2% formalin-treated sheep erythrocytes in sea saline were added to the microtiter wells.

2. Hemolysin

Test mammalian erythrocytes (1% in HBSS, 0.5 mℓ) were added to an equal volume of *Glycera* coelomic fluid, or coelomic fluid serially diluted with HBSS. This mixture was incubated 60 min at 21°C. Two milliliters of HBSS was added and the tubes were immediately centriguted for 3 min at 750 *g*. The absorbance of the supernatant was determined at 541 nm. A coelomic fluid color-control tube was prepared by following the above procedure, but substituting HBSS for the test erythrocyte suspension. All values were corrected by subtracting the absorbance of the coelomic fluid color control. Test erythrocytes, following the above procedure, were incubated in distilled water, and the resultant absorbance, representing 100% lysis, was measured. Using this value, the corrected experimental absorbance was converted to percent lysis.

Hemolysin concentration in *Mercenaria* cell-free hemolymph was determined using the same technique, but with 2% fresh rabbit erythrocytes in 0.15 *M* saline.

3. Bacterial Agglutinin

Into wells #2 to 12 of a U-bottom microtiter plate was placed 50 μℓ of HBSS. Cell-free *Mercenaria* hemolymph (100 μℓ) was placed in the first well and was serially diluted through well #11, leaving well #12 as a control. Formalin-treated *Serratia marcescens* (50 μℓ) in HBSS (A_{650} = 0.7) was then added to each well. The plate was vibrated gently to mix the wells, covered, and incubated overnight at 4°C. The next day, the plate was placed on a microtiter magnifier apparatus. The titer was established as the most dilute serum producing definite agglutination.

The method for the determination of *Glycera* bacterial agglutinin was identical to that for *Mercenaria*.

4. Lysozyme

Hemolymph (3 mℓ) was withdrawn from the adductor muscle sinus of *Mercenaria* using a syringe with an 18-gauge needle. The hemolymph was centrifuged at 750 *g* for 10 min at 4°C and the supernatant was saved for further analysis. The cells were resuspended in 0.6 mℓ of 0.2 *M* sucrose and homogenized for 6 min at 1°C with an Eberbach Con-Torque tissue grinder at maximum speed. Triton X-100, at a final concentration of 0.1%, was then added and the mixture was briefly mixed with a vortex mixer. This mixture was frozen and thawed 6 times using dry ice and 95% ethanol. Aliquots of both the cells and the supernatant were saved for Lowry protein determination.

To tubes containing 3.0 mℓ of 0.15 g *Micrococcus lysodeikticus* per milliliter 0.06 *M* potassium phosphate buffer, pH 5.5, was added to 0.1 mℓ of 10 g standard lysozyme per milliter (from egg white, Grade I, Sigma Chemical Co., St. Louis, Mo.), or 0.1 mℓ of the serum or cell solution. Using phosphate buffer as a blank and the *M. lysodeikticus* solution as a control, the optical densities of the tubes were read at 450 nm at 0 min and after 60 min at room temperature. The amount of the lysozyme in each sample was calculated using the change in absorbance. The activity of each sample tested was finally expressed as micrograms lysozyme per milligram protein in hemocytes or serum.

5. Glycera Bactericidal Activity

Two types of assay of bactericidal activity, using overnight cultures of *Serratia marcescens* as the test organism, were adapted for use with *Glycera* coelomic fluid. The first was based on the turbidometric assay for measuring complement activity. A known volume of coelomic fluid was diluted in artificial sea water to give a total volume of 0.5 mℓ. *Glycera* antibacterial factor (GAF) activity was greater in artificial sea water than in the phosphate buffer system employed in earlier studies. A bacterial suspension of *S. marcescens* (6×10^5 bacteria per 0.1 mℓ) was added to the coelomic fluid. The bacteria-coelomic fluid mixture was mixed gently and incubated at room temperature for 30 min. The incubation was terminted by dilution with 4.0 mℓ cold tryptic soy broth, and the cultures were maintained at 1 to 4°C until all portions of the experiment were complete. The bacteria were then grown up at 37°C until control tubes (without coelomic fluid) reached $A_{570} = 0.25$. The relative absorbance was a measure of the number of bacteria surviving exposure to coelomic fluid. For quantitative studies, a standard curve was used to convert the relative absorbance to the number of bacteria surviving exposure and to the amount of GAF which was present at the start of the growth period.

The other assay used was a direct plating assay. Coelomic fluid and bacteria were allowed to react at room temperature, or at 37°C, for 30 min. The suspension was diluted (1:100 in 0.9% saline) to stop the reaction, mixed thoroughly, and 100-μℓ aliquots were plated in 10.0 mℓ tryptic soy agar. The plate counts obtained were a direct measure of the number of bacteria surviving exposure to the coelomic fluid.

C. Dosing Procedures

Pollutant exposure studies in this laboratory were carried out at 12°C in marine aquaria with 30 gal of recirculated artificial sea water (Instant Ocean). PCP was dissolved in 10 mℓ ethanol, which was added directly to the water to give a nominal initial concentration of 100 to 200 ppb. HCB (150 mg) was adsorbed to ignited sand and placed in a foil-covered glass column (40 × 600 mm), through which was continuously pumped aquarium water at 17 to 20 mℓ/min. These pollutant exposures caused minimal distress during the 3-week course of the experiments. Tanks were inspected daily and "gaping" or nonresponsive clams were immediately removed. Experimental animals had no higher rates of morbidity or mortality than controls.

Exposure studies were also carried out at the University of Delaware College of Marine Studies in flow-through tanks. Groups of *Mercenaria* exposed to PCP, HCB, or BaP and controls were housed in individual 184-ℓ tanks, each receiving about 3 ℓ of sea water per minute. PCP (100 mg, as NaPCP) was added to 20 ℓ of sea water and continuously pumped into 1 tank; the flow rate was 11 to 15 mℓ/min. BaP or HCB had been adsorbed to sand columns, as described above, and the flow rate was 22 mℓ/min. Columns were replaced at the end of 2 months. Control animals received the same handling as the experimentals, but no added pollutants. All groups received sea water from the same source and hence were subjected to identical environmental conditions, such as changes in salinity (28 to 30 ‰) and temperature (18°C in June, 25°C in August); the only variable was the presence of added xenobiotics.

Table 1
HEMOLYTIC ACTIVITY OF *MERCENARIA HEMOLYMPH*

Erythrocyte donor	Hemolysis[a] (%)	Hemolysin titer[b] (range)
Polychaete	0	0
Goose	0	0
Sheep	3.0 ± 2.9	0—2
Horse	3.2 ± 3.5	0—4
Chicken	3.5 ± 3.8	0—16
Bovine	3.7 ± 4.1	0—8
Guinea pig	18.0 ± 16.2	0—16
Rabbit	55.5 ± 18.8	16—64

[a] Average hemolysis ± SD for 50% coelomic fluid. N = 100 for rabbit; N = 6 for all other species.
[b] Reciprocal of highest coelomic fluid dilution mediating spectrophotometrically detectable hemolysis.

Attempts to carry out similar exposure studies in flow-through systems with *Glycera dibranchiata* were not successful. Initial attempts failed because the worms could not be recovered after burrowing into the debris at the bottom of the tanks. Large perforated plastic holding cages were constructed to house the worms while in the exposure and control tanks, however, neither control nor experimental groups survived for much more than 1 week. However, *Glycera* survived well in aquaria with recirculated artificial sea water.

II. IMMUNOLOGICAL RESPONSES OF INVERTEBRATES

A. *Mercenaria mercenaria* (Hard Clam)

Although lacking classical immunoglobulins, *Mercenaria* hemolymph contains various humoral factors that mimic the functions of antibodies. While these factors are assumed to have immunological significance, their roles in maintaining the general health of their hosts are often rather obscure. The aim of this study was to determine the effects of several marine pollutants on immunological mechanisms which probably underly the ability of *Mercenaria* to resist infection. Defense reactions of invertebrates may lack the specificity and memory characteristic of adaptive immunity in vertebrates, but these organisms are quite capable of distinguishing between self and nonself, and can mount a variety of cellular and humoral responses to effectively resist infection and parasitism. Disruption of any of these processes is likely to produce effects as deleterious as those produced in higher animals when some aspect of their immune system is compromised.

1. Hemolytic Activity

Mercenaria hemolymph could mediate lysis of horse, sheep, chicken, guinea pig, bovine, and rabbit erythrocytes (Table 1). Lysis of rabbit erythrocytes was most pronounced. Red blood cells from geese and the polychaete annelid *Glycera dibranchiata* were not lysed. Because rabbit erythrocytes were most sensitive to this lysin, they were used as the indicator cells in all subsequent studies, unless otherwise noted.

Mercenaria hemolysin was stable for several months when stored at −80°C; it was stable for at least 3 days at 4°C. Divalent cations were required for full expression of hemolytic activity. Dose-dependent inhibition was seen at various EDTA concentrations, with about 90% inhibition by 50 m*M* EDTA.

Table 2
INDUCTION OF HEMOLYSIN IN *MERCENARIA* AT 24 HR

Inducer	Animals induced (%)	$\bar{X}$ ΔHem ± SD	N	P
Saline	75.0	17.1 ± 11.0	28	—
50% rabbit RBCs	86.2	27.0 ± 13.1	29	<0.005
Packed rabbit RBCs	71.4	50.9 ± 23.9	16[a]	<0.001

[a] In this group, 2 animals died and 7 yielded no hemolymph; therefore, data are based on N = 7.

Hemolytic activity was influenced by the temperature of incubation during the assay. Activity of 50% hemolymph samples at 30 min was 56.2 ± 13.0% at 1°C, 67.9 ± 10.9% at 22°C, and 21.2 ± 5.7% at 37°C (N = 4 for each group). Activities at 1 and 22°C were comparable, but activity at 37°C was significantly reduced ($P < 0.01$). It was necessary to equilibrate all solutions at each test temperature prior to exposing the erythrocytes to the cell-free hemolymph. Hemolysis was totally inhibited when the assay was carried out at 56°C; preheating to 56°C destroyed 95% of total activity.

Hemolytic activity could be removed from the hemolymph by adsorption with red blood cells. Hemolymph samples were mixed with 2 volumes of washed, packed erythrocytes and held for 15 min. The unlysed erythrocytes were removed by centrifugation, and the adsorption process repeated again. The adsorbed hemolymph was tested for hemolytic activity; the observed absorbance was corrected for the hemoglobin released during adsorption. In a typical experiment (N = 5 for each group), average lysis of rabbit erythrocytes was 64.0 ± 2.0% for unadsorbed serum, 3.7 ± 2.4% for rabbit erythrocyte-adsorbed serum, and 45.3 ± 2.8% for sheep erythrocyte-adsorbed serum; no lysis of sheep erythrocytes by normal or adsorbed samples was evident. Clearly, adsorption by homologous cells was effective in inhibiting hemolytic activity. Interestingly, adsorption with sheep erythrocytes, which were very minimally lysed by *Mercenaria* hemolymph, produced significant inhibition ($P < 0.001$, 2-tailed t test).

Hemocytes from *Mercenaria* were shown to contain hemolytic activity. Hemocytes from hemolymph samples pooled from five clams were harvested by centrifugation and washed with sterile sea saline (SSS). The cells were lysed with 1 mℓ deionized water and homogenized. The cellular debris was removed by centrifugation, and the supernatant adjusted to be isoosmotic for rabbit red blood cells. In assays of 2 different pools, hemocyte lysate was tested for hemolytic activity; 0.25 mℓ (lysate of hemocytes from 5 mℓ hemolymph) mediated an average of 78.2% rabbit erythrocyte lysis during 1 hr incubation, and 0.5 mℓ of cell-free hemolymph from the same animals produced 32.4% lysis under similar conditions.

Shell liquor also contained anti-rabbit erythrocyte hemolytic activity. In one study, shell liquor and anterior adductor muscle sinus hemolymph samples were withdrawn from 12 individual *Mercenaria*. There was little correlation between shell liquor and hemolymph hemolysin activity from individual to individual. However, the mean lysis (31.9 ± 18.5%) mediated by dilute (1/4) shell liquor was not significantly different from that (33.7 ± 11.5%) of dilute (1/4) hemolymph.

Results of attempts to induce hemolytic activity in *Mercenaria* hemolymph are summarized in Table 2. In this study, serum hemolysin was measured in adductor muscle hemolymph before and at various intervals after treatment. Treatments included a single insertion of a 20-gauge hypodermic needle into the muscle, injection of 0.1 mℓ of sterile sea saline, of a 50% suspension of rabbit erythrocytes in SSS, or of packed rabbit erythrocytes into the muscle hemolymph sinus. Hemolysin induction (ΔHEM) was measured as actual increase of activity (absorbance, ΔA) in responding clams.

All hemolymph samples had measurable naturally occurring hemolysin. The majority (70 to 85%) of clams injected with SSS or 50% rabbit erythrocytes showed hemolysin induction, which persisted for 3 days. Neither of the treatments produced significantly different numbers of induced clams based on analysis of percent induced by the X^2 approximation. The ΔHEM in SSS-injected clams was not significantly different from those merely wounded by the hypodermic needle. Therefore, the increment in hemolytic activity in both these groups could be ascribed to injury. Injection of packed rabbit erythrocytes was evidently quite traumatic, causing 12.5% mortality and 43.8% morbidity by 24 hr. The clams termed morbid appeared normal by all gross characteristics before and after injection and had plentiful hemolymph before treatment, but apparently contained little or no available hemolymph when sampled after treatment. Clams injected with packed erythrocytes did not survive long enough for 48- and 72-hr samples.

The data in Table 2 suggest that injection of rabbit erythrocytes caused a transient, augmented HEM induction. This effect was not seen at 48 or 72 hr postinjection, when there was no statistical difference between the average HEM induction after injection of SSS or rabbit erythrocytes. However, 24 hr postinjection of 50% rabbit erythrocytes, hemolytic activity was significantly greater than after SSS injection; for those clams that survived, the same was true 24 hr following injection of packed rabbit cells. Induction by erythrocytes was shown to be dose dependent; at 24 hr, the average ΔHEM for packed cell-injected clams was greater than that for 50% erythrocyte-injected animals ($P < 0.005$).

Several protocols using multiple injections of 50% rabbit erythrocytes were employed to further increase hemolytic activity. For example, one group received daily injections for 2 days and was assayed 24 hr after the last injection; another group received a second injection at 72 hr and was assayed 24 hr later; control clams received appropriate SSS injections. The intent of these experiments was to reinject the clams at times when the ΔHEM produced by the primary injection was maximal (24 hr) or minimal (72 hr). The average ΔHEMs measured in both experimental groups were not statistically different from each other or from those at 24 hr after a single injection of rabbit erythrocyte.

The only other data concerning hemolytic activity in *Mercenaria* were presented by Graham,[1] who also found a heat-sensitive, naturally occurring hemolysin with broad specificity in both hemolymph and shell liquor. However, his data differed from those of Anderson[2] in several respects, such as erythrocyte specificity, ability to adsorb the hemolysin with homologous erythrocytes, rate of reaction at low temperatures, and shape of the curve describing reaction kinetics. These differences may have resulted from the use of clams from different, genetically undefined, wild populations and/or methodological differences.

There are few reported attempts to induce hemolysins in invertebrates. Among the annelids, hemolysin induction has been seen in earthworms[3-5] but not in polychaetes;[6] similar attempts have failed in sipunculids.[7] In the case of *Mercenaria,* injury caused an elevated hemolysin level, which persisted for at least 3 days. Injection of erythrocytes produced an additional increment of hemolytic activity, superimposed on the injury response, which was of short duration (~24 hr). No additional hemolytic activity was induced by multiple injections of erythrocytes.

Presumably this hemolysin serves a protective role in the organism, perhaps related to cytotoxic or antimicrobial activity. Hemolysin activity was not significantly altered by short- or long-term exposure of clams to PCP, HCB, or BaP.

2. *Hemagglutination and Bacterial Agglutination*

Hemagglutinin and bacterial agglutinin were routinely measured in all clams assayed and their titers were not significantly altered in animals exposed to PCP, HCB, or BaP. Hemagglutinin titers against formaldehyde-treated rabbit erythrocytes ranged from 2^3 to 2^5. This factor was not induced by any of the erythrocyte injeciton protocols employed. Agglutination

Table 3
INDUCTION OF LYSOZYME[a] IN *MERCENARIA HEMOCYTES*

Treatment	1 week	8 weeks	15 weeks	18 weeks
PCP	4.5 ± 1.7 (NS)	7.5 ± 4.3 (<0.05)	5.4 ± 2.1 (NS)	8.4 ± 2.5 (<0.05)
HCB	— —	— —	6.6 ± 2.1 (NS)	9.8 ± 2.2 (<0.01)
BaP	8.9 ± 2.7 (<0.01)	7.0 ± 4.4 (<0.05)	7.4 ± 1.7 (<0.01)	7.2 ± 3.0 (NS)
Control	5.2 ± 1.3	3.8 ± 1.9	4.6 ± 2.0	5.4 ± 1.6

[a] Mean μg lysozyme/mg hemocyte protein ± SD. Data from experimental groups were compared to unexposed control group using t test. Difference of means considered not significant (NS) if $P > 0.05$; $N = 8$ for all groups. All animals were housed in flow-through aquarium systems.

of mammalian erythrocytes by *Mercenaria* hemolymph is strictly a laboratory phenomenon; however, it may provide insight concerning the basis of immunological recognition. A likely physiological role for such hemagglutinins is as an opsonin, rendering foreign particulates more susceptible to phagocytosis.

Mercenaria contained a natural agglutinin against formaldehyde-treated marine *Flavobacterium* strain 807098. This is the same bacteria used in the bacterial clearance assays to be described later. Bacterial agglutinins probably provide protection from infection by rapidly removing bacteria from the circulation as clumped bacteria are readily phagocytosed or encapsulated by hemocytes. Bacterial agglutinin titers routinely were measured to be about 2^5, and were not altered in clams exposed to the various pollutants studied under the PRIMA program.

3. Lysozyme

Invertebrates frequently have measurable amounts of lysozyme in their body fluids, mucus, hemocytes, and other tissues. Its bactericidal activity has long been recognized; it functions by cleaving the bond between glucosamine and muraminic acid in the walls of Gram-positive bacteria. Lysosome is present in the cell-free hemolymph of *Mercenaria,* probably deriving from blood cell degranulation. The intrahemocytic level exceeded the serum concentration and provided a better indicator of lysozyme activity in this animal. Basal intracellular concentrations ranged from about 3.5 to 5.5 μg/mg protein.

The intrahemocytic lysozyme in clams exposed to PCP, HCB, and BaP are reported in Table 3. These studies were carried out in flow-through tanks at the University of Delaware College of Marine Studies, as previously described in Section I.C. As seen in Table 3, the elevation of the intrahemocytic lysozyme lvels of the experimental groups over the control levels was statistically significant for a majority of the groups. In these systems, the amount of BaP and HCB in the water was generally below the lower level of detection (<0.001 ppb), and PCP average about 5 ppb; PCP in control water was ~0.02 ppb. Tissue concentrations in exposed animals were 200 to 500 ppb PCP, 40 to 130 ppb HCB, and 2 to 4 ppb BaP; tissue concentrations in control clams were ~8 ppb PCP, ~0.4 ppb HCB, and ~0.15 ppb BaP.

Intrahemocytic lysozyme concentrations in clams exposed to higher HCB or PCP concentrations in recirculated water systems were occasionally above control levels, but the differences were not statistically significant in 6 studies ranging from 1 to 3 weeks duration. After 2 weeks, the tissue burdens leveled off at 1000 ppb PCP and 2000 ppb HCB compared to 2 ppb PCP and 0.4 ppb HCB in the controls. BaP exposure studies were not carried out.

The increase in hemocytic lysozyme at lower pollutant levels could serve as a protective mechanism against certain Gram-positive bacteria; however, this increment was not often seen in clams with high tissue burdens. It is possible that the increased intracellular lysozyme

Table 4
BACTERIAL CLEARANCE BY *MERCENARIA* IN FLOW-THROUGH SYSTEMS

Treatment	Clearance[a] (%)	P	Failure to clear	Tissue burden (ppb)
8 weeks				
PCP	77.9 ± 16.3(7)	<0.05	3/10	498 ± 44
HCB	90.8 ± 16.1(6)	NS[b]	1/10	36.9 ± 0.07
BaP	67.8 ± 26.5(9)	<0.005	1/10	2.21 ± 0.23
Control	95.0 ± 8.4(7)	—	0/10	5.0 (PCP) 0.88 (HCB) 0.3 (BaP)
16 weeks				
PCP	81.5 ± 11.7(7)	NS	1/8	270
HCB	92.0 ± 16.7(7)	NS	1/8	131 ± 14
BaP	95.8 ± 7.0(6)	NS	2/8	3.30 ± 1.00
Control	92.9 ± 17.5(8)	—	0/8	11.7 (PCP) 0.41 (HCB) 0.17 (BaP)

[a] Average ± SD; N is given in parentheses.
[b] NS indicates $P > 0.05$.

could be released as a result of pollutant-induced lysosomal membrane destabilization,[8] mediating cytotoxic reactions in the hemocytes. We have previously documented lysozyme induction following experimental trauma in other invertebrate species.[9]

4. Bacterial Clearance

The ability of clams in flow-through systems to clear a standard dose of injected marine *Flavobacterium* is summarized in Table 4; the data are representative of the six sampling times tested. In certain pollutant-exposed animals, the numbers of bacteria per milliliter hemolymph 4 hr after injection were not lower than those present 30 min after injection; these animals were considered to be totally unable to clear. All untreated clams assayed could clear bacteria; in excess of 90% of the bacteria present at 30 min was removed from the hemolymph by 4 hr. However, in every pollutant-exposed group, one or two animals were completely unable to clear any bacteria. The percentage of bacteria cleared by those experimental animals that showed ability to clear was compared to the percentage of bacteria cleared by the controls (Table 4). After considering all the data, one may make the following generalizations: HCB and BaP at low concentrations occasionally caused decreased clearance rates, but these were usually not significantly different from control values. PCP started to produce significant inhibition of clearance at tissue levels of about 500 ppb.

The effect of higher doses of HCB and PCP on bacterial clearance was measured in clams held in recirculated water systems (Table 5). These data are representative of 6 experiments of 1 to 3 weeks exposure time. In the low-dose studies, some *Mercenaria* in every experimental group completely failed to clear bacteria. In responding clams, the rate of clearance was always significantly decreased by either PCP (about 1000 ppb tissue) or HCB (about 2000 ppb tissue). The data suggest that the observed defect in bacterial clearance becomes more pronounced after a threshold level of HCB or PCP in the tissues of *Mercenaria* has been reached.

The use of this novel method for quantifying bacterial clearance in *Mercenaria* made possible the comparison of clearance in normal clams to clearance in pollutant-exposed clams. A standard dose of marine bacteria was injected into the blood sinus of the anterior

Table 5
BACTERIAL CLEARANCE BY *MERCENARIA* IN RECIRCULATED SYSTEMS

Treatment	Clearance[a] (%)	P	Failure to clear	Tissue burden (ppb)
2 weeks				
PCP	66.7 ± 29.2(7)	<0.05	1/8	1140 ± 90
HCB	81.8 ± 17.7(6)	<0.05	2/8	1670 ± 200
Control	96.9 ± 3.9(7)	—	0/7	1.61 (PCP) 0.39(HCB)
3 weeks				
PCP	76.4 ± 11.2(5)	<0.05	3/8	820 ± 60
HCB	70.1 ± 14.2(7)	<0.01	1/8	1970 ± 190
Control	89.4 ± 5.4(8)	—	0/8	2.47 (PCP) 0.56 (HCB)

[a] Average ± SD; N is given in parentheses.

adductor muscle, and dilution equilibrium in the hemolymph was permitted to occur before taking the first sample. Samples were withdrawn from the posterior adductor muscle to eliminate the possibility of obtaining clumped or tissue-associated bacteria which might be found at the site of injection. Clearance, as measured in this study, was simply the progressive loss of viable bacteria from the circulating hemolymph. The hemolymph of normal clams, when assayed under our conditions, was virtually free of bacteria. The bacteria used was a *Flavobacterium* ssp. isolated from an oyster holding tank; this genus can also be isolated from oysters collected in the field.[10]

Bacterial clearance may depend on both cellular and humoral mechanisms. Phagocytic blood cells are probably active in clearing and killing bacteria. Among the humoral substances in *Mercenaria* are hemolysins, which might mediate bactericidal mechanisms; hemagglutinins, which might serve as recognition and/or opsonic factors; and bacterial agglutinins,[11] which could remove bacteria from circulation. The exact mechanism underlying the observed pollutant-induced impairment in bacterial clearance is unknown. In preliminary studies, we have been unable to see a clear effect of pollution exposure on the levels of any of the humoral factors listed above. In addition, no change was noted in total or differential hemocyte counts or phagocytic activity of hemocytes, as measured by uptake of erythrocytes. Possibly the changes are so small as to be obscured by the natural variation inherent in the data. The effect of pollutants on cellular bactericidal activity is currently under study.

Not only as clearance partially inhibited in certain exposed clams, but also some exposed clams were totally unable to reduce the numbers of bacteria in their hemolymph. The implications of such impairment in resistance to bacterial infection are obvious. The defect in bacterial clearance is most pronounced in clams with comparatively high tissue HCB or PCP concentrations. The defect is also sometimes seen in BaP-exposed clams, but animals with BaP concentrations above 5 ppb were not available for this study. The reduction in clearance did not seem to depend on the duration of pollutant exposure, but was determined solely by the tissue burden.

B. *Glycera dibranchiata* (Polychaete)

1. Hemagglutinins and Bacterial Agglutinins

Glycera coelomic fluid contained hemagglutinins active against both rabbit and sheep erythrocytes.[6] The highest hemagglutinin titers observed were against sheep erythrocytes pretreated with formaldehyde. The hemagglutinin was heat labile (90 to 95% loss of activity after 56°C for 30 min), did not require divalent cations for activity, and was stable after repeated freezing and thawing. Adsorption studies indicated that it was cross-reactive. Numerous attempts to induce this factor by injecting erythrocytes were unsuccessful.

Bacterial agglutinin activity against formaldehyde-treated *Serratia marcescens* was determined by adding 50 $\mu\ell$ bacterial suspension (A_{650} = 0.7) to microtiter wells containing 50 $\mu\ell$ serially diluted coelomic fluid. The microtiter plates were incubated for 18 hr at 4°C prior to reading. Although our pollutant exposure regimens in recirculated systems generated considerable body burdens of HCB and PCP, no significant effects were produced in hemagglutinin titers (6.1 ± 1.0, N = 15) or bacterial agglutinin titers (4.5 ± 0.5, N = 35).

2. Hemolysin

Sheep erythrocytes were more susceptible than rabbit erythrocytes to the lytic action of *Glycera* coelomic fluid; this was shown by differences in lytic titers and in reaction kinetics.[6] Hemolytic activity was heat labile (56°C, 30 min), required divalent cations, and was destroyed by freezing and thawing. Absorption by rabbit or sheep erythrocytes markedly reduced subsequent hemolysis of either erythrocyte by the coelomic fluid. Attempts to induce hemolysin in this polychaete by several red cell injection protocols were not successful. In most cases, hemolysin activity was significantly decreased following erythrocyte injection; the hemolysin was probably simply removed from circulation by adsorption to the membranes of the injected erythrocytes. Antisheep erythrocyte lysin titers of greater than 128 were commonly seen in microtiter plate assays; hemolysis was also followed spectrophotometrically. Exposure to HCB or PCP produced no significant alterations in hemolysin titer, activity, or rate of reaction.

It has been reporated that both hemolysins and hemagglutinins could be induced in terrestrial annelids by the injection of mammalian erythrocytes.[3,5] This reaction was not observed in *Glyceria* in these experiments; therefore, one would not predict that these humoral factors play significant roles in any anamnestic immune reactions which these marine forms may possess. This does not exclude the possibility that these factors may have roles in a more primitive immunological recognition system.

3. Cytotoxic Factor

In addition to hemolytic and bactericidal action, *Glycera* coelomic fluid exhibits cytotoxic activity against cells from a great diversity of animals (Table 6). This was demonstrated by trypan blue exclusion studies carried out 1 hr after adding coelomic fluid to suspensions or monolayers of each cell type. Such treatment caused gross lysis of red or white blood cells from many hosts, but was totally inactive against the coelomocytes of the annelid species tested.[12] The exact identity of this factor is unknown, but it shares many of the properties of *Glycera* hemolysin and bactericidin. Preliminary evidence suggests that its activity was unaltered by our pollutant exposure procedures. Its strong activity for cells from both pro- and eukaryotic organisms indicates that the cytotoxic factor may play an important role in defense against many potential pathogens.

4. Bactericidin

Anderson and Chain have described a naturally occurring bactericidin termed *Glycera* antibacterial factor (GAF).[13-15] It was first observed in simple zone-of-inhibition tests in bacteria-seeded agar plates. The factor inhibited the growth of many bacterial species, but was not lytic when added to developed bacterial lawns. Strongly inhibited species included *Serratia marcescens, Enterobacter aerogenes,* and *Arizona arizonae; Pseudomonas aeruginosa* and *Salmonella typhimurium* were moderately inhibited; and among those not inhibited were *Proteus vulgaris, P. inconstans, Alcaligenes faecalis, Flavobacterium* sp. (807098), *Bacillus cereus, Micrococcus luteus, Streptococcus lactis, S. faecalis, Staphylococcus ep-*

Table 6
CYTOTOXICITY OF *GLYCERA* COELOMIC FLUID

Donor classification	Target cell	Cytotoxic activity[a]
Bacteria	*Serratia marcescens*	+ +
Annelid	*Nereis* sp.	−
	Glycera	−
	Lumbricus sp.	−
Mollusk	*Mercenaria mercenaria*	+ +
Crustacean	*Callinectes sapidus*	+ +
Insect	*Galleria mellonella*	+
	Manduca sexta	+
Vertebrate	Sheep RBC	+ +
	Rabbit RBC	+ +

[a] + +, >90% cytotoxicity; +, 10 to 90% cytotoxicity; −, <10% cytotoxicity.

idermis, S. aureus, Citrobacter freundi, and *Spirillum serpens*. Interestingly, the 11 strains of *Escherichia coli* tested were distributed in all 3 categories of inhibition. The basis for inhibitory specificity is not known; however, it was seen that GAF was particularly active against certain Gram-negative rods. Other Gram-negative rods were not inhibited and none of the Gram-positive forms were inhibited by GAF. Because of its high sensitivity to GAF, *S. marcescens* was selected as the bacterium of choice in the development of more quantitative assay systems. The first of these was the turbidometric method,[14] based on an assay of complement activity;[16] the other was a less complicated direct plating assay.[14]

Antibacterial humoral factors have been described in many invertebrates; perhaps the best known of these is lysozyme. Basal levels of lysozyme have been reported in certain annelids,[17,18] but clearly GAF does not qualify as a lysozyme. For example, its mechanism is not lytic; it has a decided preference for Gram-negative bacteria; and it is inactive against *M. lysodiekticus,* the usual test organism in lysozyme assays. GAF was partially purified using gel chromatography and affinity chromatography on concanavalin A-Sepharose. It appears to be a glycoprotein of molecular weight 2.5 to 4.5 $\times$ 10^5 daltons. There are also strong indications that the molecule contains bound divalent cations, which are essential to its structural integrity, and at least one disulfide bridge. Because of the loss in bactericidal activity, which occurred during all chromatographic separations, and the presence of multiple components in column fractions even after purification, as shown by gel electrophoresis, it is as yet impossible to rule out the possibility that several different molecules have GAF activity, or that GAF aggregates may exist. Therefore, the results obtained may only refer to one of the major antibacterial components of the coelomic fluid. The molecular weight estimates of GAF clearly differentiate it from any other antibacterial system found in either vertebrates or invertebrates. For example, it is clearly different from bactericidal proteins recently isolated from insects[17] and oligochaetes.[18] Nor is there evidence of any relationship to components of the vertebrate complement system,[19,20] although the bactericidal mechanism of action of the two systems may show some similarities. In view of the probable cytotoxic, as well as bactericidal, activity of GAF, it may also be of interest to compare it to the lymphotoxin system found in vertebrates. The most active component(s) of lymphotoxin from both human[21] and various other animal lymphoid cells[22] has a molecular weight in the same range as GAF, and will also readily aggregate or break down into component parts with little or no biological activity.

Table 7
ACTIVITY OF ANTIBACTERIAL FACTOR IN PCP- OR HCB-EXPOSED *GLYCERA*

Exposure time (days)	Pollutant	Tissue concentration (ng/g)	Antibacterial activity[a]	N	P
5	PCP	1550 ± 691	57.5 ± 27.2	10	<0.05
	HCB	330 ± 12	68.5 ± 32.1	10	<0.01
	Control	15.3 (PCP) 1.4 (HCB)	37.2 ± 12.7	12	—
7	PCP	2560 ± 173	91.6 ± 11.9	14	<0.001
	HCB	442	62.0 ± 33.1	13	<0.05
	Control	15.2 (PCP) 2.9 (HCB)	36.4 ± 30.8	16	—

[a] Measured against *S. marcescens*, percent of test dose killed ± SD.

Analysis of the kinetics of GAF activity against *S. marcescens* indicated that the reaction consisted of two stages: initial rapid binding to the bacterial surface, followed by a slower cidal reaction.[14] By following methods of data analysis already worked out for vertebrate bactericidal complement,[23] Anderson determined that three or four hits by GAF per bacteria are required for effective killing.

Exposure to PCP (25 ng/mℓ nominal, 8 to 13 ng/mℓ actual) or HCB (<1 ng/mℓ actual) produced no lethality, morbidity, or changes in behavior, but caused a marked induction of the antibacterial factor (Table 7). The results given in the table are representative of eight different experiments. Antibacterial activity appeared to be proportional to PCP tissue levels in these studies; this effect was not seen for HCB. The pollutant-stressed worms had significantly induced levels of an antibacterial glycoprotein derived, at least in part, from circulating amebocytes. This could have been the result of a compensatory protective response similar to macrophage activation, which can take place in mammals in response to xenobiotic exposure. Such activation usually is characterized by increased production and release of lysosomal enzymes. In addition to cell activation, it is possible that xenobiotic-induced lysosomal destabilization[8] might produce elevated levels of extracellular antibacterial factor in the coelomic fluid.

5. *Lysozyme*

No lysozyme activity was detected in the cell-free coelomic fluid of control *Glycera*. Furthermore, coelomic fluid contained variable numbers of nonphagocytic cells, notably gametes, which made lysozyme determinations on cell pellets after centrifugation essentially meaningless. A cell separation technique giving high yields of relatively pure coelomocytes was not developed. Therefore, lysozyme values were not routinely determined in the *Glycera* studies.

6. *Bacterial Clearance*

Technical difficulties precluded studies of bactericidal clearance in vivo in *Glycera* similar to those reported for *Mercenaria*. The major difficulty arose because, unlike normal clam hemolymph, *Glycera* coelomic fluid normally contained a highly variable background of naturally occurring bacteria. The inability to differentiate, with certainty, injected from native bacteria after plating coelomic fluid led us to abandon this assay.

Table 8
EFFECTS OF PENTACHLOROPHENOL (PCP) AND HEXACHLOROBENZENE (HCB) ON ROSETTING OF RABBIT ERYTHROCYTES BY *GLYCERA* AMEBOCYTES

Pollutant	Exposure time (weeks)	Tissue concentration (ng/g)	Rosette-forming cells (%)	N	P
PCP	1	6640 ± 150	11.3 ± 10.4	18	<0.001
Control	1	40	33.3 ± 14.9	18	
PCP	1	7420 ± 28	10.6 ± 11.2	12	<0.005
Control	1	37 ± 1	30.2 ± 19.1	20	
HCB	1	2140 ± 157	17.2 ± 17.7	5	<0.01
Control	1	< 1	43.4 ± 4.0	5	
HCB	2	4560 ± 6	25.5 ± 15.9	10	<0.05
Control	2	< 1	40.6 ± 11.9	12	

7. In Vitro Phagocytosis

Amebocytes from untreated *Glycera* were avidly phagocytic; 40 to 70% of the cells contained the test erythrocytes after 90 min of incubation. The red blood cells were usually surrounded by conspicuous phagocytic vacuoles. Amebocytes from PCP- or HCB-exposed worms were equally phagocytically active. The phagocytic indexes from experimental groups with tissue burdens reaching ~4500 ng HCB per gram and ~7500 ng PCP per gram were not statistically different from the controls.

8. Total and Differential Coelomocyte Counts

Coelomic fluid from control *Glycera* contained 189,200 ± 56,000 erythrocytes/mm^3 and 36,800 ± 8,000 amebocytes/mm^3 (N = 10). Clearly there was considerable variation between animals; however, the nubmers of amebocytes expressed as a percentage of total coelomocytes were much less variable (16.3 ± 3.8%). In none of the HCB or PCP exposure groups were either total or differential coelomocyte counts statistically different from the controls.

9. Amebocyte-Erythrocyte Rosetting

Little or no erythrocyte rosetting (0 to 3%) was observed when horse, ox, chicken, or goose erythrocytes were used in this assay. Sheep erythrocytes gave a stronger response (9 to 13%); however, rabbit erythrocytes consistently produced 30 to 45% rosetting. In all subsequent experiments, rabbit erythrocytes were used to measure rosetting activity of the amebocytes. As indicated in Table 8, rosetting was substantially reduced in *Glycera* after 1 week PCP exposure in our recirculated systems. HCB-exposed animals showed a similar inhibition during a 2-week experiment (Table 8), but the effect was not reproducible. In another HCB study using a larger number of *Glycera* (N = 20 for both week 1 and week 2), the rosetting activities in experimental and control groups were not significantly different from each other at 1 week (tissue concentration 2000 ng HCB per gram) or 2 weeks (tissue concentration ~4500 ng HCB per gram).

C. Summary

Many environmental contaminants are synergistic to infectious agents in higher animals.[23,24] There is some evidence that invertebrates also respond in a similar manner to pollutant exposure. Resistance to viral infection is reduced in shrimp exposed to PCBs.[25] Atypical secondary bacterial infections are seen in shrimp after heavy metal exposure.[26] In clams, pollutant exposure increases the incidence of *Polydora* infestation,[27] alters hemocyte phagocytic capacity, and reduces hemocyte count due to selective cytolysis.[28] The general objective of the studies reported here was to determine if exposure to sublethal concentrations of several model pollutants would produce any reproducible alterations in the activity of cellular and/or humoral components of the immune systems of *M. mercenaria* or *G. dibranchiata*. Much of the data summarized here may be found in two recent papers.[29,30]

Clams were exposed to BaP, PCP, or HCB for up to 18 weeks at pollutant levels that produced no increased mortality or morbidity. The majority of the pollutant-exposed clams showed significantly impaired ability to clear a marine *Flavobacterium* sp. which had been injected into their hemolymph; some of these animals showed complete failure to clear. Untreated clams routinely cleared 90% of the bacteria in the same period of time; none of these controls was completely unable to clear bacteria. These findings imply that resistance to bacterial infection in this common marine organism is decreased by exposure to xenobiotics. Intrahemocytic lysozyme was elevated in certain exposed clams. Lysozyme is considered to serve a protective antimicrobial function in invertebrates; generally, it is more active against Gram-positive bacteria than against Gram-negative forms like *Flavobacterium*. It appeared that hemocyte lysozyme induction was more pronounced at lower levels of pollutant exposure; however, the deficiency in bacterial clearance was directly proportional to tissue levels of the test pollutants.

The effects of HCB and PCP on components of the defense system of a polychaete, *G. dibranchiata*, were also determined. Although both pollutants were markedly concentrated in the tissues, the levels reached in these studies produced no mortality or morbidity. Certain parameters (in vitro phagocytosis by amebocytes, total and differential coelomocyte counts, hemolysin, hemagglutinin, and bacterial agglutinin activity) were not affected. However, the activity of a bactericidal glycoprotein was consistently induced in PCP- or HCB-exposed *Glycera*. Release of this factor from coelomocytes may be increased by a mechanism similar to that of macrophage activation, a characteristic of the leukocytes of higher animals. Exposure to PCP caused an apparent reduction in the numbers of amebocytes capable of forming rosettes with formaldehyde-treated rabbit erythrocytes. This was interpreted as an impairment in the ability of the cells to recognize foreign material, a process required for the normal functioning of these potentially protective blood phagocytes.

ACKNOWLEDGMENTS

The exposure studies at the University of Delaware were made possible by M. R. Tripp and C. R. Fries. Chemical analyses were carried out at Texas A & M University, under the direction of C. S. Giam and L. E. Ray. Technical assistance was provided by J. Döös and M. L. Cook.

The project was supported by grants OCE 77-23443 and OCE 80-16186 from the National Science Foundation, grant CA-08748 from the Nation Cancer Institute, and grants from the Whitehall Foundation and the Griffis Foundation.

REFERENCES

1. **Graham, H. A., Jr.,** A Hemolytic Enzyme in the Hemolymph of the Clam *(Mercenaria mercenaria)*, M.S. thesis, University of Delaware, 1968.
2. **Anderson, R. S.,** Inducible hemolytic activity in *Mercenaria mercenaria* hemolymph, *Dev. Comp. Immunol.*, 5, 575, 1981.
3. **Cooper, E. L., Lemmi, C. A. E., and Moore, T. C.,** Agglutinins and cellular immunity in earthworms, *Ann. N.Y. Acad. Sci.*, 234, 34, 1974.

4. **Andrews, E. J. and Kukulinsky, N. E.,** Hemolysis of vertebrate erythrocytes with tissue extracts of earthworms *(Eisenia foetida), J. Reticuloendothel. Soc.*, 17, 170, 1975.
5. **Chateaureynaud-Duprat, P. and Izoard, F.,** Compared study of immunity between two genera of Lumbricians: *Eisenia* and *Lumbricus*, in *Developmental Immunobiology*, Solomon, J. B. and Horton, J. D., Eds., Elsevier/North-Holland, Amsterdam, 1977, 33.
6. **Anderson, R. S.,** Hemolysins and hemagglutinins in the coelomic fluid of polychaete annelid, *Glycera dibranchiata, Biol. Bull.*, 159, 259, 1980.
7. **Triplett, E. L., Cushing, J. E., and Durall, G. L.,** Observations on some immune reactions of the sipunculid *Dendrostomum zostericolum, Am. Nat.*, 92, 287, 1958.
8. **Moore, M. N., Lowe, D. M., and Fieth, P. E. M.,** Lysosomal responses to experimentally injected anthracene in the digestive cells of *Mytilus edulis, Mar. Biol.*, 48, 297, 1978.
9. **Anderson, R. S. and Cook, M. L.,** Induction of lysozyme-like activity in the hemolymph and hemocytes of an insect, *Spodoptera eridania, J. Invertebr. Pathol.*, 33, 197, 1979.
10. **Lovelace, T. E., Tubiash, H., and Colwell, R. R.,** Quantitative and qualitative commensal bacterial flora of *Crassostrea virginica* in Chesapeake Bay, *Proc. Natl. Shellfish Assoc.*, 58, 82, 1968.
11. **Arimoto, R. and Tripp, M. R.,** Characterization of a bacterial agglutinin in the hemolymph of the hard clam, *Mercenaria mercenaria, J. Invertebr. Pathol.*, 30, 406, 1977.
12. **Chain, B. M. and Anderson, R. S.,** A bactericidal and cytotoxic factor in the coelomic fluid of the polychaete, *Glycera dibranchiata, Dev. Comp. Immunol.*, 7, 625, 1983.
13. **Anderson, R. S. and Chain, B. M.,** Antibacterial activity in the coelomic fluid of a marine annelid, *Glycera dibranchiata, J. Invertebr. Pathol.*, 40, 320, 1982.
14. **Chain, B. M. and Anderson, R. S.,** Antibacterial activity of the coelomic fluid of the polychaete, *Glycera dibranchiata.* I. The kinetics of the bactericidal reaction, *Biol. Bull.*, 164, 28, 1983.
15. **Chain, B. M. and Anderson, R. S.,** Antibacterial activity of the coelomic fluid of the polychaete, *Glycera dibranchiata.* II. Partial purification and biochemical characterization of the active factor, *Biol. Bull.*, 164, 41, 1983.
16. **Muschel, L. H. and Treffers, H. P.,** Quantitative studies on the bactericidal actions of serum and complement. I. A rapid photometric growth assay for bactericidal activity, *J. Immunol.*, 76, 1, 1956.
17. **Hultmark, O., Steiner, H., Rasmuson, T., and Boman, H. G.,** Insect immunity. Purification and properties of three inducible bactericidal proteins from the hemolymph of immunized pupae of *Hyalophora cecropia, Eur. J. Biochem.*, 106, 7, 1980.
18. **Roch, P., Valembois, P., and Lasseques, M.,** Biochemical particularities of the antibacterial factor of two subspecies *Eisenia fetida fetida* and *Eisenia fetida andrei, Am. Zool.*, 20, 794, 1980.
19. **Müller-Eberhard, H. J.,** Chemistry and reaction mechanisms of complement, *Adv. Immunol.*, 8, 1, 1968.
20. **Müller-Eberhard, H. J. and Schreiber, R. D.,** Molecular biology and chemistry of the alternative pathway of complement, *Adv. Immunol.*, 29, 1, 1980.
21. **Harris, P. C., Yamamato, R. S., Crane, J., and Granger, G. A.,** The human LT system. X. The initial form released by T-enriched lymphocytes is 150,000 M.W., associated with small nonlytic components, and can disassociate into the smaller α, β, and γ classes, *J. Immunol.*, 126, 2165, 1981.
22. **Ross, M. W., Tiangco, G. J., Horn, P., Hiserodt, J. C., and Granger, G. A.,** The LT system in experimental animals. III. Physiochemical characteristics and relationships of lymphotoxin (LT) molecules released *in vitro* by activated lymphoid cells from several animal species, *J. Immunol.*, 123, 325, 1979.
23. **Koller, L. D.,** Effects of environmental contaminants on the immune system, *Adv. Vet. Sci. Comp. Med.*, 23, 267, 1979.
24. **Koller, L. D.,** Immunotoxicology of heavy metals, *Int. J. Immunopharmacol.*, 2, 269, 1980.
25. **Couch, J. A. and Cortney, L.,** Interaction of chemical pollutants and virus in a crustacean: a novel bioassay system, *Ann. N.Y. Acad. Sci.*, 298, 497, 1977.
26. **Nimmo, D. W. R., Lightner, D. V., and Bahner, L. H.,** Effects of cadmium on the shrimps, *Penaeus duorarum, Palaemonetes pugio* and *Palaemonetes vulgaris*, in *Physiological Responses of Marine Biota to Pollutants*, Vernberg, F. J., Calbrese, A., Thurberg, F. P., and Vernberg, W. B., Eds., Academic Press, New York, 1978, 131.
27. **Jeffries, H. P.,** A stress syndrome in the hard clam, *Mercenaria mercenaria, J. Invertebr. Pathol.*, 20, 242, 1972.
28. **Fries, C. R. and Tripp, M. R.,** Depression of phagocytosis in *Mercenaria* following chemical stress, *Dev. Comp. Immunol.*, 4, 233, 1980.
29. **Anderson, R. S., Giam, C. S., Ray, L. E., and Tripp, M. R.,** Effects of environmental pollutants on immunological competency of the clam *Mercenaria mercenaria:* impaired bacterial clearance, *Aquat. Toxicol.*, 1, 187, 1981.
30. **Anderson, R. S., Giam, C. S., and Ray, L. E.,** Effects of hexachlorobenzene and pentachlorophenol on cellular and humoral immune parameters in *Glycera dibranchiata, Mar. Environ. Res.*, 14, 317, 1984.

Chapter 6

HISTOPATHOLOGY AND HISTOCHEMISTRY

M. R. Tripp and C. R. Fries

TABLE OF CONTENTS

I. Histological Techniques 112
  A. Dissection 112
  B. Fixation and Staining 112

II. Enzymatic Assays 112
  A. Histocytochemistry 112
  B. Soluble Enzymes 113
    1. β-Glucuronidase 113
    2. Acid Phosphatase 113
    3. Alkaline Phosphatase 113
    4. Glutamic-Oxaloacetic Transaminase 114
    5. Glutamic-Pyruvic Transaminase 114

III. Polyacrylamide Gel Electrophoresis 114
  A. Standard 114
  B. Sodium Dodecyl Sulfate 114

IV. Histopathology in Invertebrates 114
  A. Normal Histology 114
  B. Morphometry of the Digestive Gland 115
  C. Histopathology of Clams Exposed to Hydrocarbons 117
  D. Summary of Histopathology 119

V. Biochemical Responses in Invertebrates 119
  A. Accumulation of Pollutant Chemicals in Clams 119
  B. Hemolymph 121
    1. Hemolymph Protein in Normal *M. mercenaria* 121
    2. Hemolymph Protein in Clams Exposed to Hydrocarbons 122
    3. Hemolymph Enzymes 125
      a. β-Glucuronidase 125
      b. Glutamic-Pyruvic Transaminase 128
      c. Glutamic-Oxaloacetic Transaminase 132
  C. Digestive Gland 133
    1. Acid Phosphatase 134
    2. β-Glucuronidase 136
    3. Glutamic-Pyruvic Transaminase 138
    4. Glutamic-Oxaloacetic Transaminase 138
  D. Summary of Biochemical Parameters 140

Acknowledgments 144

Appendix A 147

Appendix B 148

Appendix C 151

References 152

I. HISTOLOGICAL TECHNIQUES

A. Dissection

A V-shaped piece of shell was broken from the clam shell *(Mercenaria mercenaria)* with a hammer. The adductor muscles were severed with a scalpel inserted through this V-shaped hole. The hinge ligament was subsequently severed. The body of the clam was removed and cut into 4 pieces (cross-section), each approximately 1 cm thick. Pieces were fixed in either Bouin or Davidson solution[1] for 24 hr.

B. Fixation and Staining

Tissue samples were fixed in Bouin, Helley, Davidson, or formalin-sea water solution for approximately 4 to 24 hr. The compositions of these solutions[1] are given in Appendix A. Fixed tissue was stored in ethanol (70 to 95%) or in formalin-sea water.

Samples were dehydrated by washing with a series of ethanol solutions followed by xylene.[3] After dehydration, they were embedded in paraffin for sectioning.

Samples were normally stained using hematoxylin and eosin, feulgen, or alcian blue. The composition of these solutions[1] and the staining procedures are detailed in Appendix B.

II. ENZYMATIC ASSAYS

A. Histocytochemistry

Tissue samples were frozen in hexane/dry ice and stored in a Revco freezer at $-70°C$ until processed. Tissue was sectioned (5 μm) with a cryostat at $-20°C$ and then stained for acid phosphatase or β-glucuronidase activity.[4-6]

For acid phosphatase, sectioned material was fixed in iced Baker's formol-calcium solution (see below) for 15 to 30 min and rinsed in 1 *M* Na acetate, pH 5.0, for 60 sec. It was then incubated with substrate solution (with or without inhibitor) or without substrate for 1 hr at 37°C. It was next rinsed in 1 *M* Na acetate, pH 5.0 for 60 sec, placed in 2% $(NH_4)_2S$ for 60 sec, rinsed again in 1 *M* Na acetate, and finally mounted in glycerine jelly.

Baker's formol-calcium fixative solution contains 1 g $CaCl_2$, 10 mℓ 37 to 40% formaldehyde, and 90 mℓ double-distilled deionized water. Glycerine jelly is prepared by swelling 10 g gelatin in 60 mℓ cold distilled water, then heating until dissolved, and finally adding 70 mℓ glycerine and 1 mℓ phenol. Acid phosphatase substrate solution contains 2.5 mℓ 1 *M* Na acetate, pH 5.0, 0.75 mℓ 5% $Pb(NO_3)_2$, 5.0 mℓ double-distilled deionized water, 2.5 mℓ 2% β-glycerophosphate, and 1.0 mℓ 1 *M* $CaCl_2$. In the no-substrate control, β-glycerophosphate was replaced by deionized water; for the inhibitor control, 8 mg L-tartrate was added to the substrate solution.

For β-glucuronidase, sectioned material was incubated with substrate (with or without inhibitor) or without substrate for 25 min at 37°C. It was rinsed twice in 2.5% NaCl, stained for 10 min at room temperature in an azo dye (10 mg fast violet B salt or fast red violet LB salt in 10 mℓ 0.1 *M* Na phosphate, pH 7.4) solution and fixed in Baker's formal-calcium solution for 10 min. It was then rinsed in distilled water and mounted in glycerine jelly.

β-Glucuronidase substrate solution contains 2.8 mg naphthol AS-BI glucuronide dissolved in 0.12 mℓ 50 m*M* $NaHCO_3$, 10 mℓ 0.1 *M* Na Acetate, pH 4.5, containing 0.25 g NaCl and 1 g Polypep (Sigma Chemical Co.). The naphthol glucuronide was omitted from the no-substrate control and 2 g D-saccharic acid-1,4-lactone was added for the inhibitor control. The azo dye solution contained 10 mg fast violet B salt or fast red violet LB salt in 10 mℓ 0.1 *M* Na phosphate, pH 7.4.

B. Soluble Enzymes

Small pieces of clam digestive gland were weighed and homogenized in 2 mℓ double-distilled deionized water. The homogenate was centrifuged at low speed for 10 min, and the supernatant was diluted with double-distilled deionized water for assay.

To obtain hemolymph, the posterior adductor muscle of *Mercenaria* was exposed by sawing the shell adjacent to the siphon. Hemolymph was withdrawn from the muscle using a syringe with a 22-gauge, 1.5-in. needle. The hemolymph was centrifuged at low speed for 15 min to sediment the hemocytes, and the supernatant was removed for assay.

1. β-Glucuronidase

β-Glucuronidase was assayed by mixing 0.2 mℓ hemolymph or tissue extract with 0.2 mℓ 0.03 *M* phenolphthalein glucuronic acid (Sigma Chemical Co.) and 0.6 mℓ 0.2 *M* sodium acetate, pH 4.5.[6,7] The assay mixture was incubated at 56°C for 1 hr and then 5 mℓ 2-amino-2-methyl-1-propanol, pH 11.0 (Sigma Chemical Co.), was added. The absorbance at 550 nm was measured using water as the reference. Each assay included a reagent blank, with 0.2 mℓ distilled deionized water replacing the sample, and a serum blank, with 0.2 mℓ distilled deionized water replacing the substrate.

The corrected absorbance [test absorbance − (serum blank + reagent blank)] was used to determine the phenolphthalein concentration from a standard curve prepared using phenolphthalein. Serum β-glucuronidase activity (Sigma units/mℓ) was calculated as μg phenolphthalein × 30/mℓ hemolymph. Tissue β-glucuronidase activity (Sigma units/mg) was calculated as μg phenolphthalein × 30/mg wet tissue.

2. Acid Phosphatase

Acid phosphatase was assayed by adding 0.2 mℓ hemolymph or tissue extract (distilled water for reagent blank) to 0.5 mℓ 100 μ*M* disodium p-nitrophenylphosphate in 0.5 mℓ sodium citrate, pH 4.8.[6,8] After incubating for 30 min at 37°C, the reaction was stopped by adding 5 mℓ 0.1 *N* NaOH. The absorbance was measured at 400 nm using the reagent blank as reference. A serum blank was prepared by mixing 0.2-mℓ sample with 6 mℓ 0.1 *N* NaOH, and its absorbance was measured at 400 nm against water. Acid phosphatase activity was calculated from the sample absorbance corrected for the serum blank using a standard calibration curve.

3. Alkaline Phosphatase

Alkaline phosphatase activity was measured by adding 0.1 mℓ hemolymph (or distilled water for the reagent blank) to 0.5 mℓ 100 μ*M* disodium p-nitrophenylphosphate in 0.5 mℓ 1.5 *M* 2-amino-2-methyl-1-propanol, pH 10.3.[6,8] The assay was carried out at 37°C for 30 min and was stopped by adding 10 mℓ 0.02 *M* NaOH. The absorbance of the sample assay was measured at 400 nm against the reagent blank. Concentrated HCl (0.1 mℓ) was added to each tube and the absorbance was measured again. The corrected alkaline phosphatase activity was determined by subtracting the value after adding HCl from that before adding HCl.

4. Glutamic-Oxaloacetic Transaminase

Glutamic-oxaloacetic transaminase (GOT) activity was determined by adding 0.2 mℓ hemolymph to 1.0 mℓ substrate (0.2 *M* D,L-aspartic acid and 1.8 m*M* α-ketoglutaric acid in Sorenson phosphate, pH 7.5)[4] and incubating at 37°C.[6,9] After 60 min, the reaction was stopped by adding 1.0 mℓ 0.1 *M* 2,4-dinitrophenylhydrazine (Sigma Chemical Co.) and incubating at room temperature. After 20 min, 10 mℓ 0.4 *M* NaOH was added. After 5 min at room temperature, the absorbance at 530 nm was measured using water as the reference. GOT activity, in Sigma-Frankel units, was determined from a standard curve.

5. Glutamic-Pyruvic Transaminase

Glutamic-pyruvic transaminase (GPT) activity was determined as described above for GOT except that the substrate solution contained 0.2 *M* alanine and 1.8 m*M* α-ketoglutaric acid in Sorensen phosphate, pH 7.5.[4,6,9] GPT activity in Sigma-Frankel units was determined from a standard curve.

III. POLYACRYLAMIDE GEL ELECTROPHORESIS

A. Standard

Polyacrylamide gel electrophoresis (PAGE) was carried out in 7.5% gels (see Appendix C for composition of stock solutions used for preparation of gels) in 2.5 mm (I.D.) × 7.5 mm glass tubes.[10] Approximately 100 μg protein, as determined by the method of Lowry et al.,[11] as applied to each gel. Electrophoresis was performed at 3 mA per tube until the tracking dye (stock, 5 mg bromophenol blue per 100 mℓ distilled water) had reached the stacking gel-separating gel interface. Electrophoresis was then continued at 5 mA per tube until the tracking dye was within 2 mm of the end of the gel. The gels were removed, fixed in 12.5% trichloroacetic acid (TCA) for 30 min, stained in 0.05% Coomassie blue in 10% TCA for 2 hr and destained in aqueous acetic acid-methanol (45:10:45). R_f values were calculated as the ratio of the distance traveled by the protein to that traveled by the tracking dye.

B. Sodium Dodecyl Sulfate

Sodium dodecyl sulfate (SDS)-PAGE was carried out in 10% slab gels,[10] 15.5 cm × 15 cm × 0.8 mm (see Appendix C for preparation of gels). Approximately 100 μg protein[11] was added to each sample well. Electrophoresis was carried out at 10 mA per gel until the tracking dye (stock, 100 mg bromophenol blue per 100 mℓ distilled water) reached the bottom of the gel. The gel was removed, fixed in 12.5% TCA for 30 min, stained in 0.05% Coomassie blue in 10% TCA overnight, and destained in 7% acetic acid.

IV. HISTOPATHOLOGY IN INVERTEBRATES

A. Normal Histology

Mercenaria mercenaria were collected, immediately fixed, and their tissues were studied microscopically to establish normal histology. The digestive gland of *Mercenaria* is similar to that of other bivalve mollusks.[12-16] Primary ciliated tubules lead from the stomach to secondary tubules that contain absorptive cells. The digestive tubules are clustered in groups and each is round to elliptical in cross-section.[17] Typically the absorptive cells are columnar; a vesicular nucleus is in the basal third of the cell. The cytoplasm is light staining and often contains distinct yellow vesicles in the distal half. These cells are arranged regularly to comprise the elongated portion of the elliptical tubule. Basophilic apical cells ("young cells" of Yonge[18]) at each end of the ellipse are densely concentrated and small.[14] The tubule is sheathed in a thin basement membrane with connective tissue, muscle fibers, and hemocytes

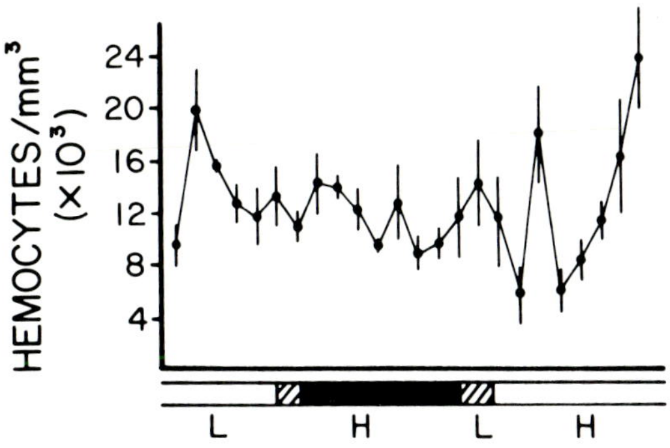

FIGURE 1. Concentration of hemocytes in the hemolymph of *M. mercenaria* sampled hourly for 24 hr: L, low tide; H, high tide; open bar, light; solid bar, dark; crossbars, dusk or dawn.

between adjacent tubules. The terminal tubules are not ciliated although, in some instances, a thickening of the distal membrane may be seen. In the lumina of some tubules, intact cells are detached and clearly visible; conventional Type IV tubules that contain cell fragments and secretions are rare.[16]

When studying possible adverse effects of environmental factors on biological systems, it is essential to be able to distinguish between normal variations and those that might be induced by a stressor. Numerous studies have found that feeding and digestion in marine bivalves are discontinuous and often cyclic. Periodic changes in digestive tubule structure as well as cyclic variations in the crystalline style, pH of the stomach, and filtration rates have been noted and suggest responses to environmental stimuli.[15,19-21] Changes associated with tidal cycles have been recorded[12,22-26] and, in some species, diurnal rhythms have been detected.[27] The importance of food as a stimulus rather than tide per se has been emphasized by Wilson and LaTouche[28] and Robinson and Langton.[16]

For most experiments, clams were held in a flow-through system supplied with Broadkill River (Lewes, Del.) water (natural food plentiful and always available, 25 to 30‰ salinity, 10 to 28°C, depending on season). These clams were normal histologically and most contained considerable food in the digestive system. *M. mercenaria* held in recirculated artificial sea water systems in the laboratory (no food added, 30‰ salinity, 21 to 24°C) for up to 17 days were not different histologically from clams freshly removed from the river. Hemocytes from clams maintained in the laboratory were normal in their ability to attach to glass, their numbers in circulation (Figure 1), and their cytological features.

B. Morphometry of the Digestive Gland

Three experiments were performed to detect possible cyclic changes in clam digestive gland tubule morphology. A total of 380 adult *M. mercenaria* of uniform size (9 to 10 cm long), sexually mature, and continuously feeding were used. In histological sections, individual tubules (Types I and II of Robinson and Langton[16]) were carefully selected for cross-sectional measurements. Cell width and height, tubule width and length, and lumen width and length were recorded (Figure 2).

In the first experiment, samples were taken every hour for 24 hr. The typical absorptive cell was 9 × 26 μm, the lumen was 12 × 61 μm, and the tubule was 65 × 89 μm (Figure 3). There was no evidence that these dimensions varied with tidal phase or with light/dark conditions.

Clams examined over a 3-day period that encompassed 6 continuous tidal cycles (samples taken at high, mid-ebb, low, and mid-flood tides) presented a similar picture. Digestive

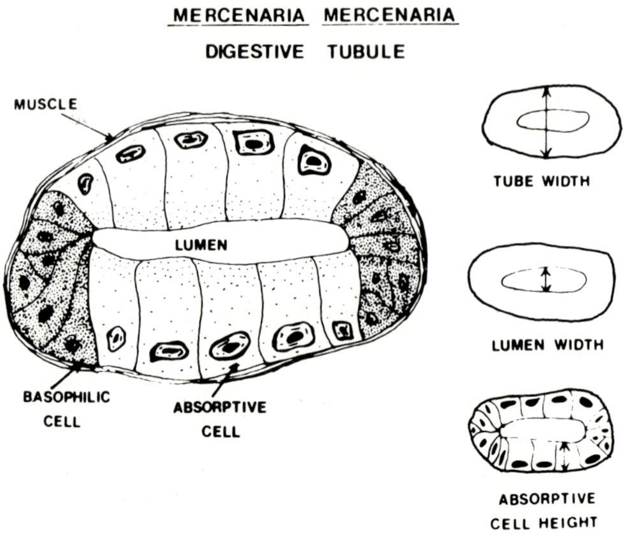

FIGURE 2. Cross-section of a digestive tubule in *M. mercenaria* showing cell types and dimensions measured.

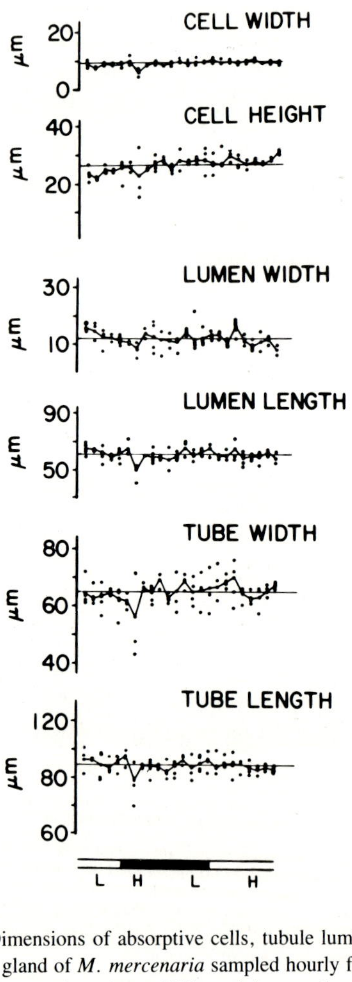

FIGURE 3. Dimensions of absorptive cells, tubule lumina, and tubules in the digestive gland of *M. mercenaria* sampled hourly for 24 hr: L, low tide; H, high tide; open bar, day; solid bar, night; crossbars, dusk or dawn.

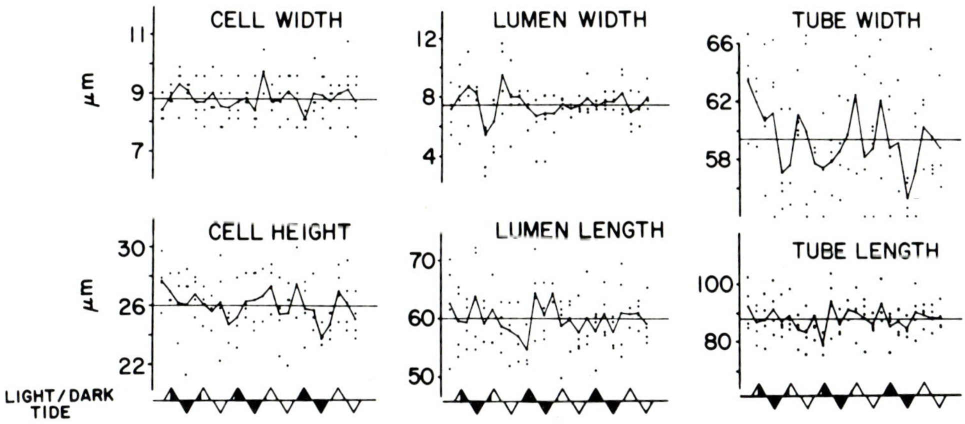

FIGURE 4. Dimensions of absorptive cells, tubule lumina, and tubules in the digestive gland of *M. mercenaria* over 6 tidal cycles: peak of triangle, high or low tide; shaded, night; unshaded, day.

TABLE 1
Mean Values (μm) for Digestive Tubule Parameters for 6 Continuous Tidal Cycles

Parameter	Overall average	Light	Dark	High tide	Low tide	High light	High dark	Low light	Low dark
Cell width	8.8 ± 1.8	8.8	8.9	8.7	8.8	8.8	8.1	8.8	8.8
Cell height	26.1 ± 3.6	26.1	26.0	26.5	25.6	26.6	25.8	25.6	25.5
Lumen width	7.6 ± 2.5	7.7	7.5	7.3	7.8	7.3	7.4	7.8	7.7
Lumen length	59.9 ± 9.5	59.5	60.7	59.0	59.5	59.3	57.8	58.3	60.7
Tubule width	59.4 ± 7.1	59.8	58.7	59.9	58.8	60.1	58.8	59.2	58.4
Tubule length	88.0 ± 12.3	87.9	88.3	87.2	86.8	87.5	85.7	86.2	87.5

tubules were remarkably constant histologically, and most clams were feeding when sacrificed. Cells averaged 8.8 × 26 μm, the lumina were 7.4 × 60 μm, and the tubules measured 59 × 88 μm (Figure 4). Mean values for these measurements are given in Table 1. These data vary both within and between samples, and represent normal fluctuations over a 3-day period not strongly influenced by either photoperiod or tide (Figure 4).

Another set of observations made on clams at approximately 6-hr intervals (at each high and low tide) over a period of 2 weeks is summarized in Figure 5. Cells averaged 8.4 × 24 μm, lumina 9.3 × 53 μm, and tubules 60 × 84 μm. There was no evidence of consistent variation with either tide or photoperiod.

The clams used were from a single supplier and of uniform size; it is presumed that they were from a single population. The cell and tubule dimensions measured are quite constant even though length of experiment, frequency of sampling, season, and holding conditions varied among these three experiments. In no case is there any indication of anything other than a constantly feeding animal with remarkably stable absorptive cells in the digestive system.

C. Histopathology of Clams Exposed to Hydrocarbons

Clams *(M. mercenaria)* injected with corn oil or corn oil plus BaP, HCB, or PCP showed similar responses. The inoculum was deposited directly into the digestive gland. Hemocytes accumulated around oil droplets at the injection site within 1 day. By 3 days, some small

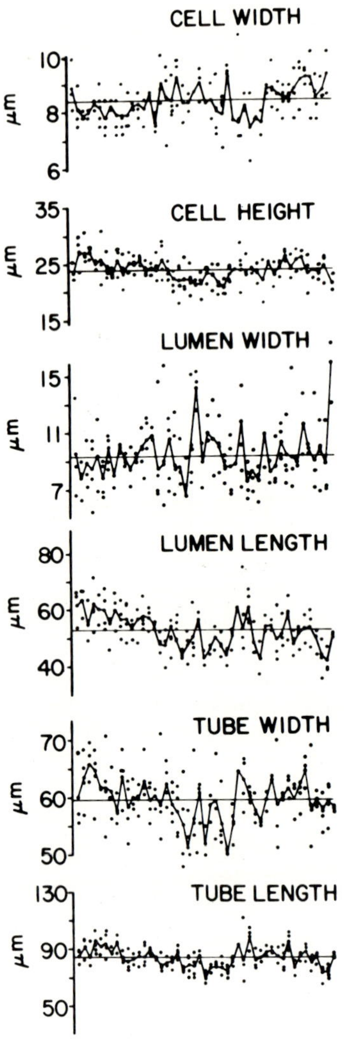

FIGURE 5. Dimensions of absorptive cells, tubule lumina, and tubules in the digestive gland of *M. mercenaria* sampled at high or low tide for 2 weeks.

droplets had begun to pass to gill sinuses and often they had a few hemocytes applied to their surfaces. After several days, they disappeared from the gill. No necrosis or other toxic effects were noted aside from those due to the mechanical trauma of injection.

Significant changes in digestive tubule dimensions are summarized in Figure 6. Tube width was decreased by injection of PCP and HCB; tube length was decreased by HCB. Lumen length was decreased by both PCP and HCB. Lumen width was not affected by injection of any of the test chemicals. Cell height was decreased by PCP and HCB, but cell width was increased by HCB. None of the morphometric parameters measured in BaP-treated clams differed significantly from the controls.

These results indicate that the three hydrocarbons tested in these experiments have very little effect on the tissues of *M. mercenaria*. When chemicals were injected there was a transient inflammatory response to the corn oil vehicle, but no clear evidence of cytotoxicity. The experiment was not designed to test possible effects of long-term exposure which might result in other types of cellular responses.

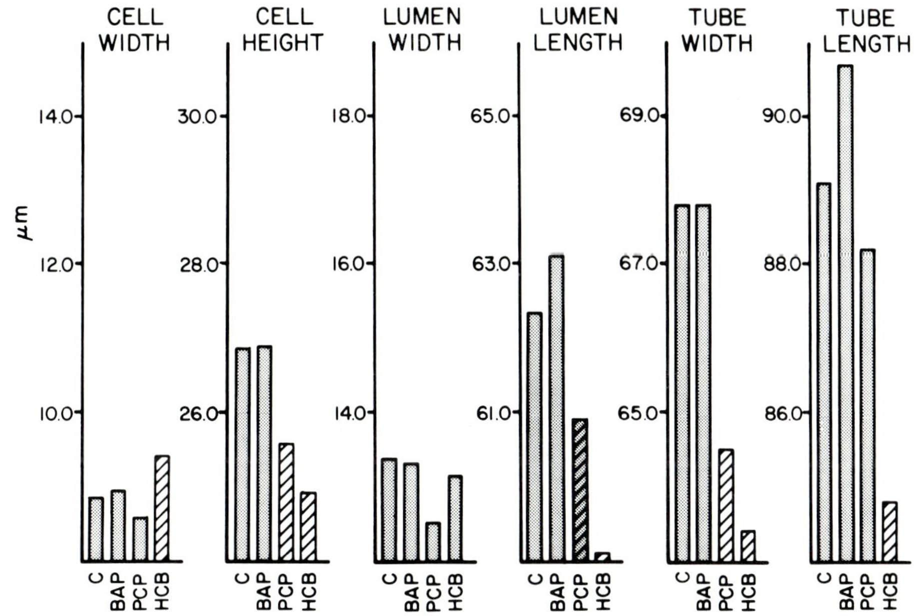

FIGURE 6. Average dimensions of absorptive cells, lumina, and tubules of the digestive gland of *M. mercenaria* injected with corn oil: C, control; BaP, corn oil plus benzo(a)pyrene; PCP, corn oil plus pentachlorophenol; HCB, corn oil plus hexachlorobenzene. Striped bars differ significantly ($P < 0.05$) from stippled bars. (From Tripp, M. R., Fries, C. R., Craven, M. A., Grier, C. E., and Gates, C. E., *Mar. Environ. Res.*, 14, 521, 1984. With permission.)

Clams exposed to BaP, HCB, or PCP in the water produced no characteristic histopathology. Exposure to BaP, HCB, or PCP in a flow-through system for 10 or 25 weeks also produced no sloughing or necrosis of absorptive cells. Digestive tubules maintained normal morphology, but some statistically significant changes in cell and tubule dimensions were observed. Clams exposed to HCB or BaP for 25 weeks had significantly narrower absorptive cells; PCP-treated animals showed narrower cells as well as decreased lumen length (Figure 7).

D. Summary of Histopathology

The normal histology of the digestive gland of *M. mercenaria* reveals no cyclic pattern of de- and regeneration of epithelial absorptive/digestive cells. Likewise, there is no evidence of tidal or photoperiod variability in structure in normal clams. Exposure to certain hydrocarbons (BaP, HCB, or PCP) in the water or by injection does not induce detectable histopathology.

V. BIOCHEMICAL RESPONSES IN INVERTEBRATES

A. Accumulation of Pollutant Chemicals in Clams

Selected hydrocarbon pollutants were added to the tanks by eluting the chemical from dosing columns (see Chapter 2, Section II D) with water as it flowed into the tanks. Thus, the clams in these tanks were continuously exposed to the chemicals. This system did not exclude the possibility that chemicals could be introduced into the tanks from exogenous sources via river water that supplied the flow-through system. The data summarized in Table 2 indicate that the sand column was not the sole source of BaP. The control tank had some

TABLE 2
Accumulation of Benzo(a)pyrene (BaP) in Water and Sediment in Flow-Through System with BaP Added via Sand Column

		BaP (ppb)	
Week	Source	H_2O	Sediment
6	Control tank	0.001	38.1
	BaP tank	0.019	597.3
	PCP tank	0.002	—
10	BaP tank	0.002	—

Note: —, no sample.

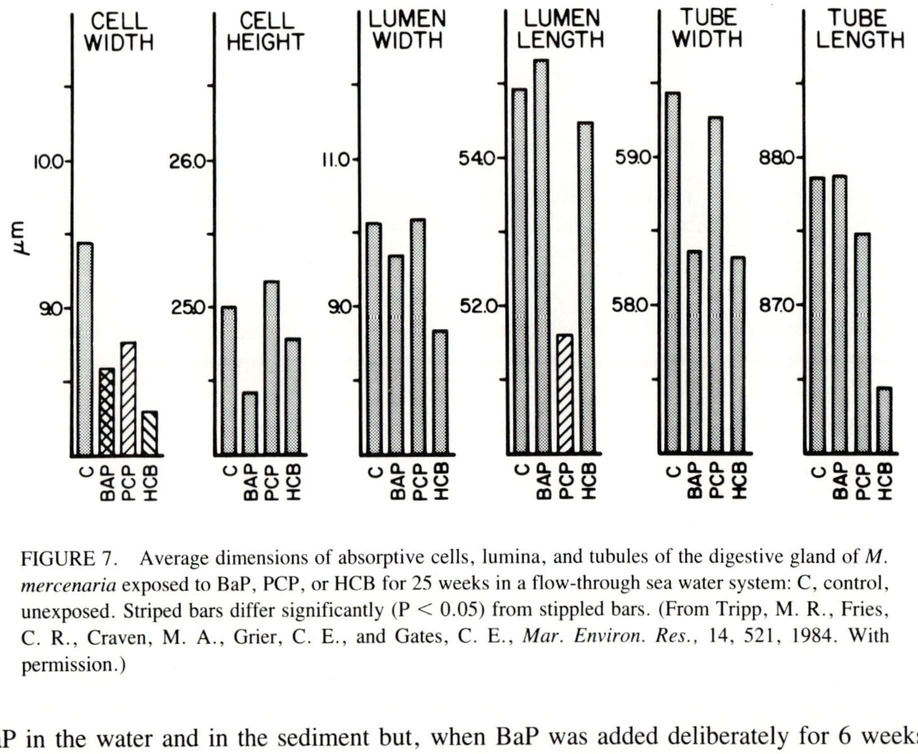

FIGURE 7. Average dimensions of absorptive cells, lumina, and tubules of the digestive gland of *M. mercenaria* exposed to BaP, PCP, or HCB for 25 weeks in a flow-through sea water system: C, control, unexposed. Striped bars differ significantly ($P < 0.05$) from stippled bars. (From Tripp, M. R., Fries, C. R., Craven, M. A., Grier, C. E., and Gates, C. E., *Mar. Environ. Res.*, 14, 521, 1984. With permission.)

BaP in the water and in the sediment but, when BaP was added deliberately for 6 weeks, the concentration in water and sediment was higher by a factor of nearly 20- and 15-fold, respectively. By 10 weeks, the sand column was exhausted and water concentrations of BaP were reduced to background values. As a result of these determinations, dosing columns were replaced at 6- to 8-week intervals in extended experiments.

It is important to know whether chemicals are transferred from the water or sediment to the experimental animals. Table 3 summarizes data indicating that this happens. BaP is accumulated up to 20-fold above ambient concentrations, and HCB may be concentrated as much as 40-fold. It was not feasible to do detailed chemical analyses of all the desirable water, sediment, and tissue samples for each experiment, but it is clear that pollutants added

TABLE 3
Accumulation of Benzo(a)pyrene (BaP) and Hexachlorobenzene (HCB) in Tissues of *Mercenaria mercenaria* in a Flow-Through System with Chemical Added via Dosing Column

Week	BaP (ppb)	HCB (ppb)
0	0.13	0.88
2	2.26 ± 0.8	23.80 ± 6.9
3	1.77 ± 0.1	24.20 ± 8.9
4	2.21 ± 0.2	36.90 ± 0.1

TABLE 4
Protein Concentration in Hemolymph of Normal *Mercenaria mercenaria*

Duration of observations	Sampling interval	Protein concentration (μg/mℓ hemolymph)		
		Mean	Min	Max
24 hr	1 hr	1253	330	2910
3 days	3 hr	400	102	816
10 weeks	1 week	688	400	940
24 weeks	3 weeks	565	396	765

to the holding system did lead to accumulation of these substances in the tissues of *M. mercenaria*.

B. Hemolymph

Hemolymph in the open circulatory systems of bivalve molluscs might be expected to contain material released from damaged cells. Therefore, hemolymph was examined for both quantitative and qualitative changes in protein content as well as the concentration of certain lysosomal enzymes (acid phosphatase and β-glucuronidase) and other metabolic enzymes (glutamic-oxaloacetic transaminase and glutamic-pyruvic transaminase).

1. Hemolymph Protein in Normal M. mercenaria

Total protein content of *M. mercenaria* hemolymph varies markedly among normal individuals sampled at one time and among groups of clams sampled at different times. The mean values obtained from four separate experiments illustrate this variability (Table 4). There is marked variability in samples taken hourly (Figure 8), at 3-hr intervals (Figure 9), weekly (Table 5), or at 2- to 3-week intervals (Table 6). There is no evidence of tidal or photoperiod influence (Figures 8 and 9). The great variability in the range of normal values of total hemolymph protein in *M. mercenaria* limits the usefulness of this measurement for evaluation of the health of populations of this clam.

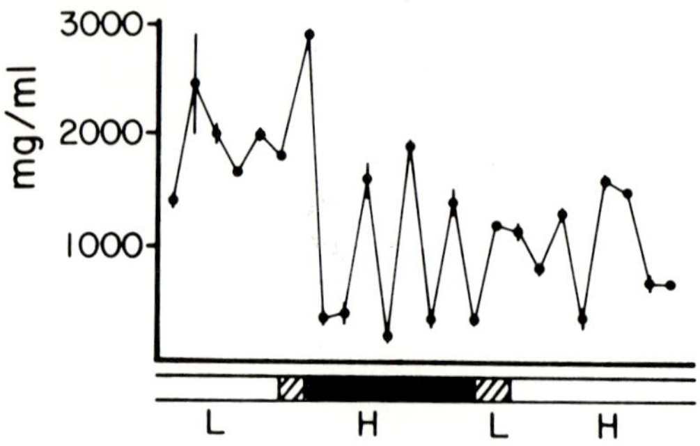

FIGURE 8. Concentration of protein in cell-free hemolymph of *M. mercenaria* sampled every hour for 24 consecutive hours: L, low tide; H, high tide; open bar, day; solid bar, night; crossbars, dusk or dawn.

FIGURE 9. Concentration of protein in cell-free hemolymph of *M. mercenaria* sampled at approximately 3-hr intervals during 3 complete tidal cycles: L, low tide; H, high tide; open bar, day; solid bar, night.

Qualitatively, PAGE of *M. mercenaria* hemolymph reveals a complex mixture of proteins including 10 major and 11 minor proteins (Figure 10). Major proteins are defined as bands detectable in the hemolymph of at least 90% of the clams examined; minor proteins are bands present in less than 90% of the hemolymph samples studied.

2. Hemolymph Protein in Clams Exposed to Hydrocarbons

Injection of sea water or potentially toxic compounds (e.g., acetone) into the hemal sinus of the posterior adductor muscle of *M. mercenaria* did not cause a change in the hemolymph electrophoretic pattern. When BaP, HCB, or PCP was injected directly into the digestive gland or when clams were exposed to those chemicals for 10 to 24 weeks, the major hemolymph proteins were essentially unaffected. However, some temporary changes in electrophoretic patterns were noted; in clams exposed to BaP for 7 and 8 weeks, and in clams exposed to HCB for 4, 5, and 8 weeks, 1 band (R_f = 0.2) disappeared. Two new bands (R_f = 0.59 and 0.96) appeared at weeks 9 and 10 in PCP-exposed clams. After 10 weeks, only normal electrophoretic patterns were found in all treated and control animals. When deletions occurred, they were among the higher molecular weight fractions (150,000 daltons), and additions occurred in the lower molecular weight fractions (10,000 to 40,000 daltons). This suggests that large molecules are degraded to smaller ones or that altered biosynthesis has changed the subunits so that they do not polymerize properly.

ELECTROPHORETIC PATTERN OF NORMAL
MERCENARIA MERCENARIA HEMOLYMPH

Identification of major bands,
Rf values, individual variation,
approximate molecular weights

Band	Rf	Distribution in population	Approx. M.W.
1	.01	100%	
2	.03	100%	
3	.04	100%	
4	.06	50%	
5	.08	87%	
6	.09	100%	
7	.10	33%	150,000
8	.13	95%	
9	.14	18%	
10	.17	100%	
11	.20	52%	
12	.21	93%	
13	.24	70%	100,000
14	.28	27%	
15	.29	100%	
16	.35	6%	
17	.36	30%	
18	.38	15%	
19	.41	77%	
20	.43	20%	
21	.47	100%	50,000
22	.53	56%	
23	.55	88%	
24	.58	68%	
25	.60	97%	
26	.63	64%	
27	.75	33%	
28	.76	22%	
29	.83	78%	25,000
30	.90	75%	
31	.95	58%	
32	.97	87%	

Slab Gel Electrophoresis:
10% polyacrylamide with SDS

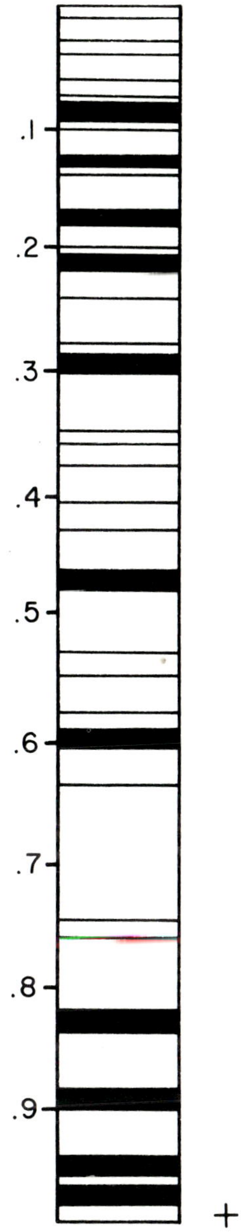

FIGURE 10. Electrophoretic pattern of *M. mercenaria* hemolymph proteins in 10% polyacrylamide with SDS.

The quantity of protein in hemolymph of clams exposed to hydrocarbons for 10 weeks is summarized in Table 5. These data are plotted as percent of control values in Figure 11. It is apparent that absolute values fluctuate from week to week and that relative values also vary irregularly over the 10-week test period. There is no consistent change in total hemolymph protein content associated with exposure to BaP, HCB, or PCP.

When a similar, but longer, experiment was conducted, hemolymph protein concentrations averaged 654 μg/mℓ over the first 10 weeks (compared to 688 μg/mℓ in the above-cited experiment), but for the whole 24 weeks the average was 565 μg/mℓ (Table 6). Protein contents of BaP-, HCB-, and PCP-treated clams varied erratically, but did not differ drast-

TABLE 5
Protein Content of Hemolymph of *Mercenaria mercenaria* Exposed to BaP, HCB, PCP, or Flowing Sea Water (Control) for 10 Weeks

	Protein concentration (μg/mℓ hemolymph)			
Week	Control	BaP	HCB	PCP
1	800	630	750	800
2	400	330	590	590
3	670	580	720	770
4	650	800	620	530
5	730	640	400	640
6	800	780	690	780
7	610	730	750	750
8	940	800	840	890
9	820	600	520	700
10	460	400	580	620
Average	688	629	646	707

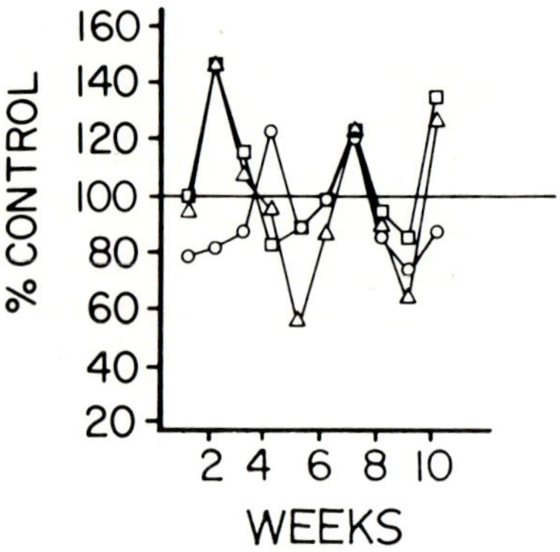

FIGURE 11. Hemolymph protein concentration of *M. mercenaria* exposed to BaP (○--○), HCB (△--△), or PCP (□--□) for 10 weeks relative to that of unexposed animals.

ically from normal values over nearly 6 months (Figure 12). Overall there is no evidence ($P > 0.05$) of any effect on hemolymph protein content, even with extended exposure to any of the 3 chemicals.

Clams injected with corn oil had depressed hemolymph protein concentration after 2 days, but by 4 days normal values were obtained (Table 7). Injection of corn oil plus any of the 3 test chemicals had very little effect on protein concentration except in PCP-treated animals at 11 and 18 days and in HCB-treated animals at 18 days; in these 3 cases, protein concentrations were less than 25% of the control values (Table 7; Figure 13). Hemolymph protein concentrations of all injected clams tended to by lower than values obtained from animals exposed to sea water plus chemical (Table 5) This suggests some loss of protein due to the mechanical trauma of the injection process itself.

TABLE 6
Protein Concentration in Hemolymph from *Mercenaria mercenaria* Exposed to BaP, HCB, PCP, or Flowing Sea Water (Control) for 24 Weeks

	Protein concentration (μg/mℓ hemolymph)			
Week	Control	BaP	HCB	PCP
1	741	840	855	735
5	765	555	510	825
8	600	540	390	420
10	510	630	540	795
13	420	495	435	405
15	396	510	555	540
17	591	606	627	510
21	660	750	735	966
24	398	780	627	651
Average	565	634	586	650

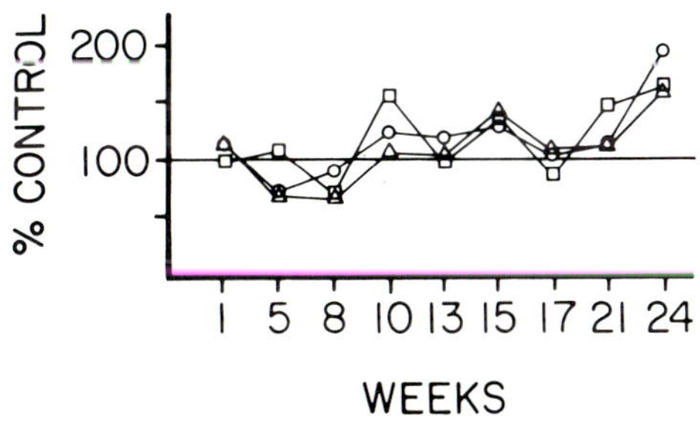

FIGURE 12. Hemolymph protein concentration of *M. mercenaria* exposed to BaP (○--○), HCB (△--△), or PCP (□--□) for 24 weeks relative to that of unexposed animals.

3. Hemolymph Enzymes

a. β-Glucuronidase

Normal clams sampled hourly (Figure 14) or at approximately 3-hr intervals (Figure 15) had serum β-glucuronidase concentrations that varied from a minimum of 20 Sigma Units/mℓ (SU/mℓ) to a maximum of 220 SU/mℓ. The activity was not related to tidal cycle or photoperiod, although it fluctuated between minimal and maximal values.

Wide variations in β-glucuronidase concentrations were noted in normal clams sampled either weekly for 10 weeks (Table 8) or at 2- to 4-week intervals for 24 weeks (Table 9). No consistent changes in hemolymph β-glucuronidase content were detectable in clams exposed to BaP, HCB, or PCP for either 10 weeks (Table 8; Figure 16) or 24 weeks (Table 9; Figure 17). There is no evidence that these chemical concentrations damage lysosomes and cause the release of lysosomal enzymes (e.g., β-glucuronidase) into the circulatory system.[29]

TABLE 7
Protein Concentration in Hemolymph of *Mercenaria mercenaria* Injected with Corn Oil (Control) or Corn Oil Plus BaP, HCB, or PCP

	Protein concentration (μg/mℓ hemolymph)			
Days	Control	BaP	HCB	PCP
2	225	750	405	381
4	642	672	435	240
9	630	510	720	420
11	450	435	342	72
15	345	300	276	357
18	381	561	87	72
Average	446	538	378	257

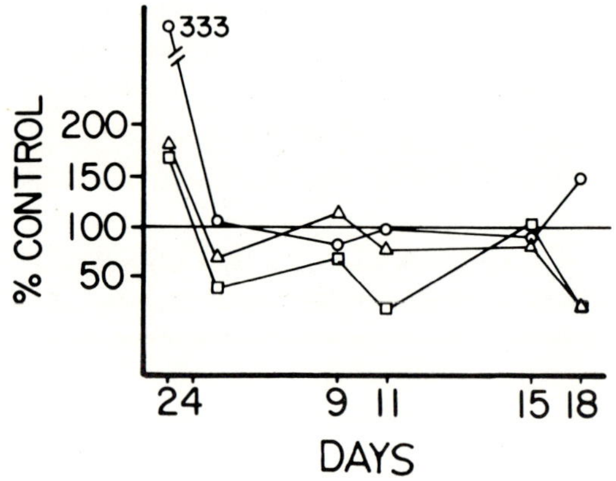

FIGURE 13. Hemolymph protein concentration of *M. mercenaria* injected with corn oil plus BaP (○--○), HCB (△--△), or PCP (□--□) relative to controls injected with corn oil only.

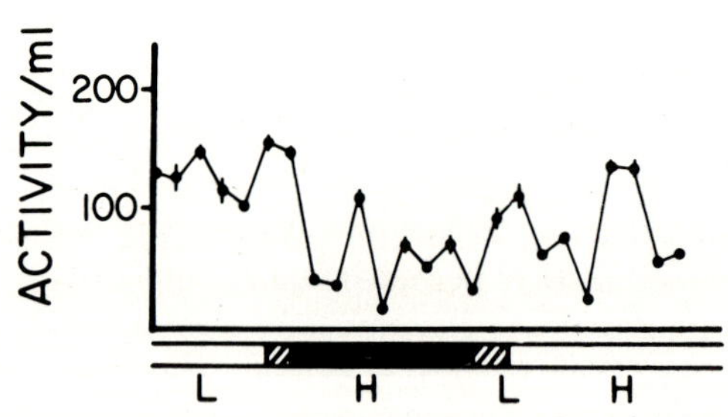

FIGURE 14. Hemolymph β-glucuronidase concentration of *M. mercenaria* sampled hourly for 24 hr: L, low tide; H, high tide; open bar, day; solid bar, night; crossbars, dusk or dawn.

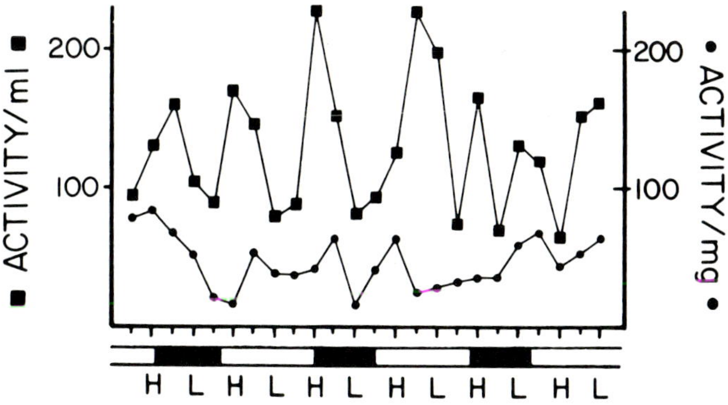

FIGURE 15. Hemolymph β-glucuronidase concentration of *M. mercenaria* sampled at approximately 3-hr intervals during 3 complete tidal cycles. Activity is expressed as both activity per milliliter hemolymph and per milligram hemolymph protein: L, low tide; H, high tide; open bar, day; solid bar, night.

TABLE 8
Concentration of β-Glucuronidase in Hemolymph of *Mercenaria mercenaria* Exposed to BaP, HCB, or PCP in Sea Water for 10 Weeks

	β-Glucuronidase [SU/mℓ (SU/mg)]			
Week	Control	BaP	HCB	PCP
1	90.4	70.5	99.0	114.0
	(113)	(113)	(132)	(143)
2	50.5	41.3	78.0	59.0
	(126)	(127)	(133)	(101)
3	73.5	67.0	67.0	78.0
	(110)	(116)	(93)	(101)
4	58.0	66.0	71.5	55.5
	(89)	(83)	(115)	(105)
5	79.0	103.5	46.5	89.5
	(108)	(162)	(116)	(140)
6	141.5	104.0	120.0	112.5
	(176)	(133)	(174)	(144)
7	66.0	80.0	75.0	77.0
	(108)	(110)	(100)	(103)
8	112.0	118.5	105.0	104.0
	(119)	(148)	(125)	(117)
9	119.5	64.0	60.0	85.5
	(146)	(107)	(115)	(122)
10	53.0	49.5	57.0	78.8
	(115)	(124)	(93)	(128)
Average	84.3	66.7	77.9	85.4
	(121)	(122)	(120)	(120)

Note: SU/mℓ, Sigma Units/mℓ hemolymph; (SU/mg), Sigma Units/mg protein.

TABLE 9
Concentration of β-Glucuronidase in Hemolymph of *Mercenaria mercenaria* Exposed to BaP, HCB, or PCP in Sea Water for 24 Weeks

	β-Glucuronidase (SU/mℓ)			
Week	Control	BaP	HCB	PCP
1	55.5	46.5	77.0	70.0
5	113.0	132.0	107.5	120.5
8	110.0	91.5	51.0	57.5
10	43.5	113.0	52.5	95.5
13	77.0	89.5	91.5	73.5
15	72.5	98.0	68.0	69.5
17	69.5	58.0	57.0	62.0
21	34.5	40.5	33.5	45.5
24	43.5	75.0	64.5	61.0
Average	68.8	82.7	66.9	72.8

Note: SU/mℓ, Sigma Units/mℓ hemolymph.

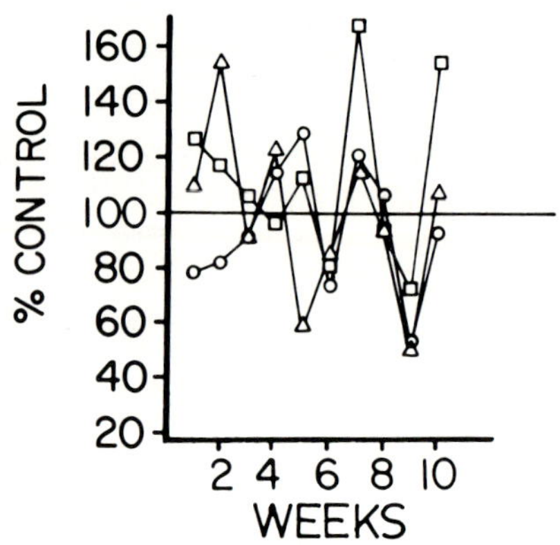

FIGURE 16. Hemolymph β-glucuronidase concentration of *M. mercenaria* exposed to BaP (○--○), HCB (△--△), or PCP (□--□) for 10 weeks relative to that of unexposed controls.

Clams injected with corn oil or corn oil plus hydrocarbon have depressed β-glucuronidase concentrations. Corn oil alone decreased the hemolymph β-glucuronidase content about 34% (Table 10); any of the hydrocarbon treatments depressed the enzyme content to 40 to 50% of that of clams receiving the same chemical from the surrounding water (Table 9). The hemolymph β-glucuronidase content of clams injected with any of the hydrocarbons tested varies irregularly relative to controls (Figure 18).

b. Glutamic-Pyruvic Transaminase

Glutamic-pyruvic transaminase (GPT) concentrations in the hemolymph of normal clams assayed at weekly intervals for 10 weeks averaged 16.2 Sigma Frankel Units/mℓ (SFU/mℓ)

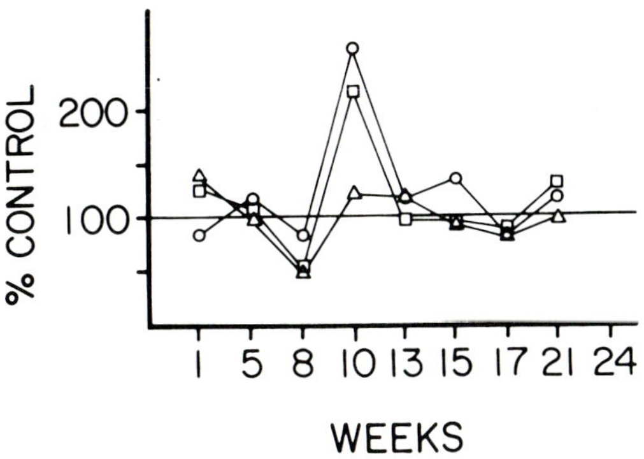

FIGURE 17. Hemolymph β-glucuronidase concentration of *M. mercenaria* exposed to BaP (○--○), HCB (△--△), or PCP (□--□) for 24 weeks relative to that of unexposed controls.

TABLE 10
Concentration of β-Glucuronidase in Hemolymph of *Mercenaria mercenaria* Injected with Corn Oil (Control) or Corn Oil Plus BaP, HCB, or PCP

	β-Glucuronidase [SU/mℓ (SU/mg)]			
Day	**Control**	**BaP**	**HCB**	**PCP**
2	50.5	78.5	38.5	61.0
	(224)	(105)	(95)	(160)
4	63.5	55.0	27.5	42.5
	(99)	(82)	(63)	(177)
9	64.5	74.5	92.0	47.5
	(102)	(146)	(128)	(113)
11	40.5	28.0	42.5	14.5
	(90)	(64)	(124)	(201)
15	28.5	18.0	20.5	27.5
	(83)	(60)	(74)	(77)
18	29.5	41.5	7.5	8.5
	(77)	(74)	(86)	(118)
Average	46.2	49.3	33.6	38.1
	(113)	(89)	(95)	(141)

Note: SU/mℓ, Sigma Units/mℓ hemolymph; (SU/mg), Sigma Units/mg hemolymph protein.

with a range of 11.6 to 23.0 SFU/mℓ (Table 11). Other clams held under the same conditions for 24 weeks averaged only 4.4 SFU/mℓ, approximately a quarter of the activity of the first group (Table 12). There is no obvious explanation for this marked difference since clams in both groups were of similar sizes, from a single source, fed actively, and appeared healthy.

Clams exposed to test chemicals for 10 weeks had avaerage GPT activities and ranges similar to those of unexposed controls whether values are expressed as activity per milliliter of hemolymph or as specific activity (SFU/mg protein) (Table 11). Similarly, clams studied

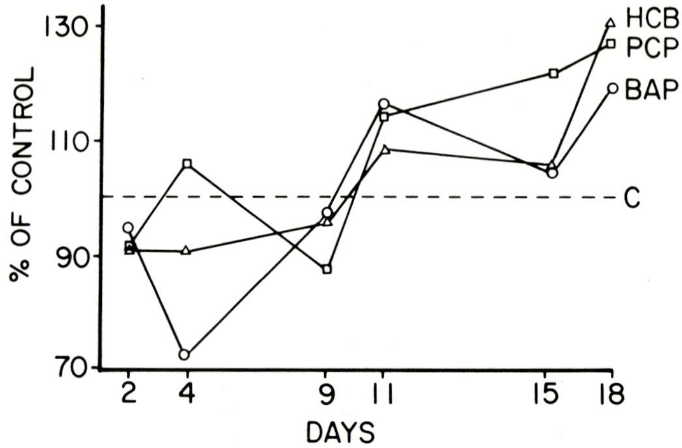

FIGURE 18. Hemolymph β-glucuronidase concentration of *M. mercenaria* injected with corn oil plus BaP (○--○), HCB (△--△), or PCP (□--□) relative to controls injected with corn oil only.

TABLE 11
Glutamic-pyruvic Transaminase Concentration in Hemolymph of *Mercenaria mercenaria* Exposed to BaP, HCB, or PCP in Sea Water for 10 Weeks

	Glutamic-pyruvic transaminase [SFU/mℓ (SFU/mg)]			
Week	**Control**	**BaP**	**HCB**	**PCP**
1	23.0	23.0	23.0	23.0
	(29)	(37)	(31)	(29)
2	11.6	11.6	13.5	11.6
	(29)	(36)	(23)	(20)
3	14.3	12.0	15.6	16.3
	(21)	(21)	(22)	(21)
4	16.3	23.6	17.0	18.0
	(22)	(28)	(36)	(34)
5	11.6	14.0	10.0	12.5
	(16)	(22)	(25)	(20)
6	15.6	16.0	11.6	16.0
	(20)	(21)	(17)	(21)
7	13.3	21.0	20.0	19.0
	(22)	(29)	(27)	(25)
8	21.6	13.6	12.0	15.0
	(23)	(17)	(14)	(17)
9	22.0	18.0	15.0	18.6
	(27)	(30)	(29)	(27)
10	12.3	16.0	23.0	22.3
	(27)	(40)	(40)	(36)
Average	16.2	16.9	16.1	17.2
	(23)	(28)	(26)	(25)

Note: SFU/mℓ, Sigma Frankel Units/mℓ hemolymph; (SFU/mg), Sigma Frankel Units/mg hemolymph protein.

TABLE 12
Glutamic-pyruvic Transaminase Concentration in Hemolymph of *Mercenaria mercenaria* Exposed to BaP, HCB, or PCP in Sea Water for 24 Weeks

	Glutamic-pyruvic transaminase (SFU/mℓ)			
Week	Control	BaP	HCB	PCP
1	3.0	1.3	5.0	0.7
5	4.0	2.3	2.0	—
8	3.3	4.2	2.5	4.3
10	4.3	1.7	2.3	5.5
13	2.0	2.0	4.7	4.0
15	—	5.0	5.5	7.3
17	6.3	10.3	5.0	4.0
21	5.3	4.3	—	7.3
24	7.0	10.0	3.3	12.7
Average	4.4	4.6	3.8	5.7

Note: SFU/mℓ, Sigma Frankel Units/mℓ hemolymph; —, no sample.

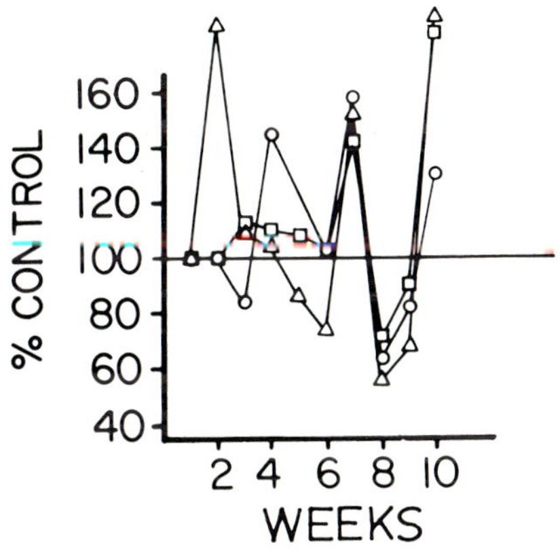

FIGURE 19. Concentration of glutamic-pyruvic transaminase in the hemolymph of *M. mercenaria* exposed to BaP (○--○), HCB (△--△), or PCP (□--□) for 10 weeks relative to that of unexposed controls.

over 24 weeks were comparable to their untreated controls, but in all cases GPT activities were consistently lower (Table 12) than in comparable animals used in the 10-week experiment. The irregular variation in GPT concentration relative to control values over time (10 or 24 weeks) can be seen in Figures 19 and 20.

Clams injected with corn oil (Table 13) had lower GPT concentrations than normal animals (Tables 11 and 12). Injections of BaP or HCB plus corn oil had little consistent effect; clams injected with PCP in corn oil had much lower GPT concentrations overall, and in some instances GPT activity was undetectable (Table 13). Concentrations of hemolymph GPT in clams injected with hydrocarbons relative to controls varied erratically (Figure 21).

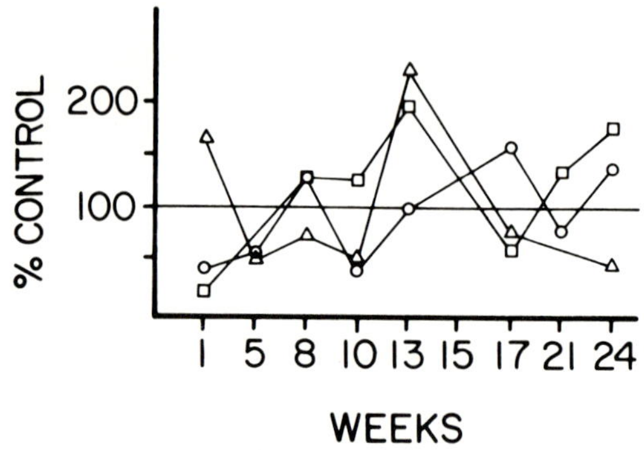

FIGURE 20. Concentration of glutamic-pyruvic transaminase in the hemolymph of *M. mercenaria* exposed to BaP (○--○), HCB (△--△), or PCP (□--□) for 24 weeks relative to that of unexposed controls.

TABLE 13
Glutamic-pyruvic Transaminase Concentration in the Hemolymph of *Mercenaria mercenaria* Injected with Corn Oil (Control) or Corn Oil Plus BaP, HCB, or PCP

	Glutamic-pyruvic transaminase [SFU/mℓ (SFU/mg)]			
Day	Control	BaP	HCB	PCP
2	0.000	7.000	4.670	1.500
	(—)	(0.009)	(0.012)	(0.004)
4	7.000	8.330	3.330	0.330
	(0.011)	(0.012)	(0.077)	(0.001)
9	4.330	2.330	2.000	6.000
	(0.007)	(0.005)	(0.003)	(0.014)
11	3.330	7.670	7.000	0.000
	(0.007)	(0.018)	(0.020)	(—)
15	5.330	5.000	2.330	0.000
	(0.015)	(0.017)	(0.008)	(—)
18	1.000	0.330	0.330	1.000
	(0.003)	(0.001)	(0.004)	(0.014)
Average	3.500	5.100	3.300	1.500
	(0.007)	(0.010)	(0.021)	(0.006)

Note: SFU/mℓ, Sigma Frankel Units/mℓ hemolymph; (SFU/mg), Sigma Frankel Units/mg hemolymph protein; —, no sample.

These data illustrate the difficulty of establishing useful ranges of GPT that are meaningful for different groups of clams. The concentration of GPT in hemolymph does not appear to be a useful diagnostic tool for *M. mercenaria*.

c. *Glutamic-Oxaloacetic Transaminase*

Glutamic-oxaloacetic transaminase (GOT) activity in the hemolymph of normal clams examined at weekly intervals for 10 weeks averaged 15.9 SFU/mℓ, with a range of 12.6 to

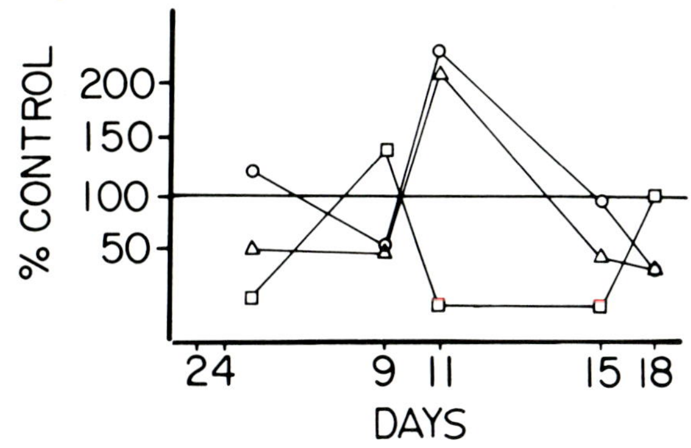

FIGURE 21. Concentration of glutamic-pyruvic transaminase in the hemolymph of *M. mercenaria* injected with corn oil plus BaP (○--○), HCB (△--△), or PCP (□--□) relative to controls injected with corn oil only.

33.0 SFU/mℓ (Table 14). Other clams examined over 24 weeks had markedly less activity, averaging 5.9 SFU/mℓ and ranging from 1.3 to 24.3 SFU/mℓ (Table 15), less than 50% of the activity of the other group.

In clams exposed to BaP, HCB, or PCP in sea water for 10 weeks, there was little difference in the average values among treatments (Table 14). Likewise, clams treated for 24 weeks had similar average values, all slightly higher than normal control values (Table 15). The variability within treatment groups is sometimes great, but occurs without apparent cause. There is no evidence that any of the test chemicals induce a significant increase or decrease in GOT activity (Figures 22 and 23). The observed changes in activity cannot be associated with environmental changes, spawning condition, or histopathology.

Clams injected with corn oil have varying amounts of GOT in the hemolymph (Table 16), but they are comparable to the values observed over a 24-week period in normal animals (Table 15). Clams injected with BaP or PCP in corn oil have irregular variations in enzyme concentrations, but overall were not different from controls (Table 16) or clams exposed to those same chemicals in sea water for 24 weeks (Table 15). HCB-treated animals had markedly lower GOT concentrations except at 9 days when the value was very high (Table 16). Hemolymph GOT values of clams injected with hydrocarbons vary erratically relative to controls (Figure 24).

Because GOT hemolymph concentration varies greatly between groups of apparently comparable healthy clams, as well as among chemically treated groups, GOT is not a useful diagnostic tool to measure chemical stress in *M. mercenaria.*

C. Digestive Gland

The digestive gland tissue of lamellibranch molluscs is exposed to pollutants in water ingested during feeding, and these materials might affect digestion and absorption by digestive gland cells. We considered the possibility that digestive lysosomes might be destabilized and enzymes (acid phosphatase and β-glucuronidase) released from them because of stress.[29] Digestive gland was analyzed for both acid phosphatase (ACP) and β-glucuronidase by counting histochemically demonstrable granules (expressed in granules/mm^2 tissue); tissue extracts were also assayed for ACP activity. The granules represent lysosomes, while extraction data represent both lysosomal and other microsomal material. β-Glucuronidase activity was found in both digestive tubules and surrounding connective tissue (Plate 2), while ACP activity was associated with tubules only (Plate 3). The ACP assay used β-

TABLE 14
Glutamic-oxaloacetic Transaminase Concentration in Hemolymph of *Mercenaria mercenaria* Exposed to BaP, HCB, or PCP in Sea Water for 10 WEEKS

	Glutamic-oxaloacetic transaminase [SFU/mℓ (SFU/mg)]			
Week	**Control**	**BaP**	**HCB**	**PCP**
1	33.0	25.0	26.0	28.3
	(41.3)	(40.0)	(34.7)	(35.4)
2	12.6	8.6	27.6	14.0
	(31.5)	(26.5)	(47.2)	(23.9)
3	18.0	11.3	15.6	14.0
	(26.9)	(19.5)	(21.7)	(18.2)
4	14.0	18.5	14.6	13.0
	(21.5)	(23.1)	(23.5)	(24.5)
5	10.6	14.0	9.3	12.0
	(14.5)	(21.9)	(23.3)	(18.8)
6	14.3	13.6	12.3	16.8
	(17.9)	(17.4)	(17.8)	(21.5)
7	11.3	15.3	20.0	15.3
	(18.6)	(21.0)	(26.7)	(20.4)
8	14.6	13.0	15.0	13.6
	(15.5)	(16.3)	(17.8)	(15.3)
9	18.0	13.3	12.3	15.3
	(21.9)	(22.2)	(23.6)	(21.9)
10	12.6	10.0	14.6	14.6
	(27.4)	(25.0)	(25.2)	(23.5)
Average	15.9	14.3	16.7	15.7
	(23.7)	(23.3)	(26.2)	(22.3)

Note: SFU/mℓ, Sigma Frankel Units/mℓ hemolymph; (SFU/mg), Sigma Frankel Units/mg Protein.

glycerophosphate as substrate, and activity was eliminated when 10 m*M* tartrate was used as inhibitor, indicating the lysosomal nature of the enzyme.[30,31]

To determine if other metabolic processes were affected by chemical stressors, alterations in concentration of GPT and GOT were sought (see below).

1. Acid Phosphatase

In normal clams sampled hourly for 1 day (2 tidal cycles), digestive gland ACP activity (extraction procedure) averaged 3.6 SU/mg of tissue and ranged from 2.2 to 4.4 SU/mg (Figure 25). When these observations were extended over 3 days (6 tidal cycles), the average value (4.06 SU/mg) and the range were only slightly greater (Table 17; Figure 26). Comparison of the histochemical and extraction procedures shows that histochemical methods yield less erratic variation; neither procedure indicates that ACP activity is correlated with tide or photoperiod (Table 17; Figures 25 and 26). Over a period of nearly 2 weeks, ACP activity tends to be relatively stable with only occasional significant variations from normal (Figure 27).

When clams are injected with chemicals (corn oil only in controls) and digestive tissues

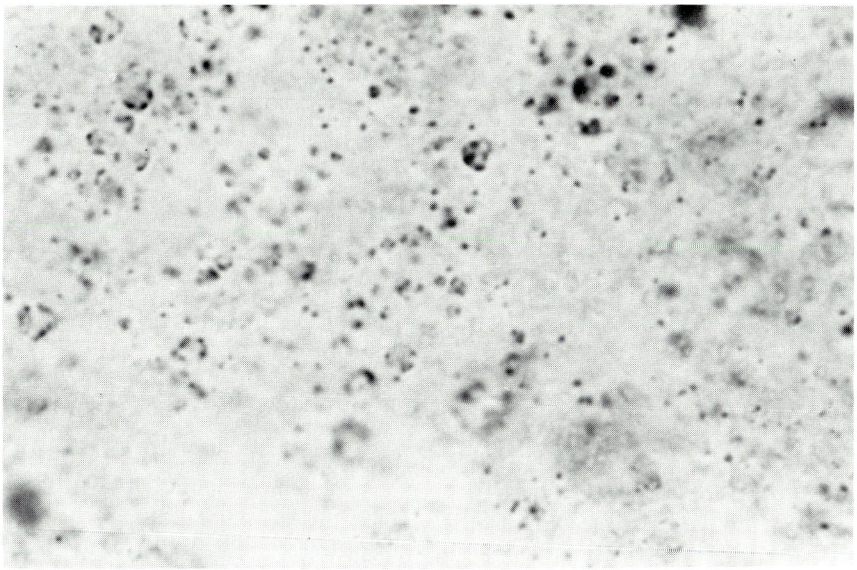

PLATE 2. Frozen section of digestive gland of *M. mercenaria* showing β-glucuronidase activity. (Magnification × 1200.)

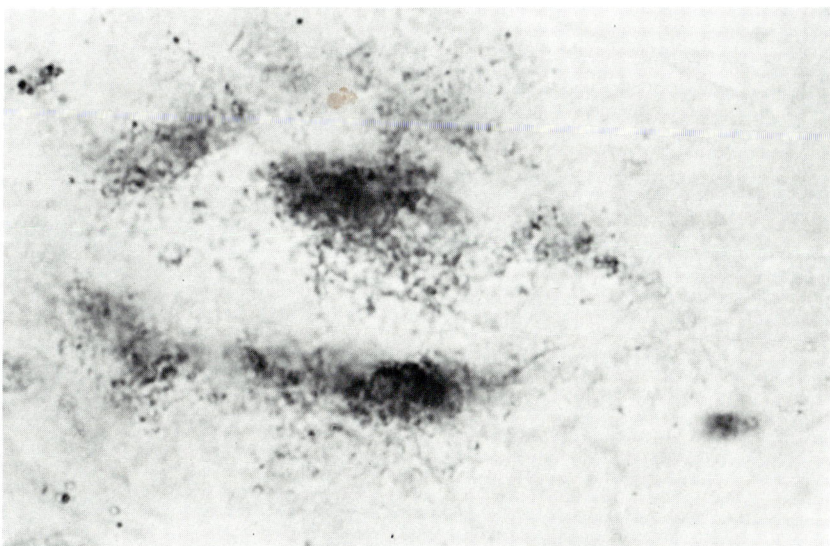

PLATE 3. Frozen section of digestive gland of *M. mercenaria* showing acid phosphatase activity (Gomori technique). (Magnification × 1200.)

TABLE 15
Glutamic-oxaloacetic Transaminase Concentration in Hemolymph of *Mercenaria mercenaria* Exposed to BaP, HCB, or PCP in Sea Water for 24 Weeks

	Glutamic-oxaloacetic transaminase (SFU/mℓ)			
Week	**Control**	**BaP**	**HCB**	**PCP**
1	5.7	7.3	3.0	5.0
5	6.0	5.7	2.0	5.3
8	1.3	4.0	3.5	1.0
10	4.0	3.8	0.5	5.0
13	3.0	6.0	6.7	13.7
15	4.3	7.3	6.7	6.3
17	3.0	16.0	5.5	15.0
21	24.3	17.3	17.0	15.0
24	1.5	0.0	20.0	0.0
Average	5.9	7.5	7.2	7.4

Note: SFU/mℓ, Sigma Frankel Units/mℓ hemolymph.

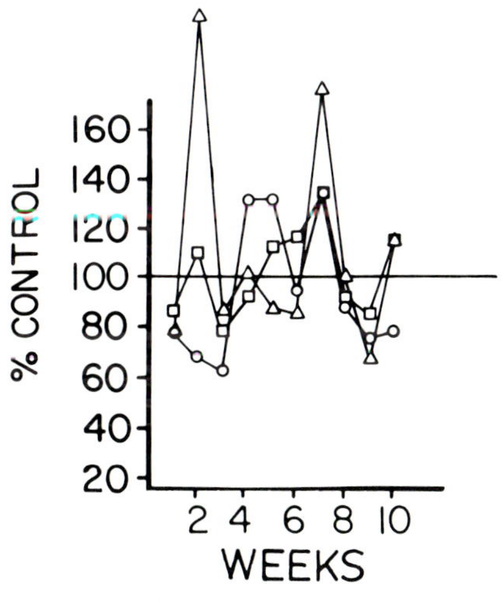

FIGURE 22. Concentration of glutamic-oxaloacetic transaminase in the hemolymph of *M. mercenaria* exposed to BaP (○--○), HCB (△--△), or PCP (□--□) for 10 weeks relative to unexposed controls.

are assayed histochemically, control values are lower than normal, but ACP values for chemically treated animals remain similar to normal ones (Table 18). There is no pattern of ACP increase or decrease that can be attributed to a unique effect of BaP, HCB, or PCP alone (Table 18). When relative values for chemical treatments are compared to controls, all these chemicals seem to induce initial depression (day 2), a rebound well above normal (day 4), and a gradual return to normal over a period of several days (Figure 28), resembling patterns often explained as hormesis.[32]

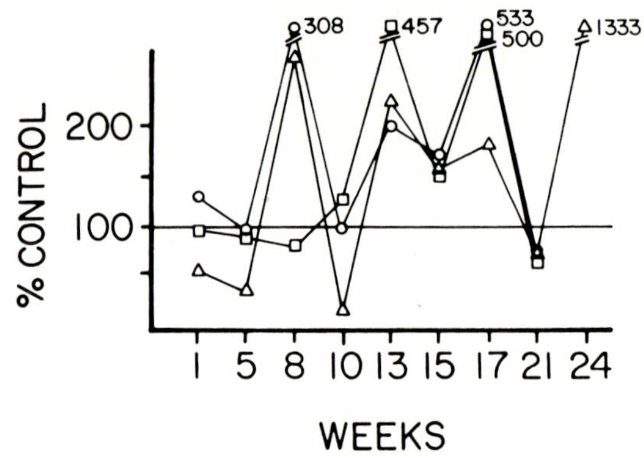

FIGURE 23. Concentration of glutamic-oxaloacetic transaminase in the hemolymph of *M. mercenaria* exposed to BaP (○--○), HCB (△--△), or PCP (□--□) for 24 weeks relative to unexposed controls.

TABLE 16
Glutamic-oxaloacetic Transaminase Concentration in the Hemolymph of *Mercenaria mercenaria* Injected with Corn Oil (Control) or Corn Oil Plus BaP, HCB, or PCP

	Glutamic-oxaloacetic transaminase [SFU/mℓ (SFU/mg)]			
Day	**Control**	**BaP**	**HCB**	**PBP**
2	0.300	7.000	4.700	0.700
	(0.001)	(0.009)	(0.002)	(0.012)
4	4.500	9.000	0.000	2.000
	(0.007)	(0.013)	(0.008)	(—)
9	3.700	5.700	14.300	20.700
	(0.006)	(0.011)	(0.020)	(0.049)
11	6.300	4.300	12.700	3.300
	(0.014)	(0.010)	(0.046)	(0.037)
16	7.000	5.000	2.300	0.000
	(0.020)	(0.017)	(—)	(0.008)
18	10.300	12.000	4.700	3.000
	(0.027)	(0.021)	(0.417)	(0.537)
Average	5.400	7.200	5.000	6.800
	(0.013)	(0.014)	(0.102)	(0.087)

Note: SFU/mℓ, Sigma Frankel Units/mℓ hemolymph; (SFU/mg), Sigma Frankel Units/mg hemolymph protein; —, no sample.

2. *β-Glucuronidase*

Digestive gland β-glucuronidase activity estimated by the histochemical procedure averaged 349 granules/mm^2 tissue over a 24-hr period (Figure 29) or 331 granules/mm^2 over a 3-day period (Table 17; Figure 30). The ranges in both cases were similar, and in neither case was there evidence of tidal or photoperiod influence (Figures 29 and 30). Normal clams

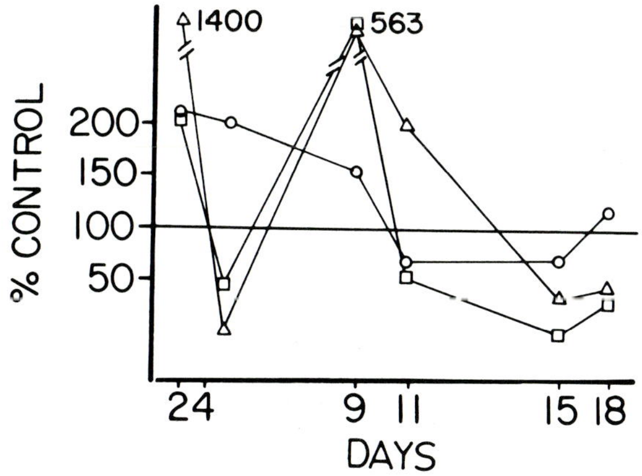

FIGURE 24. Concentration of glutamic-oxaloacetic transaminase in the hemolymph of *M. mercenaria* injected with corn oil plus BaP (○--○), HCB (△--△), or PCP (□--□) relative to controls injected with corn oil only.

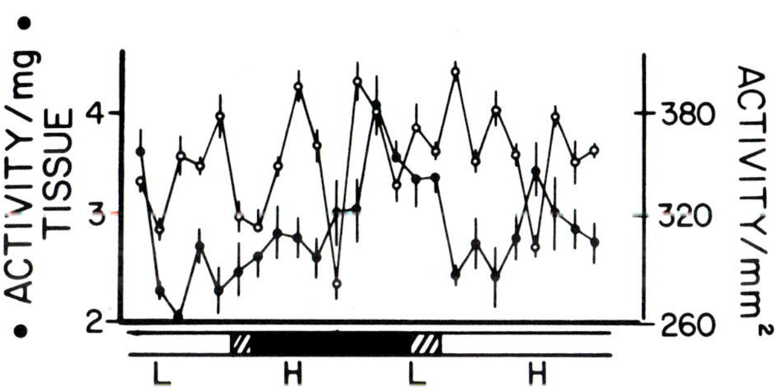

FIGURE 25. Acid phosphatase activity in the digestive gland of *M. mercenaria* sampled hourly for 24 hr. Activity expressed per milligram tissue (extraction procedure) and per square millimeter (histochemical procedure): L, low tide; H, high tide; open bar, day; solid bar, night; crossbars, dusk or dawn.

sampled at high and low tides for 2 weeks had an average activity of about 256 granules/ mm^2, with a range of values from 154 to 358 granules/mm^2 (Figure 31).

When clams were exposed to contaminants in sea water for nearly 6 months, the average values for β-glucuronidase activity (measured by the extraction procedure) for controls and the experimental treatments did not vary significantly (Table 19). When chemically treated animals are considered relative to untreated controls, BaP values tended to be elevated during the first 10 weeks and then decreased to normal; the low value at 8 weeks is unexplained (Figure 32). PCP values remained slightly below normal except at week 21; HCB values varied around normal in an irregular fashion for the duration of the experiment.

Clams that received injections of chemicals directly into the digestive gland and then were assayed by the histochemical or the extraction (Table 20) method showed irregular variations in enzyme content. When compared to control values, the lack of any clear effect of any of the three chemicals is evident whether the histochemical (Figure 33) or extraction (Figure 34) assay is used.

TABLE 17
Activities of β-Glucuronidase and Acid Phosphatase in the Digestive Gland of *Mercenaria mercenaria* during 6 Continuous Tidal Cycles

			Acid phosphatase	
Day	Cycle	β-Glucuronidase (granules/ mm^2 tissue)	Granules/ mm^2 tissue	SU/mg tissue
1	IN-L	360 ± 33	416 ± 11	4.36 ± 0.46
	HI-L	386 ± 23	452 ± 8	3.95 ± 0.72
	OUT-D	443 ± 35	429 ± 20	5.16 ± 0.70
	LO-D	348 ± 42	469 ± 8	4.92 ± 0.43
	IN-D	330 ± 36	440 ± 6	4.27 ± 0.61
	HI-L	270 ± 33	413 ± 14	3.85 ± 0.60
	OUT-L	384 ± 6	419 ± 5	3.93 ± 0.64
	LO-L	267 ± 24	416 ± 5	2.95 ± 0.32
2	IN-L	347 ± 29	437 ± 6	4.66 ± 0.21
	HI-L	245 ± 34	403 ± 13	2.10 ± 1.05
	OUT-D	293 ± 23	416 ± 6	3.83 ± 0.40
	LO-D	349 ± 17	410 ± 4	5.24 ± 0.66
	IN-L	321 ± 10	420 ± 14	4.64 ± 0.40
	HI-L	370 ± 15	418 ± 6	4.14 ± 0.23
	OUT-L	332 ± 8	404 ± 11	3.74 ± 0.50
	LO-L	418 ± 11	408 ± 8	5.63 ± 0.42
3	IN-L	288 ± 40	406 ± 9	3.66 ± 0.44
	HI-D	346 ± 20	405 ± 16	3.98 ± 0.42
	OUT-D	358 ± 9	400 ± 23	2.96 ± 0.70
	LO-D	306 ± 37	443 ± 11	3.74 ± 0.51
	IN-L	311 ± 31	398 ± 3	5.50 ± 0.27
	HI-L	282 ± 14	409 ± 12	3.61 ± 0.37
	OUT-L	292 ± 27	391 ± 14	2.58 ± 0.52
	LO-L	304 ± 15	392 ± 16	3.99 ± 0.36
Average		331 ± 10	419 ± 4	4.06 ± 0.18

Note: Values are means ± standard error of mean. IN, flooding tide; OUT, ebbing tide; HI, high tide; LO, low tide; L, light; D, dark.

There was no consistent evidence that clams exposed to or injected with BaP, HCB, or PCP are stressed enough to have detectable changes in β-glucuronidase content of the digestive gland. This may be due to a lack of toxicity of these chemicals or the resistance of *M. mercenaria* to toxic or other stressor effects of these chemicals.

3. Glutamic-Pyruvic Transaminase

GPT activity in the digestive gland of clams varied erratically over nearly 6 months of exposure (Table 21). Control values in unexposed clams ranged from 0.2 to 7.5 SFU/mg of tissue, with an average value of 5.2 SFU/mg. Extracts of digestive gland tissue from clams exposed to BaP, HCB, or PCP showed similar variability (Table 21; Figure 35). There was no detectable effect of any of those three chemicals.

4. Glutamic-Oxaloacetic Transaminase

GOT activity in control clams did not vary greatly over the nearly 6-month exposure period (Table 22). The average value was 5.1 SFU/mg of tissue, and ranged from 2.7 to

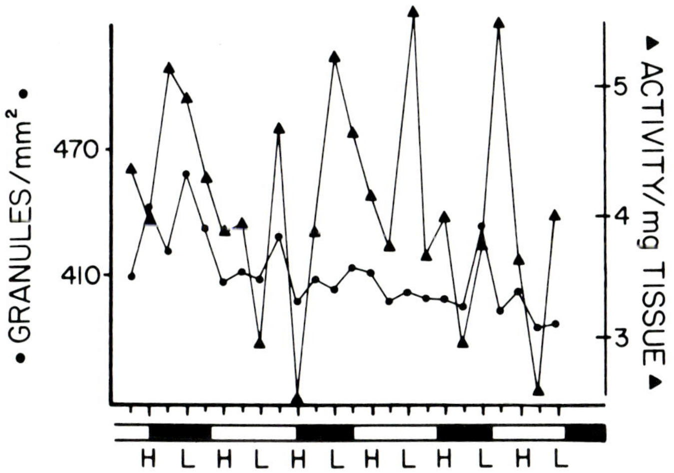

FIGURE 26. Acid phosphatase activity in the digestive gland of *M. mercenaria* sampled at 3-hr intervals over 3 tidal cycles. Activity determined by extraction and histochemical procedures: L, low tide; H, high tide; open bar, day; solid bar, night.

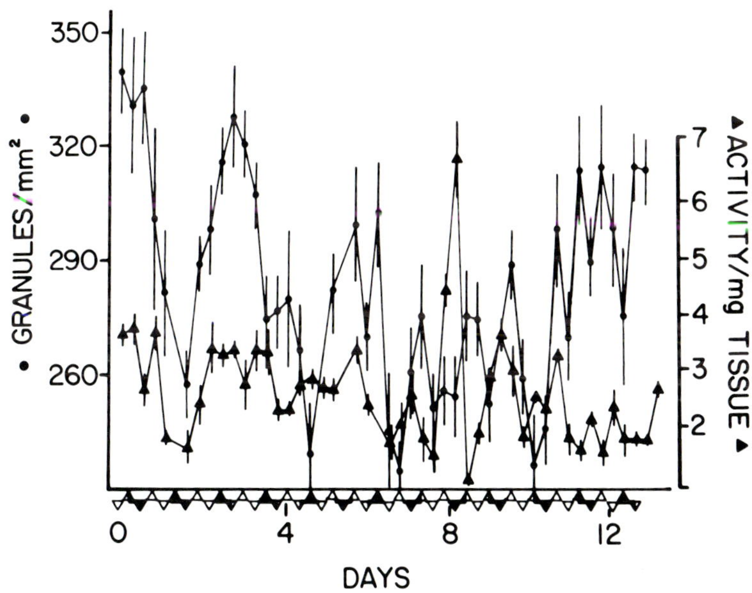

FIGURE 27. Acid phosphatase activity in the digestive gland of *M. mercenaria* sampled at high and low tides for 14 days. Activity determined by extraction (▲--▲) and histochemical (●--●) procedures; ▽, low tide; △, high tide; open △, day; solid ▲, night.

7.5 SFU/mg. Clams exposed to BaP, HCB, or PCP had similar overall averages and ranges (Table 22). At various sampling times, GOT activity in experimental animals did not differ greatly from that of control animals (Figure 36). There is no evidence that any of these three chemicals affected the concentration of GOT in the digestive gland of *M. mercenaria*.

TABLE 18
Acid Phosphatase Activity Measured by the Histochemical Technique in Digestive Gland of *Mercenaria mercenaria* Injected with Corn Oil (Control) or Corn Oil Plus BaP, HCB, or PCP

	Acid phosphatase (granules/mm² tissue)			
Days	Control	BaP	HCB	PCP
2	438	346	399	280
4	310	404	490	485
9	294	444	285	332
11	330	363	432	444
15	534	630	518	501
18	330	383	531	422
Average	372	428	443	411

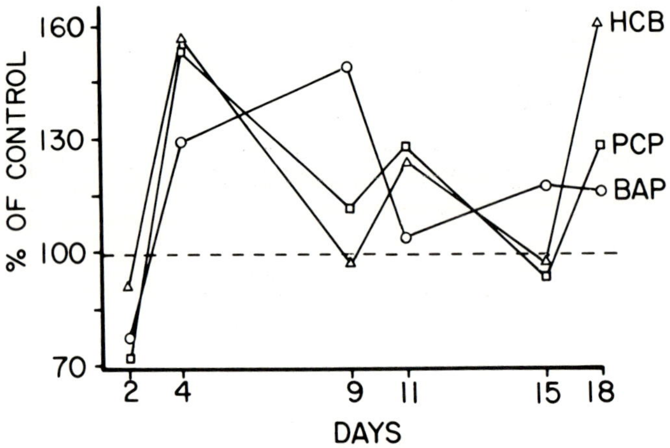

FIGURE 28. Acid phosphatase activity in the digestive gland of *M. mercenaria* injected with corn oil plus BaP (○--○), HCB (△--△), or PCP (□--□) relative to controls injected with corn oil only.

D. Summary of Biochemical Parameters

Several biochemical parameters in *M. mercenaria* hemolymph and digestive gland tissue were measured to determine normal values and to learn if significant changes in those values could be induced by exposure to three common hydrocarbon pollutants. Total hemolymph protein varies greatly in normal clams both between groups and within a single group when individuals are sampled at different times. Qualitative analysis of *M. mercenaria* hemolymph by polyacrylamide electrophoresis revealed a mixture of over 20 proteins that was relatively unvarying between individuals and between groups examined at different times. The concentrations of β-glucuronidase, glutamic-oxaloacetic transaminase, and glutamic-pyruvic transaminase varied widely among individuals and between groups. Because of this variability, it is not known if the three hydrocarbons tested stressed clams enough to cause

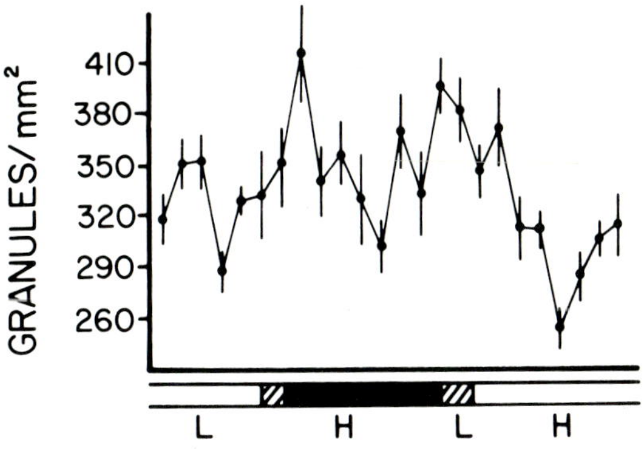

FIGURE 29. β-Glucuronidase activity in the digestive gland of *M. mercenaria* sampled hourly for 24 hr: L, low tide; H, high tide; open bar, day; solid bar, night; crossbars, dusk or dawn.

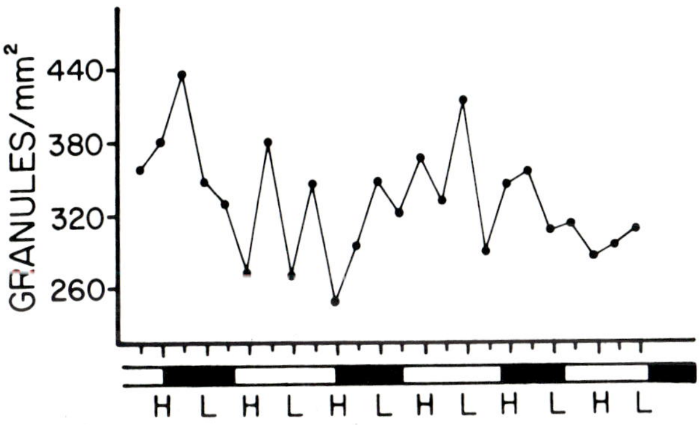

FIGURE 30. β-Glucuronidase activity in the digestive gland of *M. mercenaria* sampled at 3-hr intervals during 3 tidal cycles: L, low tide; H, high tide; open bar, day; solid bar, night.

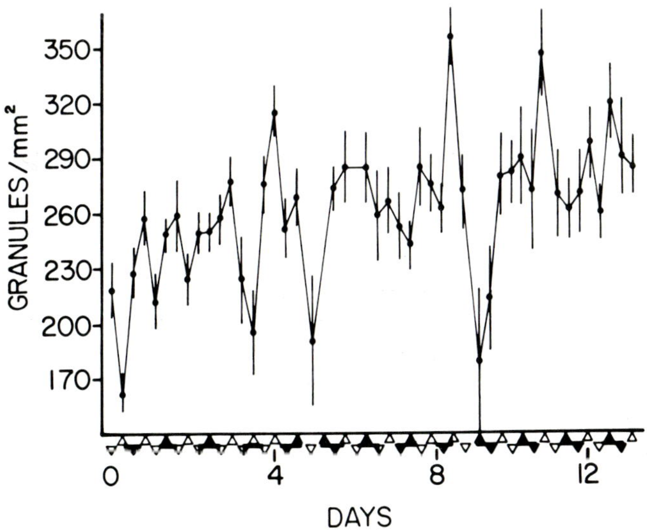

FIGURE 31. β-Glucuronidase activity in the digestive gland of *M. mercenaria* sampled at high and low tide for 14 days.

TABLE 19
β-Glucuronidase Activity in the Digestive Gland of *Mercenaria mercenaria* after Exposure to BaP, HCB, or PCP for 24 Weeks

	β-Glucuronidase (SU/mg)			
Week	Control	BaP	HCB	PCP
5	11.08	18.02	17.93	8.91
8	20.36	7.55	13.10	18.93
10	6.93	13.95	9.53	4.44
13	9.82	16.61	12.34	7.49
15	11.19	13.19	10.16	7.08
17	15.40	12.11	9.17	11.78
21	5.44	5.78	9.42	18.86
24	8.15	5.44	6.82	8.73
Average	11.05	11.58	11.06	10.78

Note: SU/mg, Sigma Units/mg tissue.

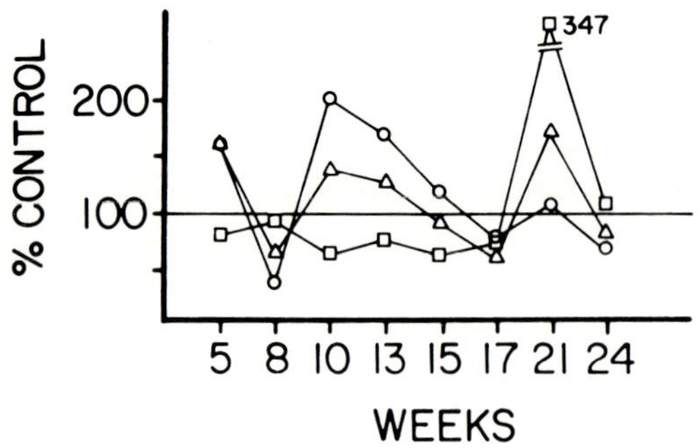

FIGURE 32. β-Glucuronidase activity in the digestive gland of *M. mercenaria* exposed to BaP (○--○), HCB (△--△), or PCP (□--□) for 24 weeks relative to unexposed controls.

detectable changes in these biochemical parameters. The only consistently positive change associated with chemical treatment was a transient change in electrophoretic banding of hemolymph protein.

Digestive gland tissue was quite stable as measured by the concentration of selected enzymes (i.e., acid phosphatase, β-glucuronidase, glutamic-pyruvic transaminase, and glutamic-oxaloacetic transaminase). When clams were injected with any of three hydrocarbons, acid phosphatase concentrations decreased initially, rebounded sharply within a few days, and then decreased to normal over several more days. The other enzymes showed no change to exposure or injection with those same chemicals.

Overall, there is little evidence to suggest that the hydrocarbons tested in these experiments (BaP, HCB, and PCP) have any significant effect on *M. mercenaria*. Neither long exposure to chemicals in the water nor injection of the chemical into digestive tissue resulted in changes in hemolymph or digestive gland tissues. Thus, there is nothing to suggest that these biochemical indexes would be useful indicators of chemical stress.

TABLE 20
β-Glucuronidase Activity in the Digestive Gland of *Mercenaria mercenaria* Injected with Corn Oil (Control) or Corn Oil Plus BaP, HCB, or PCP

	Control		BaP		HCB		PCP	
Days	SU/mg	Gran/mm²	SU/mg	Gran/mm²	SU/mg	Gran/mm²	SU/mg	Gran/mm³
2	5.71	211	4.24	200	4.85	193	5.71	193
4	4.44	187	5.09	135	3.06	169	5.56	197
9	5.71	189	5.96	184	7.63	180	5.11	165
11	7.40	165	5.95	192	5.11	179	4.21	188
15	3.62	198	6.18	207	5.80	209	5.35	242
18	5.18	117	3.68	153	3.70	140	3.38	148
Average	5.34	178	5.18	179	5.03	179	4.89	189

Note: SU/mg, Sigma Units/mg tissue; Gran/mm², Granules/mm² tissue.

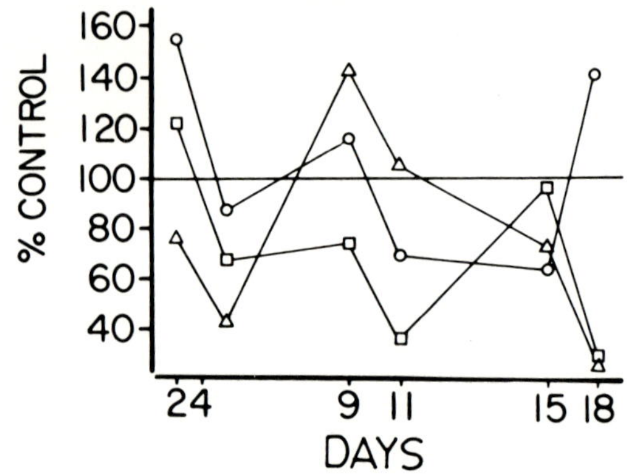

FIGURE 33. β-Glucuronidase activity measured by histochemical assay in the digestive gland of *M. mercenaria* injected with corn oil plus BaP (○--○), HCB (△--△), PCP (□--□) relative to controls injected with corn oil only.

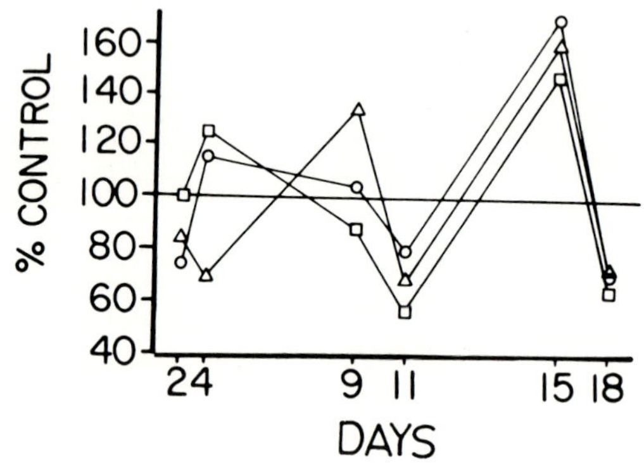

FIGURE 34. β-Glucuronidase activity measured by extraction assay in the digestive gland of *M. mercenaria* injected with corn oil plus BaP (○--○), HCB (△--△), or PCP (□--□) relative to controls injected with corn oil only.

ACKNOWLEDGMENTS

We thank M. A. Craven and C. E. Grier for excellent technical assistance, and C. E. Gates for statistical analysis. This work was supported in part by grant OCE-77 24037 from the National Science Foundation.

TABLE 21
Glutamic-pyruvic Transaminase Activity in the Digestive Gland of *Mercenaria mercenaria* Exposed to BaP, HCB, or PCP in Sea Water for 24 Weeks

	Glutamic-pyruvic transaminase (SFU/mg)			
Week	**Control**	**BaP**	**HCB**	**PCP**
5	1.9	5.1	4.8	6.9
8	6.3	1.6	2.3	4.4
10	7.5	6.3	8.7	13.0
13	7.5	2.6	3.7	3.1
15	0.2	1.2	0.4	2.2
17	6.7	5.9	3.5	4.7
21	7.3	5.8	1.9	7.7
24	3.9	6.2	4.0	7.2
Average	5.2	4.2	6.2	3.7

Note: SFU/mg, Sigma Frankel Units/mg tissue.

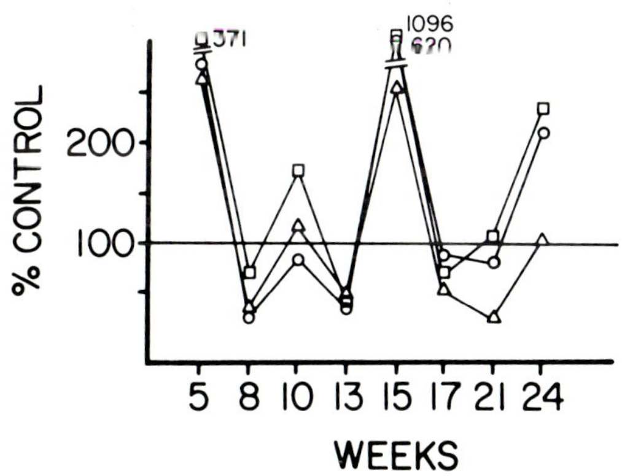

FIGURE 35. Glutamic-pyruvic transaminase activity in the digestive gland tissue of *M. mercenaria* exposed to BaP (○--○), HCB (△--△), or PCP (□--□) for 24 weeks relative to unexposed controls.

TABLE 22
Glutamic-oxaloacetic Transaminase Activity in the Digestive gland of *Mercenaria mercenaria* Exposed to BaP, HCB, or PCP in Sea Water for 24 Weeks

	Glutamic-oxaloacetic transaminase (SFU/mg)			
Week	**Control**	**BaP**	**HCB**	**PCP**
5	5.1	5.5	2.3	7.5
8	7.5	6.4	5.5	5.6
10	6.5	7.5	7.4	3.1
13	2.7	4.4	2.7	1.8
15	4.9	5.9	2.7	3.5
17	3.5	6.8	6.9	5.0
21	4.0	3.7	3.5	9.4
24	6.4	6.0	4.8	5.6
Average	5.1	5.8	4.5	5.2

Note: SFU/mg, Sigma Frankel Units/mg tissue.

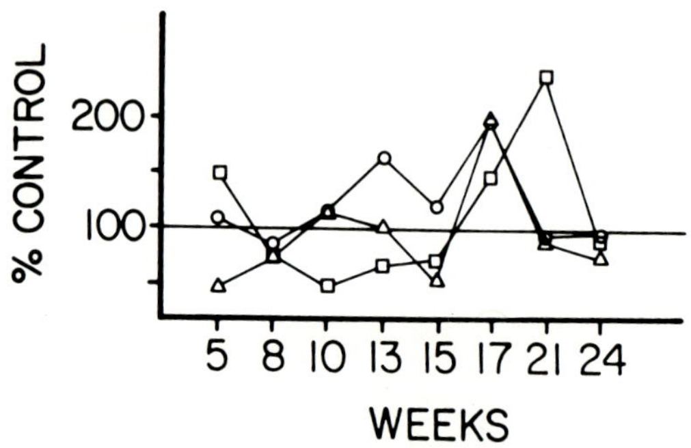

FIGURE 36. Glutamic-oxaloacetic transaminase activity in the digestive gland tissue of *M. mercenaria* exposed to BaP (○--○), HCB (△--△), or PCP (□--□) for 24 weeks relative to unexposed controls.

Appendix A
FIXATIVE SOLUTIONS AND DEHYDRATION SCHEDULE

I. Bouin's Solution

Picric acid, saturated aqueous solution	750.0 mℓ
37 to 40% formaldehyde	250.0 mℓ
Glacial acetic acid	50.0 mℓ

Fix tissue for 4 to 24 hr depending on its size. Tissue may be washed in several changes of 50% ethyl alcohol and stored in 70% ethyl alcohol.

II. Helley's Solution (Zenker-Formalin)

Mercuric chloride	50.0 g
Potassium dichromate	25.0 g
Sodium sulfate	10.0 g
Distilled water	1000.0 mℓ

Add 5 mℓ 37 to 40% formaldehyde to 95 mℓ of the above solution. Tissue may be washed in several changes of 50% ethyl alcohol and stored in 70% ethyl alcohol.

III. Davidson's Solution

37 to 40% formaldehyde	200.0 mℓ
Glycerin	100.0 mℓ
95% ethyl alcohol	300.0 mℓ
Glacial acetic acid	100.0 mℓ
Distilled water	300.0 mℓ

Fix for 24 hr. Store specimens in a solution of the same formula minus the glacial acetic acid or in 95% ethyl alcohol.

IV. Formalin-Sea Water Solution

37 to 40% formaldehyde solution	100.0 mℓ
Sea water	900.0 mℓ

Tissues may be fixed and stored in this solution.

V. Dehydration Schedule

Tissues which have been stored in 70% ethyl alcohol are processed as follows:

80% ethyl alcohol	2 hr
95% ethyl alcohol	2 hr
95% ethyl alcohol	2 hr
100% ethyl alcohol	1 hr
100% ethyl alcohol	1 hr
100% ethyl alcohol	1 hr
Xylene	1 hr
Xylene	1 hr
Paraffin	1 hr
Paraffin	1 hr

Embed in paraffin.

Appendix B
STAINING SOLUTIONS

I. Hematoxylin and Eosin

A. Solutions

1. Harris Alum Hematoxylin

Hematoxylin	5.0 g
Ammonium alum	100.0 g
Ethyl alcohol	50.0 mℓ
Distilled water	1000.0 mℓ
Mercuric oxide	2.5 g

Dissolve hematoxylin in the alcohol and the alum in water. Heat the alum solution, remove from heat, and add hematoxylin solution. Reheat for a short period of time, remove from heat, and slowly add mercuric oxide. Reheat until the solution becomes dark purple, remove from heat, and cool immediately. Add 2 to 4 mℓ glacial acetic per 100 mℓ solution. Filter before use.

2. Eosin

Eosin Y	1.0 g
Distilled water	100.0 mℓ
80% ethyl alcohol	300.0 mℓ

Dissolve eosin in distilled water. Add this solution to 80% ethyl alcohol. Bring to pH 5.4 to 5.6 with 1 *N* HCl.

3. Iodine Solution for tissues fixed with mercuric chloride

Iodine	0.5 g
70% ethyl alcohol	100.0 mℓ

4. Sodium Bisulfite Solution for tissues fixed with mercuric chloride

Sodium bisulfite	5.0 g
Distilled water	100.0 mℓ

5. 1% acid alcohol

70% ethyl alcohol	1000.0 mℓ
Hydrochloric acid (concentrated)	10.0 mℓ

6. Ammonia-alcohol

70% ethyl alcohol	100.0 mℓ
Ammonium hydroxide (concentrated)	0.2 mℓ

B. Staining procedure

Xylene	2 min
Xylene	2 min
100% ethyl alcohol	1 min
100% ethyl alcohol	1 min
95% ethyl alcohol	1 min
95% ethyl alcohol	1 min

Appendix B (continued)

70% ethyl alcohol*	1 min
50% ethyl alcohol	1 min
Tap water	4 dips
Harris hematoxylin	15 min
Tap water	4 dips
50% ethyl alcohol	3 dips
70% ethyl alcohol	3 dips
1% acid alcohol	Continuous dips until tissue with little nucleic acid (e.g., muscle) is gray and tissue with high nucleic acid content (e.g., gonad) is deep purple.
70% ethyl alcohol	4 dips
70% ammonia alcohol	Continuous dips until tissue that is purple turns blue
70% ethyl alcohol	3 min
Eosin	15 to 30 sec
95% ethyl alcohol	4 dips
95% ethyl alcohol	1 min
100% ethyl alcohol	1 min
100% ethyl alcohol	2 min
Xylene	3 min
Xylene	3 min
Mount	

II. Feulgen

A. Solutions

1. Fuchsin-Sulfurous Acid Solution

Basic fuchsin	1.0 g
Distilled water	200.0 mℓ
1 *N* hydrochloric acid	20.0 mℓ
Sodium bisulfite, anhydrous	1.0 g

Bring distilled water to the boiling point and add the dye. This should be done in a large vessel as the dye tends to be explosive on contact with the water. Cool and filter. Add HCl at 50°C, then cool to 25°C and add the sodium bisulfite. The solution should be straw color in 12 to 24 hr. Keep the solution in the dark in a tightly stoppered bottle.

2. 0.1% Fast Green

Fast Green FCF	0.1 g
80% ethyl alcohol	100.0 mℓ

* For tissues fixed with mercuric chloride:

Xylene through 70% ethyl alcohol, then 0.5% iodine in 70% ethyl alcohol	5 min
Tap water	Rinse
5% sodium bisulfite	5 sec
Tap water (running)	5 min
To hematoxylin, etc.	

Appendix B (continued)

3. Dilute sulfurous acid

Distilled water	200.0 mℓ
1 *N* hydrochloric acid	10.0 mℓ
Sodium bisulfite	1.0 g

B. Staining procedure

Remove paraffin and hydrate through a series of alcohols to distilled water.

Fuchsin-sulfurous acid solution	30 min
Dilute sulfurous acid	3 changes
Tap water	5 min
Fast green FCF	3 min
90% ethyl alcohol	1 min
100% ethyl alcohol	1 min
100% ethyl alcohol	1 min
Xylene	3 min
Xylene	3 min
Mount	

III. Alcian Blue

A. Solutions

1. Alcian Blue Solution

Alcian Blue	0.1 g
0.01 *N* hydrochloric acid (pH 2.2)	100.0 mℓ

2. Nuclear Fast Red Solution

Nuclear fast red	0.1 g
5% aluminum sulfate	100.0 mℓ

3. Alkaline Alcohol

95% ethyl alcohol	100.0 mℓ
28 to 30% ammonium hydroxide	1.0 mℓ

B. Staining procedure

Remove paraffin and hydrate through a series of alcohols to distilled water.

Alcian blue solution	30 min
Water	2 changes
Alkaline alcohol	2 hr
Running water	4 dips
Nuclear fast red solution	3 to 5 min
Dehydrate, clear, and mount.	

Appendix C
POLYACRYLAMIDE GEL ELECTROPHORESIS

I. Standard PAGE

Solution A (pH 8.9)
- 24 mℓ 1 *N* Hydrochloric acid
- 18.1 g Tris(hydroxymethyl)aminomethane
- 0.12 mℓ N,N,N′,N′-Tetramethylethylenediamine
- Double-distilled deionized water to make 100 mℓ

Solution B (pH 6.7)
- 48 mℓ 1 *N* Hydrochloric acid
- 5.98 g Tris(hydroxymethyl)aminomethane
- 0.46 mℓ N,N,N′,N′-Tetramethylethylenediamine
- Double-distilled deionized water to make 100 mℓ

Solution C
- 28 g Acrylamide
- 0.735 g N,N′-Methylenebisacrylamide
- Double-distilled deionized water to make 100 mℓ

Solution D
- 20 g Acrylamide
- 5 g N,N′-Methylenebisacrylamide
- Double-distilled deionized water to make 100 mℓ

Solution E
- 4 mg Riboflavin
- Double-distilled deionized water to make 100 mℓ

Solution F
- 40 g Sucrose
- Double distilled deionized water to make 100 mℓ

Solution G
- 140 mg Ammonium Persulfate
- Double-distilled deionized water to make 100 mℓ

Solution H (pH 8.3)
- 6 g Tris(hydroxymethyl)aminomethane
- 28.8 g Glycine
- Double-distilled deionized water to make 100 mℓ

Separating gel
- 1 Part A
- 1 Part C
- 2 Parts G

Stacking gel
- 1 Part B
- 1 Part D
- 1 Part E
- 1 Part double-distilled deionized water
- 4 Parts F

II. Sodium dodecyl sulfate PAGE

10% Sodium dodecyl sulfate
- 25 g Sodium dodecyl sulfate
- Double-distilled deionized water to make 250 mℓ

30 Acrylamide
- 60 g Acrylamide
- 1.6 g Methylenebisacrylamide
- Double-distilled deionized water to make 200 mℓ

Lower gel stock (pH 8.8)
- 90.85 g Tris(hydroxymethyl)aminomethane
- 20 mℓ 10% Sodium dodecyl sulfate
- Double-distilled deionized water to make 500 mℓ

Upper gel stock (pH 6.8)
- 12.12 g Tris(hydroxymethyl)aminomethane
- 8 mℓ 10% Sodium dodecyl sulfate
- Double-distilled deionized water to make 200 mℓ

Appendix C (continued)

Running buffer (pH 8.3)
- 12.12 g Tris(hydroxymethyl)aminomethane
- 57.64 g Glycine
- 40 mℓ 10% Sodium dodecyl sulfate
- Double-distilled deionized water to make 4 ℓ

Sample buffer (pH 6.8, prior to adding 2-mercaptoenthanol)
- 10 g Glycerol
- 3 g Sodium dodecyl sulfate
- 0.76 g Tris(hydroxymethyl)aminomethane hydrochloride
- 5 mℓ 2-Mercaptoethanol
- Double-distilled deionized water to make 100 mℓ

10% Ammonium Persulfate
- 100 mg Ammonium persulfate
- 1 mℓ Double-distilled deionized water

10% Separating gel
- 4 mℓ Lower gel stock, pH 8.8
- 5.3 mℓ 30% Acrylamide
- 6.7 mℓ Double-distilled deionized water
- 0.08 mℓ 10% Ammonium persulfate
- 0.008 mℓ N,N,N′,N′-Tetramethylethylenediamine

4.5% Stacking gel
- 1.25 mℓ Upper gel stock, pH 6.8
- 0.75 mℓ 30% Acrylamide
- 3.0 mℓ Double-distilled deionized water
- 0.015 mℓ Ammonium persulfate
- 0.005 mℓ N,N,N′,N′-Tetramethylethylenediamine

REFERENCES

1. **Luna, L. G., Ed.,** *Manual of Histologic Staining Methods of the Armed Forces Institute of Pathology,* McGraw-Hill, New York, 1968.
2. **Johnson, P. T.,** *Histology of the Blue Crab, Callinectes sapidus,* Praeger, New York, 1980.
3. **Galigher, A. E. and Kozloff, E. N.,** *Essentials of Practical Microtechnique,* Lea & Febiger, Philadelphia, 1971.
4. **Lillie, R. D. and Fullmer, H. M.,** *Histopathologic Technic and Practical Histochemistry,* McGraw-Hill, New York, 1976.
5. **Pearse, A. G. E.,** *Histochemistry: Theoretical and Applied,* Churchill Livingston, Edinburgh, 1972.
6. **Bergmeyer, H., Ed.,** *Methods of Enzymatic Analysis,* Academic Press, New York, 1965.
7. Sigma Chemical Co., β-Glucuronidase, Sigma Tech. Bull. No. 325, St. Louis, Mo., 1976.
8. Sigma Chemical Co., Phosphatases, Sigma Tech. Bull. No. 104, St. Louis, Mo., 1974.
9. Sigma Chemical Co., Transaminases, Sigma Tech. Bull. No. 505, St. Louis, Mo., 1977.
10. **Maurer, H. R.,** *Disc Electrophoresis and Related Techniques of Polyacrylamide Gel Electrophoresis,* Walter de Gruyter, New York, 1971.
11. **Lowry, O. H., Rosebrough, N. J., Farr, A. L., and Randall, R. J.,** Protein measurement with the folin-phenol reagent, *J. Biol. Chem.,* 193, 265, 1951.
12. **Morton, J. E.,** The tidal rhythm and action of the digestive system of the lamellibranch *Lasaea rubra, J. Mar. Biol. Assoc. U.K.,* 35, 563, 1956.
13. **Owen, G.,** Lysosomes, peroxisomes and bivalves, *Sci. Prog. Oxf.,* 60, 299, 1972.
14. **Owen, G.,** Feeding and digestion in the bivalvia, *Adv. Comp. Physiol. Biochem.,* 5, 1, 1974.
15. **Palmer, R. E.,** A histological and histochemical study of digestion in the bivalve *Artica islandica, Biol. Bull.,* 156, 115, 1979.
16. **Robinson, W. E. and Langton, R. W.,** Digestion in a subtidal population of *Mercenaria mercenaria* (Bivalvia), *Mar. Biol.,* 58, 173, 1980.

17. **Robinson, W. E.,** Assessment of bivalve intracellular digestion based on direct measurements, *J. Molluscan Stud.*, 49, 1, 1983.
18. **Yonge, C. M.,** The digestive diverticula in lamellibranches, *Trans R. Soc. Edin.*, 54, 703, 1926.
19. **Morton, B. S.,** The tidal rhythm and rhythm of feeding and digestion in *Cardium edule*, *J. Mar. Biol. Assoc. U.K.*, 50, 499, 1970.
20. **Mathers, N. F.,** A comparative histochemical survey of enzymes associated with the processes of digestion in *Ostrea edulis* and *Crassostrea gigas*, *J. Zool.*, 169, 169, 1973.
21. **Palmer, R. E.,** Behavioural and rhythmic aspects of filtration in the bay scallop, *Argopecten viradians* (Say) and the oyster *Crassostrea virginica* (Gmelin), *J. Exp. Mar. Biol. Ecol.*, 45, 273, 1980.
22. **Morton, B. S.,** The diurnal rhythm and tidal rhythm of feeding and digestion in *Ostrea edulis*, *Biol. J. Lin. Soc.*, 3, 329, 1971.
23. **Morton, B. S.,** The biology, ecology and functional aspects of the organs of feeding and digestion of the S.E. Asian mangrove bivalve, *Enigmonia aenigmatica* (Mollusca: Anomiacea), *J. Zool.*, 179, 437, 1976.
24. **Langton, R. W. and Gabbott, P. A.,** The tidal rhythm of extracellular digestion and the response to feeding in *Ostrea edulis*, *Mar. Biol.*, 24, 181, 1974.
25. **Mathers, N. F.,** The effects of tidal currents on the rhythm of feeding and digestion in *Pecten maximus* L., *J. Exp. Mar. Biol. Ecol.*, 24, 271, 1976.
26. **Langton, R. W.,** Digestive rhythms in the mussel *Mytilus edulis*, *Mar. Biol.*, 41, 53, 1977.
27. **Morton, B. S.,** Studies on the biology of *Dreissena polymorpha* Pall. II. Correlation of the rhythms of adductor activity, feeding, digestion and excretion, *Proc. Malacol. Soc. Lond.*, 38, 401, 1969.
28. **Wilson, J. H. and LaTouche, R. W.,** Intracellular digestion in two sublittoral populations of *Ostrea edulis* (Lamellibranchia), *Mar. Biol.*, 47, 71, 1978.
29. **Moore, M. N.,** Cytochemical demonstration of latency of lysosomal hydrolases in digestive cells of the common mussel, *Mytilus edulis*, and changes induced by thermal stress, *Cell. Tissue Res.*, 175, 279, 1976.
30. **Neil, M. W. and Horner, M. W.,** Acid p-nitrophenyl phosphate phosphohydrolases of adult guinea pig liver, *Biochem. J.*, 92, 217, 1964.
31. **Neil, M. W. and Horner, M. W.,** Studies on acid hydrolases in adult and fetal tissues. II. Acid phenyl phosphomonoesterases of adult mouse liver, *Biochem. J.*, 93, 220, 1965.
32. **Stebbing, A. R. D.,** Hormesis-stimulation of colony growth in *Campanulusia flexuosa* (Hydrozoa) by copper, cadmium and other toxicants, *Aquat. Toxicol.*, 1, 227, 1981.
33. **Tripp, M. R., Fries, C. R., Craven, M. A., Grier, C. E., and Gates, C. E.,** Histopathology of *Mercenaria mercenaria* as an indicator of pollutant stress, *Mar. Environ. Res.*, 14, 521, 1984.

Chapter 7

BIOCHEMICAL STRESS RESPONSES AND ALTERATIONS OF TISSUE ASCORBIC ACID AND GLUTATHIONE CONTENT

P. Thomas, R. S. Carr, and J. M. Neff

TABLE OF CONTENTS

I. Measurement of Biochemical Responses 156
A. Animals 156
1. Striped Mullet and Atlantic Croaker 156
2. Sandworms 156
B. Exposure to Pollutants 157
1. Striped Mullet and Atlantic Croaker 157
2. Sandworms 158
C. Analytical Methods 158
1. Plasma Cortisol 158
2. Glucose 159
3. Glycogen 159
4. Osmolality 159
5. Lipids 159
6. Biogenic Amines 159
7. Amino Acids 159
8. Tissue Ascorbic Acid and Ascorbic Acid-2-Sulfate 160
9. Acid-Soluble Thiols 160
10. Glutathione 160
11. Cysteine 160

II. Biochemical Stress Responses and Ascorbic Acid Fluctuations in Invertebrates 161
A. Results with Sandworms 161
B. Results with Blue Crabs 163
C. Discussion 163

III. Biochemical Stress Responses in Teleosts 163
A. Results 163
B. Discussion: Plasma Corticosteroid Stress Responses 166
C. Discussion: Secondary Metabolic Disturbances 170

IV. Alterations in the Ascorbic Acid and Glutathione Status of Fish Tissues 170
A. Results 170
B. Discussion: Tissue Ascorbic Acid, Glutathione, and Acid-Soluble Thiols 172

V. Field Studies: *N. virens* 174

VI. Field Studies: Winter Flounder 175

VII. Biological Significance of Biochemical Alterations in Fish 176

Acknowledgment 177

References 177

I. MEASUREMENT OF BIOCHEMICAL RESPONSES

A. Animals

1. Striped Mullet and Atlantic Croaker

Juvenile striped mullet *(Mugil cephalus),* 10 to 20 cm long, were collected with a seine near Port Aransas, Texas. Juvenile Atlantic croaker *(Micropogonias undulatus)* were collected in the same vicinity with a trawl. The fish were acclimated for 2 weeks in the laboratory in large recirculating marine aquaria. Fish were maintained at 33‰ salinity (S) and at 20 ± 0.5°C with lights on from 07:00 to 19:00 daily. Fish were fed once daily with Tetramin algal flake.

Six fish were transferred to each exposure aquarium containing 40 ℓ of artificial sea water (33‰ S) and 2 charcoal filters. The fish were undisturbed for 16 hr prior to sacrifice, which was between 10:00 and 11:00 on the 14th day after transfer. Fish were removed with a dip net and blood samples were collected in syringes by cardiac puncture. All the fish in each tank were bled within 2 min. The aquaria were positioned so that fish in other exposure tanks were not disturbed by the sampling procedure. This protocol prevented nonspecific activation of the interrenal glands. The blood was centrifuged in heparinized tubes and the plasma was removed for biochemical analysis. Glucose, cholesterol, and osmolality were measured on the same day. The remaining sample was stored at −80°C until analyzed for cortisol. Immediately after blood collection, the fish were sacrificed and placed on ice. Pieces of liver, gill, kidney, and brain tissue were removed within a few minutes of death, weighed, and frozen in liquid nitrogen. Samples were subsequently stored at −80°C for up to 2 weeks until analyzed for glutathione, acid-soluble thiols, ascorbic acid, and glycogen. Fish carcasses were stored at −80°C for up to 2 months until analyzed for pollutant accumulation.

2. Sandworms

Sandworms *(Neanthes (Nereis) virens)* were obtained from a commercial bait supplier in Maine and delivered via Express Air Package in insulated containers provided with frozen blocks of foam refrigerant to minimize temperature stress during shipment. The polychaetes were carefully separated from the moist *Thalassia* sp., in which they were shipped, and acclimated to Instant Ocean sea water (32 to 35‰ S; Aquarium Systems, Inc., Eastlake, Ohio) in 100 ℓ fiberglass tanks equipped with Ehiem filters in a 10°C cold-room in the dark for at least 2 weeks before they were used in experiments. Although 100% survival of the animals received was not uncommon, the 2-week acclimation allowed for the detection and removal of individuals which had been unduly stressed during collection or transport. The worms were fed fresh minced *Rangia cuneata* which had been depurated before use. *N. virens* has been maintained in our laboratory for periods exceeding 2 years on this sole food source. Neither holding nor experimental tanks contained any sediment, but oyster valves were strewn about the bottom of the tanks to provide some cover for the animals.

Unless otherwise mentioned, all polychaete experiments in this study were conducted at

10°C in the dark at 32 to 35‰ S with Instant Ocean sea water to which trace element blocks (Jungle Lab. Corp.) were added as necessary. Sea water salinities were measured weekly with an American Optical temperature-compensated refractometer, and adjusted when necessary with distilled water. The ammonia content of control and experimental tanks was also monitored periodically by means of an ion-specific ammonia electrode (Orion Research, Inc.) because it was suspected that elevated ammonia levels might induce a nonspecific response during experiments.

In the long-term PCP experiment and the acute and chronic cadmium experiments, all animals were sampled in a similar fashion. Twelve worms were randomly selected from either experimental or control tanks, except at the termination of the long-term studies, when less than 12 worms remained. The animals were transferred to individual 15-cm finger bowls after recording wet weights and descriptive information concerning the state of sexual maturity, presence and location of wounds, number of regenerated segments, etc. Replicate coelomic fluid samples were then taken with a microsyringe (26 gauge) and transferred to nonheparinized capillary tubes. Tissue samples from posterior segments (and parapodia in the long-term PCP experiment) were blotted dry and transferred to tared weight boats on ice. The animals were then frozen in labeled plastic bags in liquid nitrogen and subsequently stored at −60°C for further analyses.

B. Exposure to Pollutants

1. Striped Mullet and Atlantic Croaker

Saturated aqueous solutions of PCP and BaP were made using a sand column. The column was charged with 25 gm PCP or with 50 mg BaP (Aldrich, Gold Label, 99+% pure) dissolved in 50 mℓ diethyl ether and evaporated with a stream of nitrogen on approximately 500 g of coarse sand. The charged sand was subsequently packed between layers of clean sand in a glass column. The sand was kept in place with glass wool plugs. Aluminum foil was wrapped around the column to exclude light. Saturated solutions of PCP and BaP were produced by pumping water into the bottom of the column at the rate of 60 and 36 mℓ/hr, respectively. The concentration of the pollutants in the eluate stabilized after 2 days operation. The PCP concentrations in the exposure tanks were monitored by the spectrophotometric techniques of Carr et al.[1] BaP was measured by the method of Giam et al.[2]

One hundred percent water-soluble fractions (WSF) of a No. 2 fuel oil were prepared by stirring a mixture of 1 part oil and 9 parts artificial sea water for 20 hr.[3] Freshly prepared WSF was diluted with sea water and added to the exposure aquaria to produce final concentrations of 1, 5, 10, and 20% WSF.[4] Disappearance of naphthalenes from the exposure aquaria was monitored spectrophotometrically.[5]

Mullet were exposed to dibenzofuran (750 μg/ℓ), carbazole (100 μg/ℓ), and fluorene (100 μg/ℓ) (Aldrich Chemical Co.), which were added to the exposure aquaria as a concentrated stock solution dissolved in acetone (0.5 mℓ acetone per aquarium). Acetone was also added to control aquaria. Concentrations of these compounds in water were measured spectrophotometrically by a modification of the technique of Neff and Anderson.[5] The exposure aquaria were redosed daily.[6]

A concentrated stock solution of naphthalene (Aldrich Chemical Co.) was made up in ethanol (80 mg/mℓ). Aliquots (1 mℓ) of the stock solution were added to the exposure aquaria to give a final concentration of 2.0 mg/ℓ.[7] An equal volume of ethanol was added to the control aquaria at the same time. Napththalene concentrations in the aquaria were monitored spectrophotometrically.[5]

Cadmium chloride ($CdCl_2 \cdot 2H_2O$) was prepared as a concentrated stock solution (40 g Cd per milliliter) and added to the exposure aquaria to give final concentrations of 0.1 to 10 mg Cd per liter.[8,9] The aquarium water was replaced with fresh sea water every 3 days and redosed with cadmium. Cadmium concentrations in the water were monitored by atomic absorption spectrophotometry.

Mercuric chloride was added to the exposure tanks to give a final concentration of 200 μg Hg per liter. Loss of mercury from the aquarium water was monitored with ^{203}Hg. Mercury was added every 3 days to the exposure aquaria to maintain water concentrations of the metal.

2. *Sandworms*

Both acute and long-term experiments at different exposure levels were conducted with two toxicants: PCP and cadmium. Two short-term experiments (96 hr of less) were conducted with PCP (Aldrich Chemical Co., 99+% purity) in a static system; the concentration of PCP was monitored daily during the exposure. These two acute exposures were performed in 35-ℓ plastic tanks. Gentle aeration was supplied to both control and exposure tanks, but the organisms were not fed.

Protracted PCP experiments employed a sand column dosing system, as described previously for the fish experiments, to maintain the exposure concentrations at a constant level. Used in connection with an Ehiem filter containing activated charcoal, this system supplied a continuous flow of eluate from the PCP-saturated sand column to the exposure tank. This system was utilized in two chronic experiments. The first experiment lasted 12 days and was conducted in 35-ℓ plastic tanks. The second long-term PCP exposure was conducted in 100-ℓ fiberglass tanks for 63 days. The animals were fed during these chronic experiments, and the control tanks were also equipped with Ehiem filters containing activated charcoal.

From the results of a preliminary bioassay study, 40 mg Cd (as $CdCl_2$) per liter (ROC/RIC, anhydrous 99% purity) was selected for the acute exposure concentration. This experiment was conducted in 100-ℓ fiberglass tanks equipped with Ehiem filtration; food was supplied. The cadmium concentration was monitored daily during this 7-day exposure. A long-term (8 week) exposure to 10 mg Cd per liter was also conducted under the same conditions. During the chronic experiment, the cadmium concentration was measured and the tank redosed weekly in order to maintain the exposure concentration at 10 mg/ℓ. A control group was also sampled on the same day as the Cd-exposed group throughout this study.

C. Analytical Methods

1. *Plasma Cortisol*

Plasma cortisol was measured by radioimmunoassay using antiserum obtained from New England Nuclear. This antiserum cross-reacts 25.7% with cortisone, 21.2% with 17 α-hydroxyprogesterone, 17% with corticosterone, and 2.8% with 11-deoxycortisol so that preliminary clean-up of samples by chromatography on Sephadex LH20 columns may be necessary. The radioimmunoassay for cortisol determination in striped mullet plasma was validated by thin-layer chromatography and high-performance liquid chromatography.[10] Corticosterone was also found, but in insufficient concentrations to significantly affect the radioimmunoassay for cortisol. Preliminary purification on a Sephadex LH20 column was therefore unnecessary for mullet plasma. Plasma samples (50 μℓ) were extracted in 5 mℓ methylene chloride (analytical grade). The aqueous layer was aspirated and the organic layer was evaporated to dryness under a stream of air. The sides of the tube were washed three times with a small volume of methanol (analytical grade). The three washings were transferred to a clean tube and the methanol was evaporated under a stream of air. This procedure prevented possible interference in the assay by plasma proteins when the sample was redissolved in assay buffer. The concentration of charcoal was increased fourfold to prevent nonspecific interference by lipids which are present in high concentrations in fish plasma. Three dilutions of each sample were assayed. Recovery was always above 80%. The assay typically had a sensitivity of 10 pg, a within-assay variance of 6.8% (mean of 10 determinations, 32.5 ± 2.2 ng/mℓ), and a between-assay variance of 7.7% (mean of 8 determinations, 35.4 ± 2.7 ng/mℓ).

2. Glucose

Plasma and coelomic fluid glucose were analyzed with a YSI model 23 A glucose analyzer (Yellow Springs Instrument Co., Yellow Springs, Ohio) which measures the H_2O_2 released upon enzymatic oxidation of glucose. The assay has a sensitivity of 1 mg/dℓ in 25 μℓ plasma and a within-assay variance of 1.6% (mean of 10 determinations, 53.45 ± 0.86 mg/dℓ).

3. Glycogen

Glycogen was measured in hepatic tissue of striped mullet and Atlantic croaker and in several tissues of blue crabs by the technique of Montgomery.[11]

Glycogen concentrations in sandworm and winter flounder tissues were quantified by the method of Carr and Neff.[12] Briefly, this method employs the enzymatic hydrolysis of glycogen in a tissue homogenate to glucose. This glucose concentration of the homogenate supernatant is then determined with the aid of a glucose analyzer. This method is specific and highly reproducible.

4. Osmolality

Plasma osmolality was measured with a vapor pressure osmometer (model 5100, Wescor Inc.).

5. Lipids

Total plasma cholesterol and percent high-density lipoprotein cholesterol in striped mullet and Atlantic croaker were measured colorimetrically after enzymatic conversion using a commercial kit (Pierce Chemical Company). A commercial kit (Dade) was also used for colorimetric assay of plasma triglycerides.

Total lipids in liver tissue of winter flounder were determined in triplicate by the method of Holland and Gabbott.[13]

6. Biogenic Amines

Brain norepinephrine, dopamine, and serotonin concentrations in striped mullet were determined by the fluorometric procedure of Jacobowitz and Richardson[14] without modification.

Catecholamines and indoleamines in winter flounder were determined by ion pairing high-performance liquid chromatography (HPLC) with electrochemical detection. Frozen brains were homogenized in 250 m*M* $HClO_4$. The homogenate was centrifuged at 27,000 g for 30 min and the supernatant analyzed without further extraction or clean-up. An isocratic Waters HPLC system was employed with an amperometric detector (Bioanalytical Systems, LC-4) using a thin-layer glassy electrode (+800 mV vs. Ag/AgCl). A Partisil SAX guard column in conjunction with a Resolve C_{18} column provided excellent separation and resolution of the components of interest. The mobile phase was 0.1 *M* KH_2PO_4 to methanol, 85:15 (pH 3.76), with heptane sulfonic acid (Waters Assoc.) added as the ion-pairing reagent. At a flow rate of 0.5 mℓ/min, 5-hydroxytryptamine (the last component of interest) eluted approximately 15 min after injection. Dihydroxybenzylamine (DHBA) was used as the internal standard.

7. Amino Acids

Free amino acid concentrations in winter flounder and sandworm tissues were determined by gradient HPLC with postcolumn derivitization (o-pathaldehyde) and fluorometric detection. The tissue samples were homogenized in 7% trichloroacetic acid (TCA) as described by Roesijadi et al.[15] TCA supernatants were extracted three times with diethyl ether to remove TCA and lipids, and then lyophilized. Lyophilized samples were redissolved in 0.1 *N* HCl just prior to analysis. A Waters gradient HPLC with a WISP autosampler, system controller, data module, and fluorometric detector was employed. The chromatographic conditions have been described previously.[16] The cation exchange column with a sulfonated polymeric resin was maintained at 62°C with a column heater. Norleucine was used as the internal standard.

8. Tissue Ascorbic Acid and Ascorbic Acid-2-Sulfate

Two techniques were used to assay ascorbic acid in striped mullet and Atlantic croaker tissues. The α,α′-dipyridyl method[17] was usually satisfactory for the determination of ascorbic acid in hepatic tissues. However, a poor correlation was found between the values measured in certain tissues by the α,α′-dipyridyl technique[18] and those obtained by the dinitrophenylhydrazine (DNPH) technique[19] and an HPLC method with electrochemical detection.[20] The α,α′-dipyridyl assay was therefore validated for each tissue by the dinitrophenylhydrazine technique. All blue crab tissues were analyzed by the DNPH technique.

Both techniques were modified to permit assay of less than 50 mg of tissue. Ascorbic acid-2-sulfate and ascorbic acid were assayed by the modification of the DNPH technique.[8] The assay had a sensitivity of 0.05 μg ascorbic acid and 1.25 μg ascorbic acid-2-sulfate per tube. A close correlation was found between the values obtained for mullet tissue by this assay and an HPLC technique with electrochemical detection.[18]

Ascorbic acid was determined in perchloric acid extracts of winter flounder tissue homogenates or in sandworm tissues by a HPLC technique with electrochemical detection.[20] This technique can detect 2 μg of ascorbic acid in a 10-μℓ sample. The α,α′-dipyridyl method[17] was found to be unsatisfactory for the determination of ascorbic acid in invertebrate tissues.[18]

9. Acid-Soluble Thiols

The acid-soluble thiol (AST) content of tissues was determined using dithiobis 2-nitrobenzoic acid (DTNB).[21] DTNB reacts with reduced thiols to produce a yellow product which can be quantitated spectrophotometrically. Oxidation of glutathione or cysteine standards was less than 2%. Recovery of glutathione added to liver homogenates was approximately 96%.

10. Glutathione

The total glutathione content of striped mullet tissues was determined enzymatically by a modification of the method of Owens and Belcher[22] which permits measurement in small tissue samples.[9] The assay is highly specific for reduced and oxidized glutathione. Other thiols, such as cysteine and methionine, do not interfere. Oxidized glutathione gives changes in absorbance identical to that of reduced glutathione.[22] Recovery of glutathione added to liver homogenates was approximately 100%.

Hepatic glutathione in winter flounder was determined by isocratic HPLC with amperometric detection using a Au/Hg thin-layer electrode (+15 mV vs. Ag/AgCl). Frozen tissue was briefly homogenized in 250 m*M* $HClO_4$ and centrifuged at 27,000 g for 30 min. The supernatant was injected without further clean-up. A Partisil SAX guard column was used in conjunction with a Resolve C_{18} analytical column. Monochloroacetic acid ([MCAA], 9.45 g MCAA + 3 g NaOH per liter, pH 3.2) was used as the mobile phase. At a flow rate of 0.5 mℓ/min, glutathione (GSH) eluted in approximately 7 min. External standard area quantitation was used in this assay as no suitable internal standard was found.

11. Cysteine

Cysteine was measured by the acid ninhydrin technique of Gaitonde.[23] The color reaction in this procedure is specific for cysteine. An equimolar amount of glutathione gives less than 2% interference. Tissue concentrations obtained by this method (0.37 to 0.46 mmol/g) agree well with those obtained with an amino acid analyzer (0.36 to 0.43 mmol/g).

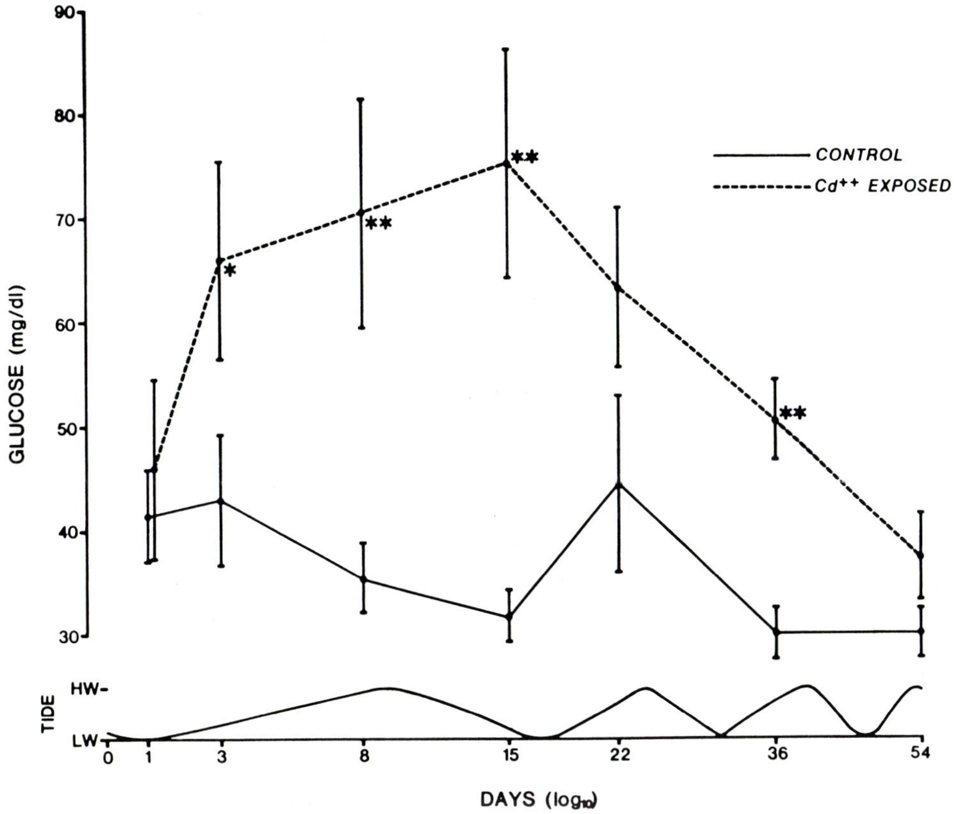

FIGURE 1. Coelomic fluid glucose concentrations in *N. virens* during long-term exposure to Cd. Mean glucose (± SE) and predicted tidal height in Maine at the time of sampling are shown. Asterisks denote significant increases compared to the controls determined by the U test (*, $\alpha < 0.05$; **, $\alpha < 0.005$). N = 12 except for 8- and 54-day Cd-exposed groups where N = 11 and 8, respectively, and for the 22- and 54-day control groups where N = 10 and 9, respectively.

II. BIOCHEMICAL STRESS RESPONSES AND ASCORBIC ACID FLUCTUATIONS IN INVERTEBRATES

A. Results with Sandworms

Several biochemical parameters including coelomic fluid glucose levels and tissue ascorbic acid and glycogen concentrations were evaluated in the sandworm *Neanthes virens* (Sars) during long-term exposures to pentachlorophenol (PCP)[24] and cadmium.[25] After 3 days of exposure to cadmium, a significant hyperglycemic response was observed which persisted throughout the 8-week experiment (Figure 1). During the same experiment, a statistically significant depletion of water-soluble glycogen was observed after 8 days of exposure (Figure 2), and most dramatically after 8 weeks of exposure, at the same time that the coelomic fluid glucose levels were rapidly decreasing. A similar decrease in the mean tissue glycogen level of PCP-exposed worms was observed after a chronic exposure. The mean ascorbic acid levels were remarkably similar between control and cadmium-exposed animals until the 8th week when the cadmium-exposed worms exhibited a statistically significant increase in tissue ascorbic acid (Figure 3). During the long-term PCP experiment, the mean ascorbic acid level of exposed worms remained elevated above the control levels throughout the experiment.[24]

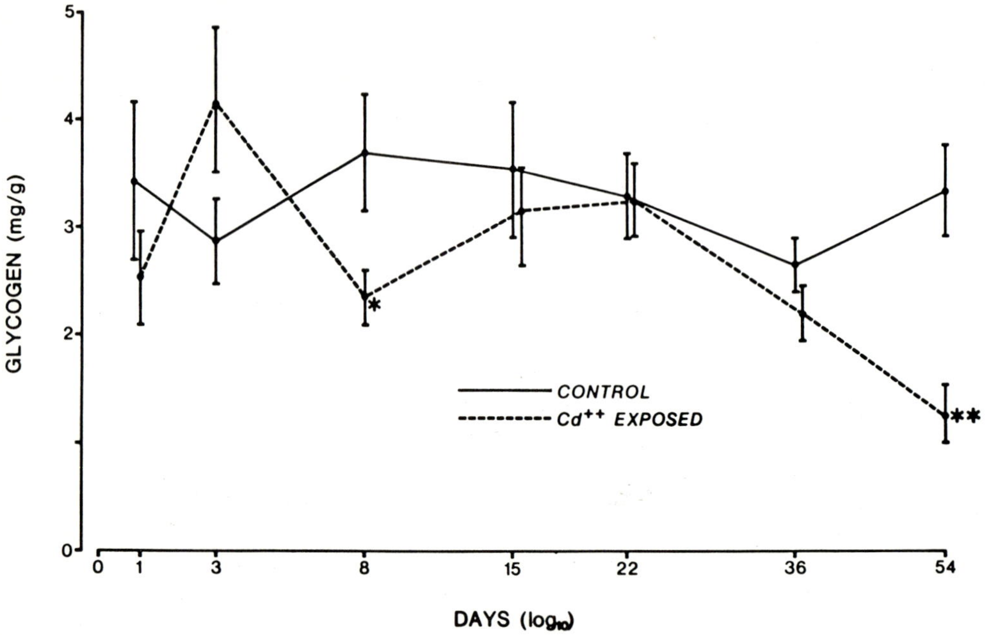

FIGURE 2. Water-soluble glycogen content (mean ± SE) of *N. virens* during long-term exposure to Cd. Asterisks denote significant decreases compared with the controls determined by the U test (*, $\alpha < 0.05$; **, $\alpha < 0.005$). N = 12 except for 54-day Cd-exposed and control groups where N = 8 and 10, respectively.

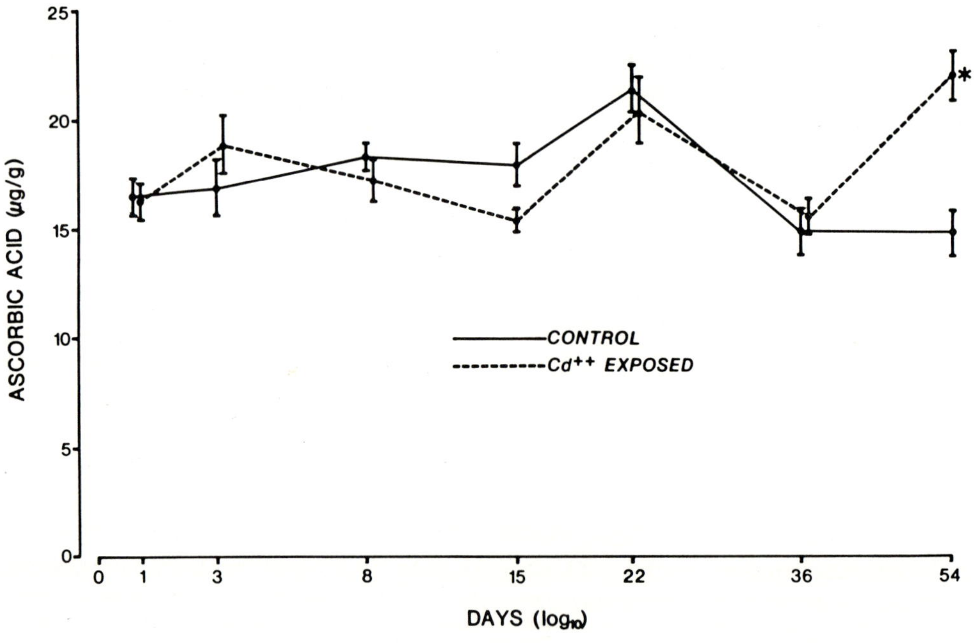

FIGURE 3. Ascorbic acid content (mean ± SE) of posterior segments of *N. virens* during long-term exposure to Cd. Asterisk denotes significant increase ($\alpha < 0.001$) compared with the control determined by the t test. N = 12 except for 54-day control group where N = 10 and the 15- and 54-day Cd-exposed groups where N = 11 and 8, respectively.

B. Results with Blue Crabs

Blue crabs *(Callinectes sapidus)* were exposed to sublethal concentrations of PCP for 7 days.[26] This exposure resulted in a rapid elevation in hemolymph glucose concentrations, accompanied by a gradual depletion of hepatopancreatic and gonadal glycogen reserves. Hemolymph total cholesterol and percent high-density lipoprotein cholesterol were decreased in PCP-exposed crabs. Hepatopancreatic reserves of ascorbic acid and ascorbate-2-sulfate also decreased in these crabs.

C. Discussion

The pattern of changes observed in the coelomic fluid glucose levels of pollutant-exposed worms appeared to be similar to the pattern of endocrine changes described by Selye in mammals exposed to a wide variety of noxious stimuli which he named the General Adaptation Syndrome (GAS).[27,28] Interestingly, similar fluctuations in glucose concentrations were observed in the hemolymph of blue crabs during PCP exposure.[26] Alterations of carbohydrate reserves have also been reported in other invertebrate species after exposure to noxious stimuli.[29] Thus, although the hormones and mechanisms by which polychaetes, crustaceans, and vertebrates regulate carbohydrate metabolism appear to differ considerably,[30-32] similarities exist in the glucose responses of these organisms to adverse stimuli. The GAS in mammals consists of three distinct phases: the alarm reaction, the stage of resistance, and the stage of exhaustion.[27,28] It is conceivable that a generalized pattern of biochemical responses to noxious agents, analogous to the GAS, also exists in some invertebrate species. A significant hyperglycemia was observed in *N. virens* after 3 days exposure to cadmium which persisted for over a month (Figure 1). A hyperglycemic response was also observed after 24 hr of exposure to an initial PCP concentration of 365 $\mu g/\ell$. In both instances in which a hyperglycemic response was elicited, the subsequent decrease in the coelomic fluid concentration coincided with a decrease in glycogen levels in the pollutant-exposed animals as the carbohydrate reserves became exhausted. Therefore, the maintenance of coelomic fluid glucose concentrations in *N. virens* appears to be dependent upon adequate glycogen reserves. These studies suggest that tissue glycogen content may be a useful measure of the metabolic reserves available in this species for resistance to adverse environmental conditions.

Of the biochemical parameters investigated, coelomic fluid glucose levels and tissue glycogen content were the most sensitive indicators of pollutant exposure. Neither coelomic fluid osmolality nor pH was a sensitive index of pollutant exposure.The hypothesis that tissue ascorbic acid would become depleted after chronic pollutant-induced stress was not supported by the results of these studies. Because the tissue ascorbic acid levels of pollutant-exposed *N. virens* increased significantly after chronic contamination, this response could conceivably be used as an index of pollutant stress for this species. Ascorbic acid does not appear to be a limiting factor for survival of this polychaete during pollutant exposure.

III. BIOCHEMICAL STRESS RESPONSES IN TELEOSTS

A. Results

Biochemical stress responses of teleosts to a variety of pollutants are summarized in Table 1. Of the nine pollutants investigated, only cadmium failed to elicit a corticosteroid stress response in mullet. Pollutant-induced elevations of plasma cortisol were accompanied by secondary metabolic disturbances except after exposure to BaP. Four secondary biochemical stress parameters — plasma glucose, cholesterol, triglycerides, and osmolality — were compared after pollutant exposure. A rise in circulating glucose titers was the most consistent pollutant response; only BaP did not cause a marked hyperglycemia. On the other hand, plasma cholesterol concentrations were largely unaffected by pollutant exposure. Certain compounds also depleted hepatic glycogen and brain catecholamine stores. Mortality was

Table 1
SUMMARY OF POLLUTANT EFFECTS ON BIOCHEMICAL STRESS RESPONSES IN TELEOSTS

Pollutant	Concentration/ℓ	Species	Plasma					Liver	Brain	Mortality	Ref.
			Cortisol	Glucose	Cholesterol	Triglycerides	Osmolality	glycogen	catecholamines		
Pentachlorophenol	100—200 µg	*M. cephalus*	+[a]	+	0		+	−		+	39
	50 µg	*P. americanus*	0	0				−			
Benzo(a)pyrene	0.2—0.3 µg	*M. cephalus*	+	0	0		0	0		0	7
		M. undulatus	+	0	0		0	0		0	7
Fluorene analogs[b]	100 µg	*M. cephalus*	+	+	0	+	0			0	6
Dibenzofuran	750 µg	*M. cephalus*	+	+	0	0	0		−	0	6
Naphthalene	2 mg	*M. cephalus*	+	+						0	7
WSF[c] no. 2 fuel oil	5—20%	*M. celphalus*	+	+	+		+			0	4
Cadmium	10 mg	*M. cephalus*	0	+	0		0			+	7, 54
	10 mg	*T. mossambicus*	+	+						+	

[a] +, significant increase; 0, no change; −, significant decrease.
[b] Fluorene, carbazole, and dibenzothiophene.
[c] WSF, water-soluble fraction.

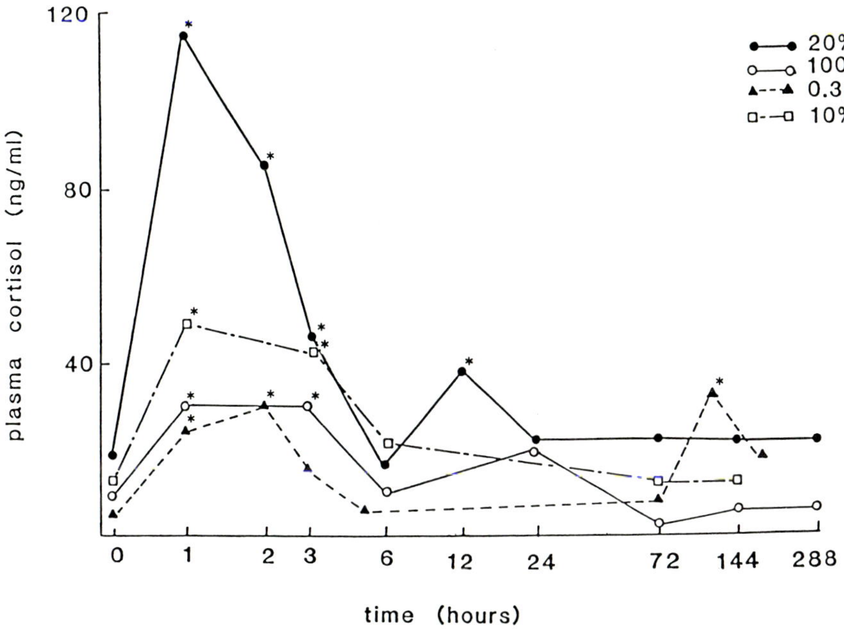

FIGURE 4. Plasma cortisol fluctuations in striped mullet exposed to 10% (□--□) and 20% (●--●) water-soluble fractions of No. 2 fuel oil, 100 μg fluorene per liter (○--○), and 0.3 μg benzo(a)pyrene per liter (▲--▲). Asterisks denote means significantly different from controls ($P < 0.05$, $N = 6$).[4,6,7]

significantly increased only by the highest concentration of PCP (200 μg/ℓ) or after chronic exposure to cadmium (6 weeks).

The fluctuations of plasma cortisol in mullet exposed to several pollutants are shown in Figure 4. Exposure to 20% WSF of No. 2 fuel oil, 10% WSF, 100 μg fluorene per liter, and 0.3 μg BaP per liter caused a similar pattern of plasma cortisol changes. Cortisol concentrations increased rapidly to reach maximum levels 1 hr after the beginning of exposure. Circulating cortisol levels subsequently declined to reach minimum levels 4 to 5 hr later. A secondary rise in cortisol titers was observed 12 hr after exposure to 20% WSF oil and after continuous exposure to BaP for 72 hr (Figure 4). The corticosteroid dynamics in mullet during continuous exposure to dibenzofuran were different from those observed after exposure to the other pollutants (Figure 5). An acute transient elevation of plasma cortisol was also observed after exposure to 750 μg dibenzofuran per liter. In addition, there was a marked secondary rise in plasma cortisol after 1 day exposure to the hydrocarbon. Plasma concentrations had declined to control levels by 3 days, but cortisol titers increased for a third time after 6 days of pollutant exposure.

Plasma cortisol concentrations in mullet 1 hr after exposure to various concentrations of oil, and after 1 week exposure to dibenzofuran and fluorene, are shown in Figure 6. There was a dose-response relationship in plasma cortisol concentrations after exposure to oil and dibenzofuran, but not after exposure to fluorene (Figure 6). Cortisol levels were not significantly altered by 2% WSF of oil and 10 μg dibenzofuran per liter.

Plasma glucose fluctuations in mullet exposed to oil, dibenzofuran, and fluorene are shown in Figure 7. Plasma glucose concentrations did not rise as rapidly as plasma corticosteroids after exposure to oil. Maximum glucose titers occurred 3 hr after the addition of oil, but the hyperglycemia was longer lasting and persisted for more than 24 hr (Figure 7). A persistent hyperglycemia was also observed during repeated exposure to 100 μg fluorene per liter. In contrast, glucose titers were not significantly elevated until after 24 hr of exposure to 750 μg dibenzofuran per liter.

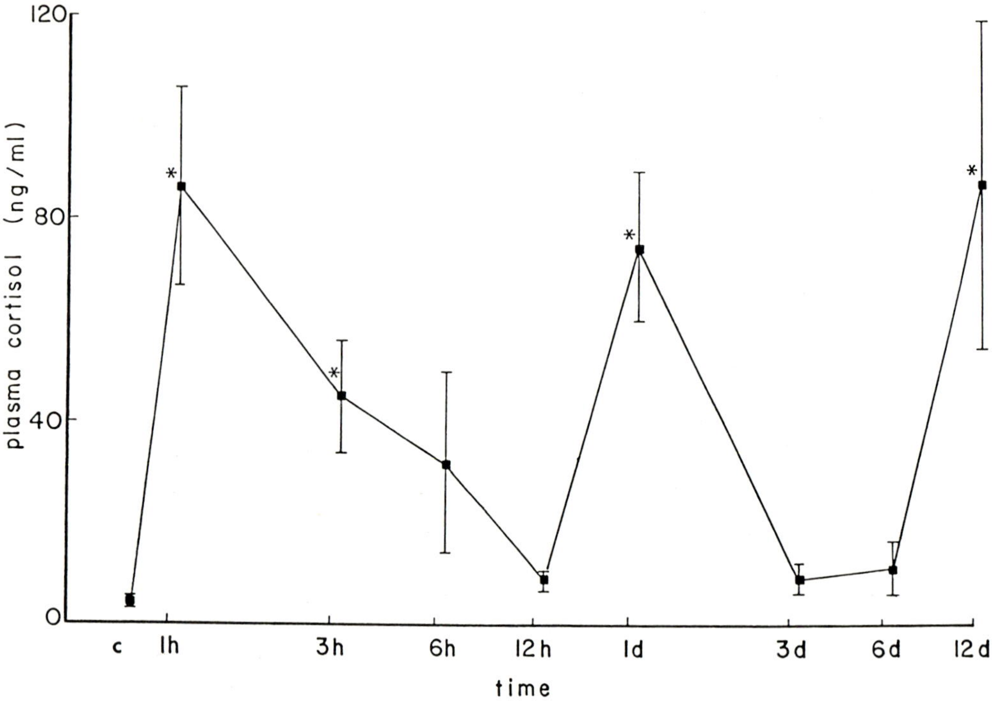

FIGURE 5. Plasma cortisol fluctuations in striped mullet exposed to 750 μg dibenzofuran per liter. Asterisks denote means significantly different from control ($P < 0.05$, $N = 5$). (From Thomas, P., Wofford, H. W., and Neff, J. M., *Aquat. Toxicol.*, 1, 332, 1981. With permission.)

Circulating levels of plasma glucose 3 hr after exposure to various concentrations of oil and after 1 week exposure to dibenzofuran and fluorene are shown in Figure 8. There was a close correlation between plasma glucose levels and cortisol titers after exposure to these pollutants (Figures 6 and 8). A dose-response relationship of glucose concentrations was observed after exposure to dibenzofuran and oil, but not after fluorene exposure (Figure 8).

The biochemical stress responses of fish to pollutants were compared to those elicited by physical stressors. The severe physical trauma of restraint in a dip net for 30 min with exposure to air (hypoxia) for 1 min every 5 min caused a marked elevation (30 min later) of plasma cortisol, glucose, and osmolality, but not cholesterol (cortisol, 218 ± 35 and 5.9 ± 4.2 ng/mℓ; glucose, 80.6 ± 7.5 and 57.4 ± 2.8 mg/dℓ; osmolality, 452 ± 24 and 344 ± 9 mOs/ℓ, for restrained and control animals, respectively; $P < 0.05$). These biochemical stress parameters were not significantly different from controls 12 hr after transfer from 30 to 5‰ S sea water.

Mullet were also exposed to 20% WSF of No. 2 fuel oil, considered to be moderately stressful, to investigate whether they were still responsive to stressful stimuli after 1 month in the laboratory. Plasma cortisol, glucose, and cholesterol rose to 5, 4, and 3 times control levels, respectively, after 1 to 3 hr exposure to 20% WSF (cortisol: 1 hr exposure to WSF, 115.2 ± 34.2 ng/mℓ; glucose: 3 hr exposure to WSF, 167.6 ± 39.7 mg/dℓ; cholesterol: 3 hr exposure to WSF, 195.2 ± 27.4 mg/dℓ; $N = 6$).

B. Discussion: Plasma Corticosteroid Stress Responses

The present results provide further evidence that the hypothalamo-pituitary-interrenal (HPI) axis of teleosts is responsive to a wide variety of adverse environmental stimuli, including

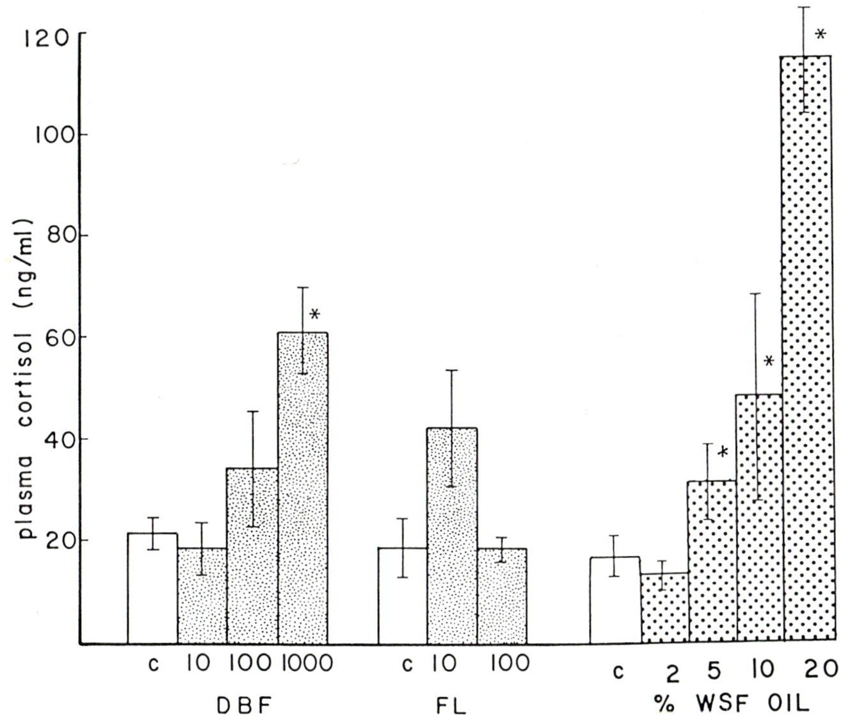

FIGURE 6. Mean plasma cortisol levels in striped mullet exposed to different concentrations ($\mu g/\ell$) of dibenzofuran (DBF) and fluorene (FL) for 1 week or to 2, 5, 10, or 20% WSFs of a No. 2 fuel oil for 1 hr. Asterisks denote means significantly different from (C) controls ($P < 0.05$, $N = 5$).[4,6]

pollutants (see review by Donaldson[33]). Elevations of plasma cortisol were observed in mullet exposed to seven organic pollutants and a complex mixture (Table 1). Other researchers have demonstrated elevated circulating corticosteroid levels in teleosts exposed to endrin,[34] naphthalene,[35] and the herbicide 2,4-D.[33] Chromium[36] and copper[37,38] also stimulate corticosteroid secretion in fish.

Although cadmium caused plasma cortisol levels to rise in tilapia *Tilapia mossambicus* (Table 1), even lethal concentrations of the metal failed to elicit a corticosteroid response in coho salmon[38] and mullet.[7] It has been suggested that only stimuli which cause fright, discomfort, or pain elevate cortisol titers in fish.[38] Thus, a deleterious environmental change may have to be detected by a fish in order to elicit a rise in plasma corticosteroids. It is possible that mullet and coho salmon could not detect cadmium at the concentrations tested in these studies.[7] As our knowledge of the mechanisms by which chemicals activate the HPI axis of teleosts is still rudimentary, it is uncertain why cadmium failed to stimulate cortisol production in these species.

The HPI axis of fish appears to be particularly sensitive to low concentrations of certain pollutants. The threshold dose of BaP for stimulation of cortisol secretion in mullet was 0.2 $\mu g/\ell$, less than one tenth the concentration of fluorene, naphthalene, or PCP required to elicit a corticosteroid response in this species.[7] In general, pollutant concentrations of 10 to 100 $\mu g/\ell$ or greater are necessary to cause a rise in plasma cortisol levels in teleosts.[6,35-39]

The magnitude of the plasma corticosteroid response to pollutants is often dose dependent (Figure 6), thus permitting an assessment of the degree of trauma associated with a particular dose of the pollutant. For example, the maximum cortisol titer recorded after exposure to

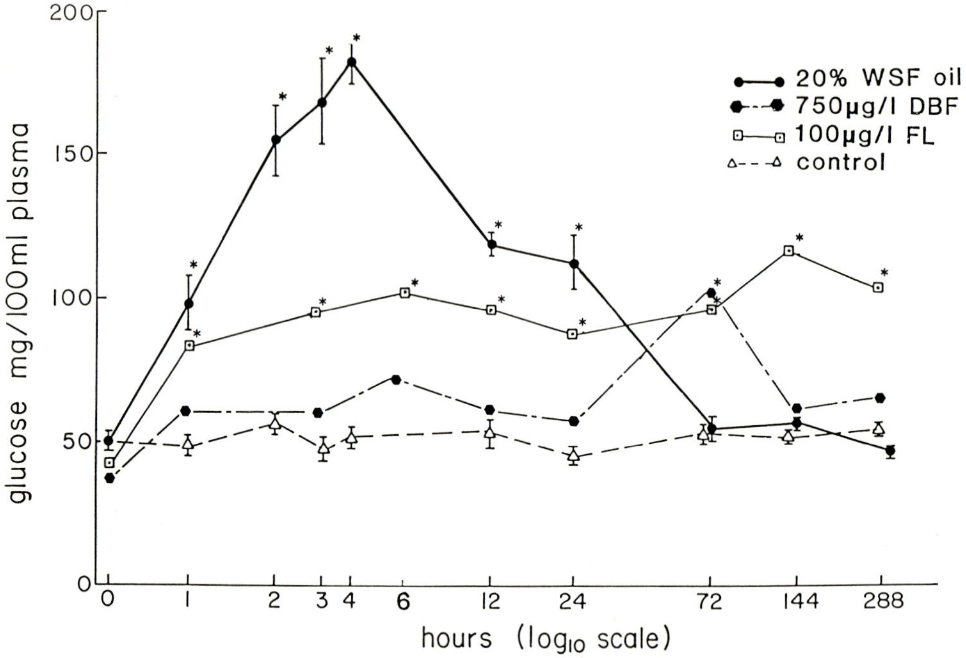

FIGURE 7. Plasma glucose fluctuations in striped mullet exposed to 20% WSF of a No. 2 fuel oil (●--●), 750 µg dibenzofuran (⬢ -- ⬢), and 100 µg fluorene per liter (□--□). Asterisks denote means significantly different from controls (△--△) (controls for 20% WSF oil shown in Figure 8, $P < 0.05$, $N = 5$).[4,6]

oil (120 ng/mℓ, Figure 4) was approximately half that observed after exposure to severe handling stress.[6] Thus, exposure to 20% WSF oil probably represents a moderate stressor in mullet. In contrast, exposure to a lethal concentration of PCP caused plasma cortisol concentrations in mullet to rise to maximum levels (300 ng/mℓ) for several days prior to death.[39] This marked and persistent elevation of plasma cortisol levels in pollutant-exposed fish may be an indication that death is imminent.

The time course of plasma corticosteroid fluctuations after pollutant exposure varies with the compound under investigation and also the exposure concentration (see Figures 4 and 5 and References 6, 37, and 39). Characteristic plasma corticosteroid responses of mullet to sublethal concentrations of pollutants are shown in Figure 4. A minor rise in plasma cortisol occurred during the first hour of exposure to the toxicant. It has been suggested that this minor elevation of plasma corticosteroids after low-level pollutant exposure may represent a fright reaction to this novel stimulus.[7,38] Cortisol concentrations declined rapidly after 1 hr of exposure and reached control levels by 6 hr (Figure 4). A secondary elevation of plasma cortisol was often observed between the 12th and 24th hr of exposure to moderate pollutant concentrations (Figures 4 and 5). This brief decline in circulating cortisol titers may be due to the presence of a corticosteroid-sensitive negative-feedback system in fish which temporarily blocks cortisol secretion.[4] The secondary cortisol rise was also transient, and circulating hormone concentrations are often at control levels 1 day later (see Figures 4 and 5 and References 37 and 38). This stage, when the HPI axis is no longer activated by the pollutants, may be equivalent to the resistance or acclimation phase of the GAS described by Selye in mammals.[40]

No further increases in plasma cortisol concentrations are usually seen in fish during chronic exposure to low sublethal pollutant concentrations. However, a further rise in plasma cortisol occurred in mullet after 6 days exposure to 750 µg dibenzofuran per liter (Figure

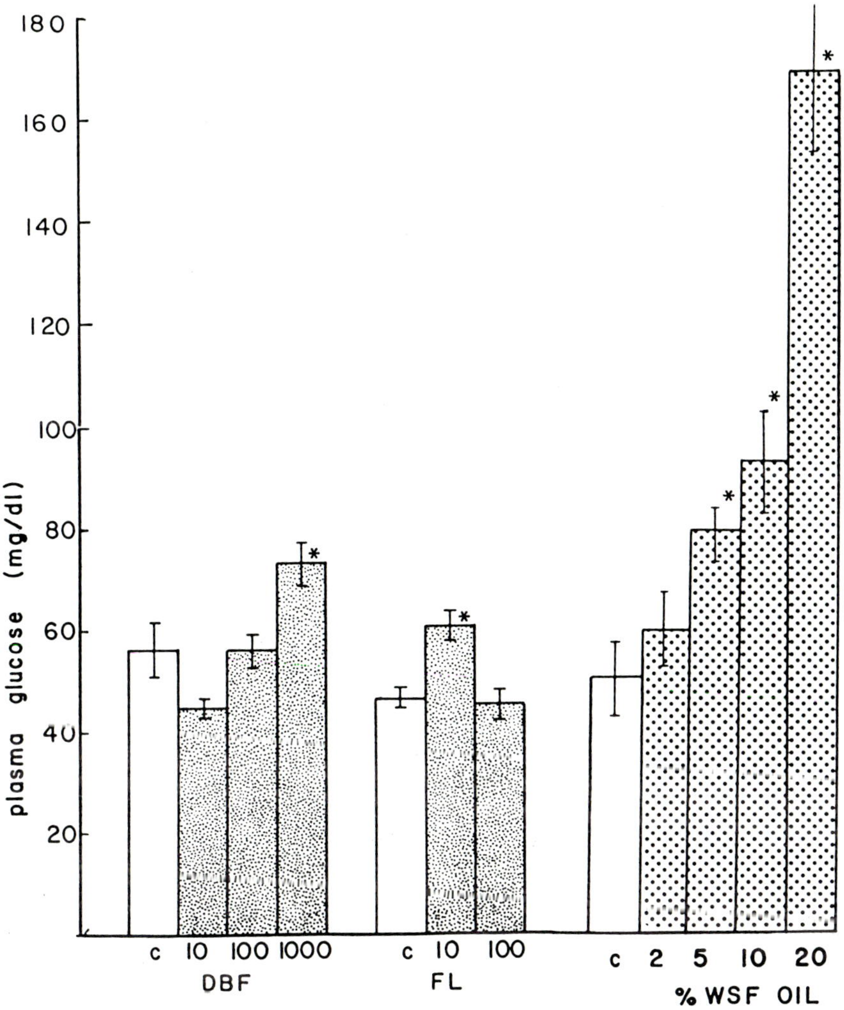

FIGURE 8. Mean plasma glucose levels in striped mullet exposed to different concentrations ($\mu g/\ell$) of dibenzofuran (DBF) and fluorene (FL) for 1 week or to 2, 5, 10, or 20% WSFs of a No. 2 fuel oil for 1 hr. Asterisks denote means significantly different from (C) controls ($P < 0.05$, $N = 5$).[4,6]

5). Similar increases in circulating corticosteroids have been reported in coho salmon during chronic exposure to doses of copper which, although sublethal, appeared to distress the fish.[38] These chronic elevations in plasma corticosteroids may be related to accumulation of the pollutants in the tissues and the onset of toxic effects and physiological dysfunction.[7,39] Adverse environmental stimuli have been shown to trigger corticosteroid stress responses in mammals by changing certain physiological parameters beyond their normal limits.[41] There is also evidence that some chemicals stimulate the secretion of corticosteroids in fish by causing physiological dysfunction.[39,42] The rise in plasma cortisol in mullet after 2 days exposure to 100 μg PCP per liter was accompanied by marked disturbances in several physiological parameters.[39] It is probable that failure to maintain homeostasis in these fish was a major cause of the increased mortality observed 2 days later.

Plasma cortisol concentration is not always a reliable indicator of HPI activity, particularly when fish are chronically exposed to a mild noxious stimulus.[4,43] Pollutants may also alter plasma cortisol concentrations by exerting direct inhibitory effects on steroidogenesis[44] or by enhancing steroid catabolism.[45,46] An increase in the nuclear diameter of interrenal cells may be a more reliable indicator of HPI hyperactivity under conditions of chronic exposure or increased catabolism,[33] although the utility of this parameter has been questioned.[47]

C. Discussion: Secondary Metabolic Disturbances

Elevations of circulating corticosteroids and catecholamines in fish exposed to adverse environmental stimuli result in secondary metabolic disturbances.[48] The present data show that these metabolic disturbances also occur after exposure to pollutants (Table 1). Hyperglycemia was the most consistent response to pollutants, although disturbances of lipid metabolism and osmoregulatory dysfunction were also occasionally observed. Hepatic glycogen may also become depleted during exposure to pollutant concentrations which cause a marked and prolonged hyperglycemia.[39] Other studies have demonstrated marked hyperglycemia in fish exposed to pollutants[35,49-51] and also alterations of lipid metabolism and impairment of ionic and osmotic regulation.[52,53]

Typical plasma glucose responses of mullet to pollutants are shown in Figure 7. The rise in glucose concentrations was more gradual than that of cortisol after exposure to oil and naphthalene, but the hyperglycemia persisted for several days after plasma cortisol had returned to control levels.[4,7] Comparison of the effects of different pollutant concentrations on plasma cortisol and glucose concentrations (Figures 6 and 8) reveals a close correlation between the changes in these two parameters. However, plasma glucose levels did not correlate with cortisol values in mullet and croaker after exposure to cadmium or BaP.[7] The hyperglycemia in mullet exposed to cadmium is probably due to the pancreatoxic effects of the metal, resulting in impaired insulin secretion.[54] It is possible that BaP did not cause sufficient increases in the plasma concentrations of catecholamines and corticosteroids to elicit a threshold hyperglycemic response in these species.

Hyperglycemia is not always a reliable indicator of metabolic disturbance in fish. Circulating glucose titers and the hyperglycemic stress response vary with the nutritional state[50] and ambient water temperature.[55] Further, plasma glucose levels may decline during chronic pollutant exposure due to exhaustion of glycogen reserves,[39] and low glucose levels are frequently found in fish close to death.[39]

IV. ALTERATIONS IN THE ASCORBIC ACID AND GLUTATHIONE STATUS OF FISH TISSUES

A. Results

The effects of a variety of pollutants on tissue reserves of ascorbic acid (ASA), glutathione, and acid-soluble thiols (AST) are summarized in Table 2. PCP and BaP caused a transient elevation of hepatic ASA concentrations, whereas chronic exposure (more than 7 days) to naphthalene, oil, and cadmium depleted hepatic ASA reserves (Table 2). The depletion of liver ASA by oil and cadmium was accompanied by a decline of ASA stores in gill and kidney tissues. Brain ASA was depleted by exposure to cadmium, but not by oil (Table 2).

All the pollutants tested (PCP, fluorene, dibenzofuran, oil, cadmium, and mercury) elevated AST concentrations in fish livers (Table 2). The largest increases in AST were observed after exposure to cadmium and mercury. The increase in AST after exposure to cadmium was due to elevations in glutathione and a metal binding protein, probably metallothionein. In contrast, glutathione accounted for all changes in AST after exposure to oil. AST concentrations were also elevated in the kidney tissues of mullet after exposure to cadmium, but brain levels were unaffected by cadmium exposure (Table 2).

The time course of ascorbic acid fluctuations in mullet hepatic tissues after exposure to

Table 2
SUMMARY OF POLLUTANT EFFECT ON TISSUE ASCORBIC ACIDS, GLUTHATHIONE, AND ACID-SOLUBLE THIOL CONCENTRATIONS

			Ascorbic acid				Acid-soluble thiols				
Pollutant	**Concentration/ℓ**	**Species**	**Liver**	**Brain**	**Gill**	**Kidney**	**Liver**	**Brain**	**Kidney**	**Glutathione liver**	**Ref.**
Pentachlorophenol	100—200 μg	*Mugil cephalus*	+				+				39,63
	50 μg	*Pseudopleuronectes americanus*					+				
Benzo(a)pyrene	0.2—0.3 μg	*M. cephalus*									
		M. undulatus	+								
Fluorene	100 μg	*M. cephalus*					+				63
Dibenzofuran	750 μg	*M. cephalus*					+				63
Naphthalene	500 μg	*M. cephalus*	−								
WSF No. 2 fuel oil	20%	*M. cephalus*	−	0	−	−	+			+	63,77
Cadmium	0.1—10 mg	*M. cephalus*	−	−	−	−	+[a]	0	+	+	8,9,63,64
Mercury	200 μg	*M. cephalus*					+				63

Note: +, significant increase; 0, no change; −, significant decrease; WSF, water-soluble fraction.

[a] Includes metallothionein.

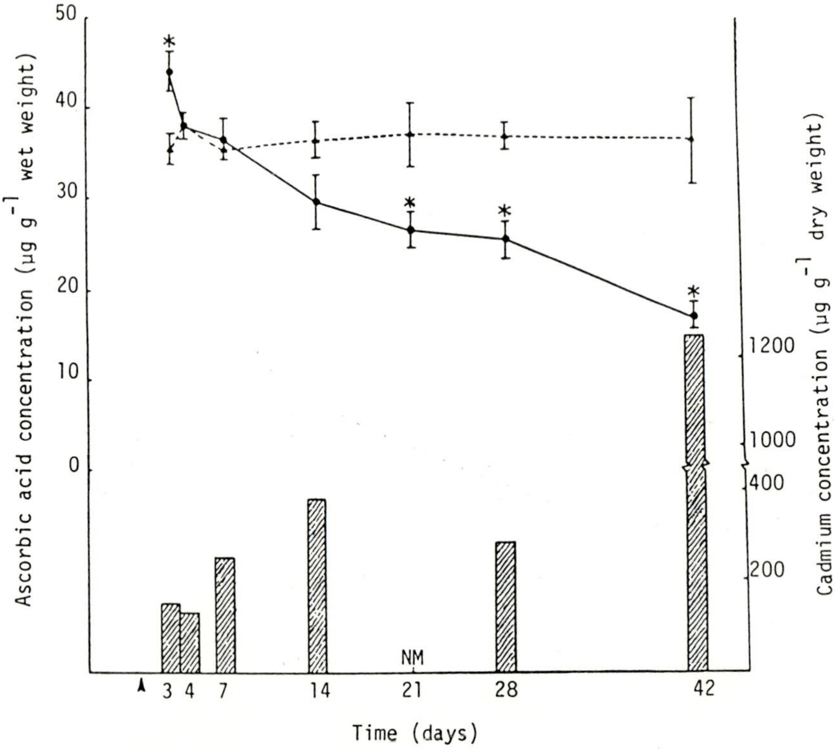

FIGURE 9. Hepatic ascorbic acid fluctuations in striped mullet exposed to 10 mg cadmium per liter (●—●). Asterisks denote means significantly different ($P < 0.005$) from controls (▲--▲). Cadmium accumulations are shown by shaded bars. NM, not measured ($N = 4$).

10 mg Cd per liter is shown in Figure 9. There was a significant depletion of hepatic ASA stores after 3 weeks exposure to 10 mg Cd per liter and tissue reserves were only 40% of original levels after 6 weeks exposure. The depletion of liver ASA was accompanied by considerable accumulation of cadmium (Figure 9). Exposure to 1 mg Cd per liter caused a similar depletion of ASA in mullet livers. There was a dose-response relationship in AST levels in mullet livers after 6 weeks exposure to cadmium (Figure 10). AST concentrations were elevated more than 3-fold after this period of exposure to 10 mg Cd per liter, whereas one hundredth of this dose of cadmium (0.1 mg/ℓ) only elevated hepatic AST concentrations 50% above control levels (Figure 10).

Neither hepatic AST nor ASA concentrations were persistently altered by a variety of physical stressors including capture and restraint, exposure to elevated water temperatures and a hypoosmotic environment, and short-term starvation.[62,63]

B. Discussion: Tissue Ascorbic Acid, Glutathione, and Acid-Soluble Thiols

Ascorbic acid, glutathione, and other acid-soluble thiols perform important functions in detoxifying foreign compounds in mammals,[57,58] and alterations in the tissue reserves of these protective agents have been observed after exposure to toxicants.[59,60] The present data show that toxicants also alter the concentrations of ascorbic acid, glutathione, and acid-soluble thiols in fish tissues. Hepatic ascorbic acid content did not show a consistent response to pollutant exposure. Both PCP and BaP caused a transient elevation in ascorbic acid, whereas other chemicals depleted hepatic reserves (see Table 2 and References 8, 39, and 61). Exposure of mullet to cadmium caused a gradual depletion of hepatic ascorbic acid

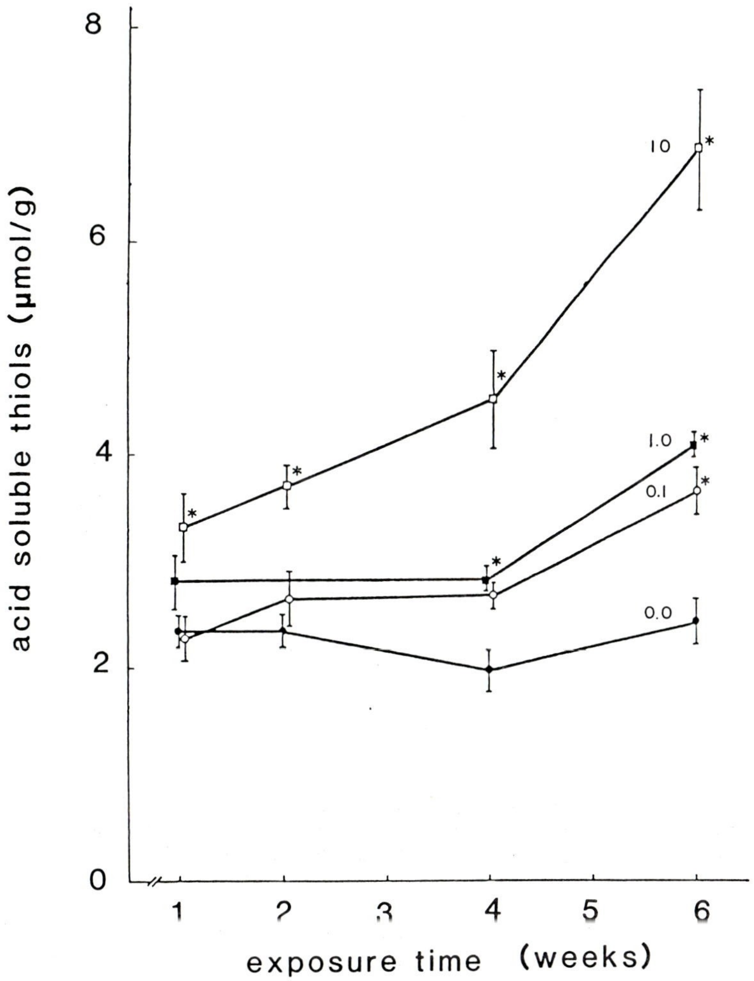

FIGURE 10. Hepatic acid-soluble thiol fluctuations in striped mullet exposed to 0, 0.1, 1, and 10 mg cadmium per liter for up to 6 weeks. Asterisks denote means significantly different ($P < 0.05$) from controls (0) ($N = 8$). (From Thomas, P., Wofford, H. W., and Neff, J. M., *Physiological Mechanisms of Marine Pollutant Toxicity*, Vernberg, F. J., Calabrese, A., Thurberg, F. P., and Vernberg, W. B., Eds., Academic Press, Orlando, Fla., 1982, 113. With permission.)

(Figure 9) which was not dose dependent. In contrast, a dose-dependent depletion of liver ascorbic acid was observed 1 week after exposure to water-soluble fractions of a fuel oil.[77] Depletion of hepatic ascorbic acid appears to be a specific response to certain chemicals, since a variety of physical environmental variables failed to significantly alter this parameter.[62] However, seasonal data showed a depletion of hepatic ascorbic acid reserves in mullet during the period of decreased food availability at the end of the winter.[56] Thus, the available data suggest that the dietary intake of the vitamin, as well as exposure to certain environmental contaminants, can influence the ascorbic acid status of fish hepatic tissues.

All the chemicals investigated except naphthalene elevated the acid-soluble thiol content of fish livers. The pattern of changes in specific thiols differed among the various treatments. The elevation in acid-soluble thiols after 10 days exposure to oil was largely due to a rise in glutathione content.[63] In contrast, a metal-binding thiol (apparent molecular weight, 9400 daltons; probably a metallothionein) accounted for most of the increase in hepatic acid-soluble thiols observed after 10 days exposure to cadmium. However, after 3 weeks exposure to cadmium, hepatic glutathione levels were also elevated.[64] Neither physical trauma nor starvation for 2 days significantly altered hepatic acid-soluble thiol levels.[63] These results suggest that the increase in hepatic acid-soluble thiol content is a specific response to chemical insult. Preliminary evidence indicates that some pollutants may elevate glutathione content by enhancing the hepatic uptake of amino acid substrates and also the activity of biosynthetic enzymes.[63]

V. FIELD STUDIES: *N. VIRENS*

A study was conducted to evaluate the feasibility of using biochemical indexes to assess the relative "health" of field populations of *N. virens*. Two field populations in Maine were selected for study. The first was at the mouth of the Fore River in Portland, a site which has been contaminated chronically with refined oil products. The reference site, approximately 15 miles south of Portland, was at the mouth of the Nonesuch River in Pine Point, an area in which soft-shelled clams are harvested commercially. A number of biochemical parameters were measured in animals from the two sites to determine whether significant differences could be detected between the two populations and to evaluate the utility of this approach for diagnosing stressed animals in the field.

Animals were collected from the two sites on consecutive days on a bimonthly basis at low tide within a day of the full moon from September 1981 through March 1982. An attempt was made to select animals of the same size from the two sites as glycogen content has been shown to be positively correlated with weight for this species.[24,25] (Sandworms were also collected in September 1981 from Deer Island, in an area adjacent to the Boston, Mass. sewage outfall; due to the limited population at this site, it was not sampled subsequently.) The animals were kept in sea water drawn from the site and changed as required to maintain the temperature within 2°C of the ambient temperature at the collection site. Twenty-four animals were sampled within 2 hr of the time of collection. Each animal was weighed before replicate coelomic fluid samples were taken with a microsyringe and transferred to nonheparinized capillary tubes. After the coelomic fluid samples were obtained, each polychaete was placed in a labeled plastic bag and frozen in liquid nitrogen for transport back to the laboratory.

Frozen tissue samples from posterior segments were analyzed for glycogen, lipids, and free amino acids, and, for one collection period, total acid-soluble proteins. Tissue glycogen concentrations were determined by the enzymatic procedure of Carr and Neff.[12] Lipid and protein determinations were performed according to the methods of Holland and Gabbot[13] and Peterson,[65] respectively.

Significant differences between populations for the various parameters were determined by Student's t test or by the Mann-Whitney U test when the data did not conform to analysis by parametric methods. The descriptive information and biochemical data were also analyzed by Spearman correlation analysis with the aid of the Statistical Analysis System[66] in order to determine the degree of association among the different parameters for the two populations. A highly significant difference in the mean coelomic fluid glucose concentration between the two groups was observed for the March 1982 collection.[67]

There was a consistent trend for both the glycogen and lipid concentrations for all samples. The glycogen concentration was always lower and the lipid content always higher in the Portland animals than in animals from the reference site. The difference was statistically significant for two separate collections for the total combined data for the lipid levels, but not for the glycogen concentrations. In September 1981 when sandworms were also collected from Deer Island, Mass., the glycogen content of animals from this site was 50% of the concentration in the animals from the reference site in Maine, a highly significant difference. No difference was observed for the acid-soluble protein content between the two populations.

A number of significant differences in the tissue free amino acid concentrations between the two populations were also observed. The concentration of tissue alanine, the primary tissue free amino acid for this species, was significantly decreased in the Portland *N. virens*, compared with the Pine Point animals. The levels of a number of other free amino acids including lysine, threonine, arginine, valine, phenylalanine, tyrosine, and isoleucine were significantly elevated in the Portland population when compared with reference animals.

There was no significant difference between the two populations for the tissue/glycine (T/G) ratio for any of the collections, but the ratio tended to be higher for the Portland *N. virens*. The threonine + serine concentration, which has been suggested for use as a stress index in mussels,[68] tended to be higher in the Portland animals than in those from Pine Point, and this difference was significantly different for one of the three collections and for the combined data from all three collections. The concentration of alanine in Portland sandworms was always less than in animals from the reference site, and for two of the three sampling periods and for the combined data, this difference was highly significant statistically.

VI. FIELD STUDIES: WINTER FLOUNDER

The purpose of this investigation was to determine the feasibility of using biochemical parameters and histopathological examinations to assess the relative health of winter flounder *(P. americanus)* collected near the discharge of a sewage treatment plant, compared to a population from a nearby reference site. Results of a 10-year tagging study showed that the movement of animals from these two populations was localized[69] and that winter flounder return to the same area to spawn each year.[70] The Deer Island sewage treatment plant discharges both primary effluent and sewage sludge into the outer Boston Harbor. Although early year class winter flounder are very abundant near the discharge area, an increased incidence of fin erosion has been observed among individuals collected at this site,[71] with increasing incidence tending to parallel the increasing water temperature. An increased incidence of fin erosion in winter flounder has also been reported for animals collected in the New York Bight Apex and the Sandy Hook-Raritan Bay areas, both of which are impacted by domestic and industrial pollution.[72]

The biochemical parameters that were investigated in this study included hepatic and pectoral fin ascorbic acid concentrations, hepatic glutathione, lipid, and glycogen concentrations, plasma glucose and cholesterol levels, muscle free amino acid concentrations, and brain indoleamine and catecholamine concentrations. Gill and liver tissues were examined histopathologically. We have evaluated these parameters in several laboratory and field investigations and have found them to be useful for diagnosing pollutant stress in marine teleosts.[4,6,8,9,39,73,74] In this investigation, the relative sensitivity of the various indexes was evaluated by correlating differences in the biochemical parameters between the two populations with any observed incidence of tissue lesions, parasitic infection, other histological abnormalities, or obvious external indications of fin erosion disease.[75]

Winter flounder were collected by otter trawl in July 1981 and June 1982 near the Deer Island sewage treatment plant outfall and at a reference site off Plymouth Beach, Mass. The results of ascorbic acid determinations for pectoral fins and liver and hepatic glycogen concentrations are presented in Figure 11 for the July 1981 collections. The hepatic glycogen and lipid concentrations in animals from the Boston Harbor site were significantly lower than in animals from the reference site for the June 1982 sampling also. The plasma glucose level of the Boston Harbor animals showed a highly significant decrease compared to the reference animals. The plasma cholesterol level of animals from Boston Harbor was elevated compared to the reference site animals, but not significantly. The neurotransmitters present

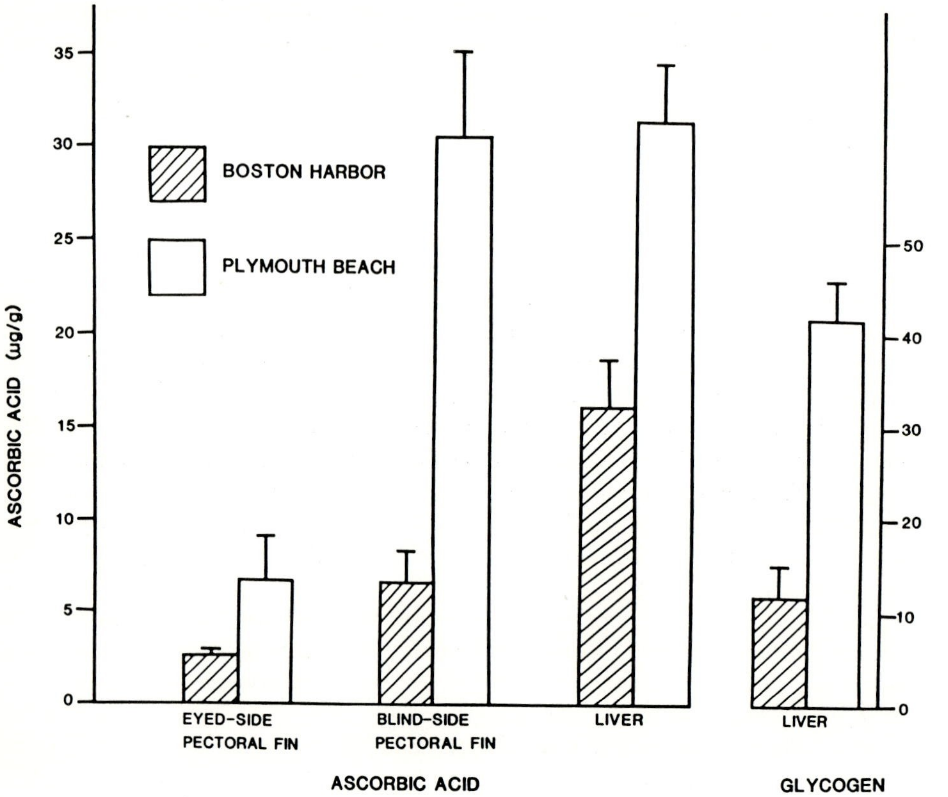

FIGURE 11. Mean (±SEM, N ≥ 13) pectoral fin ascorbic acid and liver ascorbic acid and glycogen for winter flounder collected in July 1981. The difference between the two populations was highly significant ($\alpha < 0.005$, U test) for all parameters except eyed-side pectoral fin.

in greatest concentrations were norepinephrine and serotonin (5-hydroxytryptamine). The level of both these neurotransmitters was significantly lower in the brains of winter flounder from Boston Harbor compared with animals from Plymouth Beach. In general, the concentration of muscle free amino acids was higher in the Boston Harbor animals. This elevation was statistically significant for only glycine and serine. The serine + threonine index and the taurine/glycine ratio were both highly significantly different between the two populations. Based on the histopathological examination, liver and gill tissues were assigned a pathology index rating. There was a highly significant difference between the two populations for the liver pathology index. A number of highly significant associations between glycogen and tissue ascorbic acid levels and the histopathological conditions and pathology index were also determined by Spearman correlation analysis.

VII. BIOLOGICAL SIGNIFICANCE OF BIOCHEMICAL ALTERATIONS IN FISH

Many of the biochemical and physiological changes observed in fish after pollutant exposure, such as the rise in plasma cortisol and catecholamine titers, are also characteristic responses of teleosts to a wide variety of other adverse stimuli.[48] These generalized stress responses, which are necessary for tolerance to changing environmental conditions, are transitory except when the environmental stressor is severe and homeostasis is not completely restored. Stress responses have been classified as primary and secondary by Mazeaud et al.,[48] and the classification has been extended by Wedemeyer and McLeay[76] to include tertiary effects. Noxious stimuli elicit corticosteroid and catecholamine stress responses (primary effects), which in turn cause a wide variety of effects including metabolic, osmoregulatory, hematological, immunological disturbances (secondary effects). These disturbances, if they persist, lead to tertiary effects such as increased disease susceptibility and reduced growth.[43,76] Thus, persistent alterations of hypothalamo-pituitary-interrenal and catecholaminergic activities in fish exposed to pollutants are likely to have serious ecological consequences.[4]

In contrast, hepatic ascorbic acid, acid-soluble thiol, and glutathione concentrations are largely unaffected by physical environmental variables so that the alterations in these parameters observed after pollutant exposure probably represent specific responses to chemical insult. Although the possible involvement of ascorbic acid and glutathione in detoxication processes in fish has been discussed,[8,9,39] the precise mechanisms by which pollutants alter these parameters remain to be elucidated. A better understanding of the biological significance of chronic ascorbic acid deficiency in fish has emerged from feeding studies conducted over the past several years, and the possible ecological implications of such a deficiency in fish populations have been reviewed.[8] In contrast, there is a paucity of knowledge on the interactions between glutathione and chemicals in fish tissues. Many chemicals cause an acute depletion of hepatic glutathione reserves in mammals, which is often accompanied by lipid peroxidation and tissue necrosis.[60] However, the biological significance of pollutant-induced increases in the glutathione content of fish hepatic tissues is unknown.

ACKNOWLEDGMENT

This work was supported in part by grants OCE-8308448 and OCE-8310504 from the National Science Foundation.

REFERENCES

1. **Carr, R. S., Thomas, P., and Neff, J. M.,** A simple spectrophotometric technique for the determination of pentachlorophenol in water, *Bull. Environ. Contam. Toxicol.*, 28, 477, 1982.
2. **Giam, C. S., Trujillo, D. A., Kira, S., and Hrung, Y.,** Simplified monitoring procedures for benzo(a)pyrene, hexachlorobenzene and pentachlorophenol in water, *Bull. Environ. Contam. Toxicol.*, 25, 824, 1980.
3. **Anderson, J. W., Neff, J. M., Cox, B. A., Tatum, H. E., and Hightower, G. M.,** Characteristics of dispersions and water-soluble extracts of crude and refined oils and their toxicity to estuarine crustaceans and fish, *Mar. Biol.*, 27, 75, 1974.
4. **Thomas, P., Woodin, B., and Neff, J. M.,** Biochemical responses of striped mullet *Mugil cephalus* to oil exposure. I. Acute responses — interrenal activation and secondary stress responses, *Mar. Biol.*, 59, 141, 1980.
5. **Neff, J. M. and Anderson, J. W.,** An ultraviolet spectrophotometric method for the determination of naphthalene and alkyl naphthalenes in the tissue of oil contaminated marine animals, *Bull. Environ. Contam. Toxicol.*, 14, 122, 1975.
6. **Thomas, P., Wofford, H. W., and Neff, J. M.,** Biochemical stress responses of striped mullet (*Mugil cephalus* L.) to fluorene analogs, *Aquat. Toxicol.*, 1, 329, 1981.
7. **Thomas, P. and Neff, J. M.,** Plasma corticosteroid and glucose responses to pollutants in striped mullet: different effects of naphthalene, benzo(a)pyrene and cadmium exposure, in *Marine Pollution and Physiology: Recent Advances*, Calabrese, A., Thurberg, F. P., Vernberg, F. J., and Vernberg, W. B., Eds., University of South Carolina Press, Columbia, 1985, 63.
8. **Thomas, P., Bally, M., and Neff, J. M.,** Ascorbic acid status of mullet (*Mugil cephalus* Linn.) exposed to cadmium, *J. Fish Biol.*, 20, 183, 1982.

9. **Thomas, P., Wofford, H. W., and Neff, J. M.,** Effect of cadmium on glutathione content of mullet *(Mugil cephalus)* tissues, in *Physiological Mechanisms of Marine Pollutant Toxicity,* Vernberg, F. J., Calabrese, A., Thurberg, F. P., and Vernberg, W. B., Eds., Academic Press, New York, 1982, 109.
10. **Thomas, P. and Wofford, H. W.,** High-performance liquid chromatography of corticosteroids in vertebrate plasma: assay of cortisol in mullet and corticosterone in the rat, *Comp. Biochem. Physiol.,* 78, 473, 1984.
11. **Montgomery, R.,** Determination of glycogen, *Arch. Biochem. Biophys.,* 67, 378, 1957.
12. **Carr, R. S. and Neff, J. M.,** Quantitative semi-automated enzymatic assay for tissue glycogen, *Comp. Biochem. Physiol.,* 778, 447, 1984.
13. **Holland, D. L. and Gabbot, P. A.,** A micro-analytical scheme for the determination of protein, carbohydrate, lipid and RNA levels of marine invertebrate larvae, *J. Mar. Biol. Assoc. U.K.,* 51, 659, 1971.
14. **Jacobowitz, D. M. and Richardson, J. S.,** Method for rapid determination of norepinephrine, dopamine and serotonin in the same brain region, *Pharmacol. Biochem. Behav.,* 8, 515, 1978.
15. **Roesijadi, G., Anderson, J. W., and Giam, C. S.,** Osmoregulation of the grass shrimp *Palaemonetes pugio* exposed to polychlorinated biphenyls (PCBs). II. Effect on free amino acids of muscle tissues, *Mar. Biol.,* 38, 357, 1976.
16. **Pfeiffer, R., Karol, R., Korpi, J., Burgoyne, R., and McCourt, D.,** Practical application of HPLC to amino acid analysis, *Am. Lab.,* 15, 78, 1983.
17. **Zannoni, V., Lynch, M., Goldstein, S., and Sato, P.,** A rapid micro-method for the determination of ascorbic acid in plasma and tissues, *Biochem. Med.,* 11, 41, 1974.
18. **Carr, R. S., Bally, M. B., Thomas, P., and Neff, J. M.,** Comparison of methods for determination of ascorbic acid in animal tissues, *Anal. Chem.,* 55, 1229, 1983.
19. **Terada, M., Watanabe, Y., Kunitomo, M., and Hayashi, E.,** Differential rapid analysis of ascorbic acid and ascorbic acid 2-sulfate by dinitrophenylhydrazine method, *Anal. Biochem.,* 84, 604, 1978.
20. **Carr, R. S. and Neff, J. M.,** Determination of ascorbic acid in tissues of marine animals by liquid chromatography with electrochemical detection, *Anal. Chem.,* 52, 2428, 1980.
21. **Moron, M. S., DePierre, J. W., and Mannervik, B.,** Levels of glutathione, glutathione reductase and glutathione-S-transferase activities in rat lung and liver, *Biochim. Biophys. Acta,* 582, 67, 1979.
22. **Owens, C. W. I. and Belcher, R. V.,** A colorimetric micro-method for the determination of glutathione, *Biochem. J.,* 94, 705, 1965.
23. **Gaitonde, M. K.,** A spectrophotometric method for the direct determination of cysteine in the presence of other naturally occurring amino acids, *Biochem. J.,* 104, 627, 1967.
24. **Carr, R. S. and Neff, J. M.,** Biochemical indices of stress in the sandworm *Neanthes virens* (Sars). I. Responses to pentachlorophenol, *Aquat. Toxicol.,* 1, 313, 1981.
25. **Carr, R. S. and Neff, J. M.,** Biochemical indices of stress in the sandworm *Neanthes virens* (Sars). II. Sublethal responses to cadmium, *Aquat. Toxicol.,* 2, 319, 1982.
26. **Coglianese, M. P. and Neff, J. M.,** Biochemical responses of the blue crab, *Callinectes sapidus,* to pentachlorophenol, in *Physiological Mechanisms of Marine Pollutant Toxicity,* Vernberg, F. J., Calabrese, A., Thurberg, F. P., and Vernberg, W. B., Eds., Academic Press, New York, 1982.
27. **Selye, H.,** A syndrome produced by diverse nocuous agents, *Nature (London),* 138, 32, 1936.
28. **Selye, H.,** *The Physiology and Pathology of Exposure to Stress,* Medical Publishers, Montreal, 1950.
29. **Sinderman, C. J.,** Keynote address: notes of a pollution watcher, in *Marine Pollution and Physiology: Recent Advances,* Vernberg, F. J., Thurberg, F. P., Calabrese, A., and Vernberg, W. B., Eds., University of South Carolina Press, Columbia, 1985, 11.
30. **Olive, P. J. W.,** Endocrine adaptations in Annelida, in *Hormones and Evolution,* Vol. 1, Barrington, E. J. W., Eds., Academic Press, London, 1979, 73.
31. **Kleinholz, L. H. and Keller, R.,** Endocrine regulation in Crustacea, in *Hormones and Evolution,* Vol. 1, Barrington, E. J. W., Ed., Academic Press, London, 1979, 160.
32. **Felig, R., Sherwin, R. S., Sorion, V., Wahren, J., Hendler, R., Sacca, L., Eigler, N., Goldberg, D., and Wasesky, M.,** Hormonal interactions in the regulation of blood glucose, *Recent Prog. Hormone Res.,* 35, 501, 1979.
33. **Donaldson, E. M.,** The pituitary-interrenal axis as an indicator of stress in fish, in *Stress and Fish,* Pickering, A. D., Ed., Academic Press, London, 1981, 11.
34. **Grant, B. F. and Mehrle, P. M.,** Endrin toxicosis in rainbow trout *(Salmo gairdneri), J. Fish. Res. Board Can.,* 30, 31, 1973.
35. **DiMichele, L. and Taylor, M. H.,** Histopathological and physiological responses of *Fundulus heteroclitus* to naphthalene exposure, *J. Fish. Res. Board Can.,* 35, 1060, 1978.
36. **Hill, C. W. and Fromm, P. O.,** Response of the interrenal gland of rainbow trout *(Salmo gairdneri)* to stress, *Gen. Comp. Endocrinol.,* 11, 69, 1968.
37. **Donaldson, E. M. and Dye, H. M.,** Corticosteroid concentrations in sockeye salmon *(Oncorhynchus nerka)* exposed to low concentrations of copper, *J. Fish Res. Board Can.,* 32, 533, 1975.
38. **Schreck, C. B. and Lorz, H. W.,** Stress response of coho salmon *(Oncorhynchus kisutch)* elicited by cadmium and copper and potential use of cortisol as an indicator of stress, *J. Fish. Res. Board Can.,* 35, 1124, 1978.

39. **Thomas, P., Carr, R. S., and Neff, J. M.,** Biochemical stress responses of mullet *Mugil cephalus* and polychaete worms *Nereis virens* to pentachlorophenol, in *Biological Monitoring of Marine Pollutants*, Calabrese, A., Thurberg, F. P., Vernberg, F. J., and Vernberg, W. B., Eds., Academic Press, New York, 1981, 73.
40. **Selye, H.,** Stress and the general adaptation syndrome, *Br. Med. J.*, 1, 1383, 1950.
41. **Vigas, M.,** Contribution to the understanding of the stress concept, in *Catecholamines and Stress: Recent Advances*, Usdin, E., Kvetnansky, R., and Kopin, I. J., Eds., Elsevier/North-Holland, New York, 1980, 573.
42. **Tomasso, J. R., Davis, K. B., and Simco, B. A.,** Plasma corticosteroid dynamics in channel catfish *(Ictalurus punctatus)* exposed to ammonia and nitrite, *Can. J. Fish. Aquat. Sci.*, 38, 1106, 1981.
43. **Robertson, L., Thomas, P., Arnold, C. R., and Trant, J. M.,** Plasma corticosteroid and secondary stress responses of red drum *(Sciaenops ocellatus)* to common culture procedures, in *Proc. 2nd Int. Conf. on Aquaculture Finfish, Warm Water* Brigham Young University, Provo, Utah, 1985, 583.
44. **Ilan, Z. and Yaron, Z.,** Suppression by organochlorines of the response to adrenocorticotrophin of the interrenal tissue in *Sarotherodon aureus (Teleostei), J. Endocrinol.*, 87, 185, 1980.
45. **Sivarajah, K., Franklin, C. S., and Williams, W. P.,** The effects of polychlorinated biphenyls on plasma steroid levels and hepatic microsomal enzymes in fish, *J. Fish. Biol.*, 13, 401, 1978.
46. **Hansson, T.,** Effects of treated, municipal waste water on the hepatic metabolism of 4-androstene-3, 1-dione in rainbow trout *Salmo gairdneri* Richardson, in *Stress and Fish*, Pickering, A. D., Ed., Academic Press, London, 1981, 339.
47. **Scott, D. B. T. and Rennie, S. E.,** Nuclear diameter as a criterion of adrenocortical activity in a teleost fish *Coregonus laveretus* (L.), *J. Fish. Biol.*, 17, 83, 1980.
48. **Mazeaud, M. M., Mazeaud, F., and Donaldson, E. M.,** Primary and secondary effects of stress in fish: some new data with a general review, *Trans. Am. Fish. Soc.*, 106, 201, 1977.
49. **Silbergeld, E. K.,** Blood glucose: a sensitive indicator of environmental stress in fish, *Bull. Environ. Contam. Toxicol.*, 11, 20, 1974.
50. **McLeay, D. J.,** Development of a blood sugar bioassay for rapidly measuring stressful levels of pulpmill effluent to salmonid fish, *J. Fish. Res. Board Can.*, 34, 477, 1977.
51. **Swift, D. J.,** Changes in selected blood component concentrations of rainbow trout *(Salmo gairdneri* Richardson) exposed to hypoxia or sublethal concentrations of phenol or ammonia, *J. Fish. Biol.*, 19, 45, 1981.
52. **Holmberg, B., Jensen, S., Larsson, A., Lewander, K., and Olsson, M.,** Metabolic effects of technical pentachlorophenol (PCP) on the eel *Anguilla anguilla* L., *Comp. Biochem. Physiol.*, 43B, 171, 1972.
53. **Englehardt, F. R., Wong, M. P., and Duey, M. E.,** Hydromineral balance and gill morphology in rainbow trout, *Salmo gairdneri*, acclimated to fresh and sea water, as affected by petroleum exposure, *Aquat. Toxicol.*, 1, 175, 1981.
54. **Thomas, P.,** Effect of cadium exposure on plasma cortisol levels and carbohydrate metabolism in mullet *(Mugil cephalus), J. Endocrinol.*, 94(Suppl.), 1982.
55. **Umminger, B. L.,** Osmoregulatory role of serum glucose in fresh water-adapted killifish *(Fundulus heteroclitus)* at temperatures near freezing, *Comp. Biochem. Physiol.*, 109, 229, 1971.
56. **Thomas, P., Bally, M. B., and Neff, J. M.,** Influence of some environmental variables on the ascorbic acid status of mullet *(Mugil cephalus* Linn.) tissues. II. Seasonal fluctuations and biosynthetic ability, *J. Fish. Biol.*, 27, 47, 1985.
57. **Zannoni, V. G. and Sato, P. H.,** Effects of ascorbic acid on microsomal drug metabolism, *Ann. N.Y. Acad. Sci.*, 258, 119, 1975.
58. **Reed, D. J. and Beatty, P. W.,** Biosynthesis and regulation of glutathione: toxicological implications, in *Reviews in Biochemical Toxicology*, Hodgson, E., Bend, J. R., and Philpot, R. M., Eds., Elsevier, Amsterdam, 1980, 213.
59. **Chatterjee, G. C., Banerjee, S. K., and Pal, D. R.,** Cadmium administration and L-ascorbic acid metabolism in rats: effect of L-ascorbic acid supplementation, *Int. J. Vit. Nutr. Res.*, 43, 370, 1973.
60. **Chasseaud, L. F.,** Conjugation with glutathione and mercapturic acid excretion, in *Glutathione: Metabolism and Function*, Arias, I. M. and Jakoby, W. B., Eds., Raven Press, New York, 1976, 77.
61. **Mayer, F. L., Mehrle, P. M., and Crutcher, P. L.,** Interactions of toxaphene and vitamin C in channel catfish, *Trans. Am. Fish. Soc.*, 107, 326, 1978.
62. **Thomas, P.,** Influence of some environmental variables on the ascorbic acid status of mullet *(Mugil cephalus* Linn.) tissues. I. Effect of salinity, capture-stress and temperature, *J. Fish. Biol.*, 25, 711, 1984.
63. **Thomas, P. and Wofford, H. W.,** Effects of metals and organic compounds on hepatic glutathione, cysteine, and acid-soluble thiol levels in mullet *(Mugil cephalus* L.), *Toxicol. Appl. Pharmacol.*, 76, 172, 1984.
64. **Wofford, H. W. and Thomas, P.,** Interactions of cadmium with sulfhydryl-containing compounds in striped mullet *(Mugil cephalus* L.), *Mar. Environ. Res.*, 14, 119, 1984.

65. **Peterson, G. L.,** A simplification of the protein assay method of Lowry et al. which is more generally applicable, *Anal. Biochem.*, 83, 346, 1977.
66. SAS Institute Inc., *SAS User's Guide: Basics,* 1982, ed., SAS Institute Inc., Cary, N.C., 1982.
67. **Carr, R. S. and Neff, J. M.,** Field assessment of biochemical stress indices for the sandworm *Neanthes virens* (Sars), *Mar. Environ. Res.*, 14, 267, 1984.
68. **Livingstone, D. R., Widdows, J., and Firth, P.,** Aspects of nitrogen metabolism of the common mussel *Mytilus edulis.* I. Adaptation to abrupt and fluctuating changes in salinity, *Mar. Biol.*, 53, 41, 1979.
69. **Howe, A. B. and Coates, P. G.,** Winter flounder movements, growth and mortality off Massachusetts, *Trans. Am. Fish Soc.*, 104, 13, 1975.
70. **Saila, S. B.,** A study of winter flounder movement, *Luminol. Oceanogr.*, 6, 292, 1981.
71. **Williams, J. W.,** The occurrence of fin erosion in winter flounder, *Pseudopleuronectes americanus,* from the vicinity of Boston Harbor, Massachusetts, Internal Tech. Rep., Battelle New England Marine Research Laboratory, Duxbury, Mass., 1981.
72. **Ziskowski, J. and Murchelano, R.,** Fin erosion in winter flounder, *Mar. Pollut. Bull.*, 6, 26, 1975.
73. **Haensly, W. E., Neff, J. M., Sharp, J. R., Morris, A. C., Bedgood, M. F., and Boehm, P. D.,** Histopathology of *Pleuronectes platessa* from Aber Wrach'h and Aber Benoit, Brittany, France: long-term effects of the AMOCO CADIZ crude oil spill, *J. Fish Dis.*, 5, 365, 1982.
74. **Neff, J. M. and Haensly, W. E.,** Long-term impact of the *AMOCO CADIZ* crude oil spill on oysters *Crassostrea gigas* and plaice *Pleuronectes platessa* from Aber Benoit and Aber Wrac'h, Brittany, France. I. Oyster histopathology. II. Petroleum contamination and biochemical indices of stress in oysters and plaice, in Ecological Study of the AMOCO CADIZ Oil Spill, Report of the NOAA-CNEXO Joint Commission, USDC/NOAA, Washington, D.C., 1982, 269.
75. **Carr, R. S., Hillman, R. E., and Neff, J. M.,** Field assessment of the impact of an ocean sewage outfall on the health of winter flounder *Pseudopleuronectes americanus,* submitted.
76. **Wedemeyer, G. A. and McLeay, D. J.,** Methods for determining the tolerance of fishes to environmental stressors in *Stress and Fish,* Pickering, A. D., Ed., Academic Press, London, 1981, 247.
77. **Thomas, P.,** Influence of some environmental variables on the ascorbic acid status of mullet *(Mugil cephalus)* Linn.) tissues. III. Effects of exposure to oil, *J. Fish Biol.*, in press.

INDEX

A

Accumulation, see also Bioaccumulation; Bioaccumulation factors
  of pollutants
    in clams, 119—121
    in marine systems, 2, 8
  rates of, 4
Acid phosphatase
  assay for, 112—113
  in *Mercenaria mercenaria* digestive gland, 133—135, 137—140
Acid-soluble thiols, 160, 170—174
Adrenocortical stress response, 4, 10—11
Adrenocorticotropic hormone, 10
Alabama estuaries, pollutants measured in, 37
Alcian blue stain, 112, 150
Alkaline phosphatase, 113
Amine, biogenic, 159
Amino acids, 159—160, 174—176
Aminopyrine N-demethylase, 79—80
ANF, see α-Naphthoflavone
Anguilla anguilla L., 34
Animals, test, 6—8, 28
Annelida, 3, 6—8, 100
Anthracene, 11
Antibodies, polyclonal, 86
Arene oxides, 9
Aroclor 1242, 35, 38
Aroclor 1254, 54—55
Aroclor 1260, 37—38, 43
Aromatic hydrocarbons, 7, 87
Arthropoda, 3, 6—7
Aryl hydrocarbon hydroxylase (AHH)
  activities
    during development, 84
    in fish, 86—87
  in crustaceans, 10
  in fish liver, 3, 9
  in *Fundulus* liver, 70—71
  inhibition by antibodies against cytochrome P-450, 86
  microassay for, 78—79, 82
  in oyster, 8
  as substrate, 76—79
Ascorbic acid, 10—12, 160—163, 170—177
Ascorbic acid-2-sulfate, 160, 163
D,L-Aspartic acid, 114
Atlantic Shelf, 40

B

Back Cove, Maine, 35—40
Bacterial agglutinins, 96, 100—101, 103—104
Bacterial clearance, 95, 102—103, 106
Bactericidal activity, 97
Bactericidin, 104—106
Bahamas, 37
Baker's formol-calcium solution, 112
Benthic organisms, 2, 4—8, 41—42
Benz(a)anthracene, 5, 58
5,6-Benzoflavone (β-napthoflavone, BNF), 77, 82—88
7,8-Benzoflavone (α-napthoflavone, ANF), 82—83, 85—86
Benzo(a)pyrene (BaP)
  clams
    accumulation in, 119—121
    effects on, 117—119, 122—144
  in coal tar, 5
  in croaker, 85
  dosing procedures with, 97—98, 157
  in field studies, 40—42
  in fish, 163—165, 167, 170—172
  invertebrate responses to, 100—103, 108
  metabolism of, 66—67
  metabolites
    analysis of, 82
    carcinogenic, 9, 66—67
    profile of, 88
  as model pollutant, 5
  preparation of, 78
  quantitation of, 24—30
  in worms, 54—55
Benzo(a)pyrene (BaP) hydroxylase, 9—10
  in estuarine crabs, 56—58
  in induced systems, 85
  from *Neanthes virens*, 54
  subcellular distribution of, 70
Benzo(a)pyrene (BaP) monooxygenase, see Aryl hydrocarbon hydroxylase
Bioaccumulation, see also Accumulation, 8
  measurement of, 28, 30—35
Bioaccumulation factors, 30—35, 39—40
Biochemical changes
  as indicators of pollutant exposure, 163, 174
  in pollutant-stressed animals, 2, 11—12, 163—170, 176—177
  secondary, in response to stress, 163, 170, 176—177
Biogenic amines, 159
Biological indexes, of pollution-induced damage, 2, 8—12
Biomagnification, 8
Biotransformation, 9—10, 66
Blood cells, 94—95
Blood worm, 6
Blue crab (*Callinectes sapidus*)
  aryl hydrocarbon hydroxylase activity in, 10
  collection and handling of, 52
  cytochromes P-450 in, 52—54, 56—59, 62
  field studies of, 39, 41
  microsomes, *n*-octylamine binding to, 62—63
  mixed-function oxygenase activity in, 56—58
  pentachlorophenol effects in, 163

as test animal, 6—7
Bluegill (*Lepomis macrochirus*), 33
Boston Harbor, Massachusetts, 175—176
Bouin solution, 112, 147
Bromophenol blue, 114
Brown shrimp, 42

C

Cadmium
dosing with, 157—158
in fish, 163—165, 167, 170—174
in sandworms, 161—162
California, southern, 37
Callinectes sapidus, see Blue crab
Canada, 39, 42
Capitella capitata, 7
Carassius auratus (goldfish), 33—34
Carbazole, 157
Carcinogenic metabolites, 9, 66—67
Carcinogens, chemical, 3
Casco Bay, Maine, 40
Catalytic function, probes for, 76—77
Catecholamines, 10, 159, 163—164, 177
Catfish (*Galeichthys felis; Ictalurus punctatus*), 31, 39
Chaetodipterus faber (spade fish), 39
Chemicals, representative of pollutant types, 3—6
Chlordane, 36, 38—39, 43—46
Chlorinated hydrocarbons, 4, 9
Chlorinated phenols, 5—6
Cholesterol, 159, 163—164, 166, 175
Chordata, 3, 6—7
Clams
aromatic hydrocarbon accumulation by, 8
collection and handling of, 28
field samples, pollutant concentrations in, 37—38, 40—41
hard-shell, see *Mercenaria mercenaria*
hydrocarbon-exposed, histopathology of, 117—119
immunosuppression in, 12, 108
Clam worm, see *Neanthes (Nereis) virens*
Coal, as PAH source, 5
Cod, 87
Coelomocyte counts, 107
Coho salmon, 167, 169
Combustion, as PAH source, 5
Coomassie blue, 114
Coos Bay, Oregon, 41
Copper, 169
Corpus Christie Bay, Texas, 43—47
Corticosteroid stress response, 10—12, 163—170, 176—177
Cortisol
analysis for, 158
in stress response, 10—11, 163—170, 176
Crabs, see also Blue crab; Spider crab; Stone crab
cytochromes P-450 from, 56—63
mixed-function oxygenase activity in, 9—10, 56—58
Crassostrea virginica, see Oyster
Crayfish, 39
Croaker (*Micropogon undulatus*)
biochemical responses to pollutants, 164, 171
collection and maintenance of, 156
field studies of, 37, 39
hepatic microsomal electron transport components in, 85
pollutant exposure of, 157—158
as test animals, 7
Crude oil, 5, 7
Crustaceans, 9—11
Crustecdysones, 10
Cynoscion nothus (sand trout), 39
Cysteine, 160
Cytochrome b_5, 73—74, 76
Cytochrome P-420, 56, 68
Cytochrome P-450, see also Mixed-function oxygenase system, 51—63
carcinogen activation by, 66
catalytic functions of, 76
from crabs and polychaetes, 52—54
fish liver, induced in, 9
immunochemical analysis of, 84
of insect microsomes, 59
in mollusks, lack of, 8
monooxygenases, environmental induction of in fish, 84—89
multiple forms of, 10, 53, 58—60
purification of, 53, 58—60, 74—75
reconstitutions of, 83
sex difference in, 10
spectral studies of, 53—56, 59—63, 73
steroid metabolism by, 88
Cytochrome P-450A, 88
Cytochrome P-450E, 75, 86—88
Cytosolic factors, 70—71
Cytotoxic factor, 104

D

Davidson's solution, 112, 147
DDT, 35—41, 43—46
Deer Island, Massachusetts, 174
Dehydration schedule, 147
N-Demethylase, 77, 79—80
Depuration, measurements of, 28, 30, 34
Detection limits, 29—30, 81—82
Detergent, nonionic, 75—76
Detoxification systems, 9, 172
Dibenzofuran, 157, 164—171
Di(n-butyl)phthalate (DBP), 36—41, 43—47
Dieldrin, 43—46
Diet
for crabs and polychaetes, 52
sufficiency, effect on xenobiotic metabolism, 68
Di(2-ethylhexyl)phthalate (DEHP), 36—41, 43—47
Diethylphthalate (DEP), 41

Digestive gland, *M. mercenaria*
enzymes of, 133—144
histopathology of, 114—119
Disodium *p*-nitrophenylphosphate, 113
Dissection, 68, 72, 112
Distribution, of pollutants in body, 4
Diurnal rhythms, see Light-dark cycle
Dopamine, 159
Dose-response relationship, 165—167, 172
Dosing procedures, 97—98, 157—158

E

Earthworms, 100
Ecdysterone, 11—12
Eel, yellow (*Anguilla anguilla* L.), 34
Electrophoresis
in monooxygenase studies, 83
polyacrylamide gel, 83, 114
of *M. mercenaria* hemolymph, 122—123, 140—142
solutions for, 151—152
Embryos, frationation of, 71—72
Emulgen 911, 75—76
Enzymes
assay methods for, 77, 112—114
digestive gland, of *M. mercenaria*, 133—144
hemolymph, in *M. mercenaria*, 125—133
isolation of, 69—71
marker, 77
purification of, 72—76
soluble, 113—114
Epoxides, 9
Ethoxycoumarin, 77
Ethoxycoumarin *O*-deethylase, 80
Ethoxyresorufin, 77
Ethoxyresorufin *O*-deethylase (EROD)
assay for, 79
in induced systems, 85—86
as substrate, 77
Extraction procedures, analytical, 24—27

F

Fathead minnow (*Pimephales promelas*), 31
Feulgen stain, 112, 149—150
Fiddler crab (*Una pugnax*), 56—58
Fin rot, 11, 175
Fish, see also specific types
collection and handling of, 28
freshwater, pollutant bioaccumulation in, 31—35
histopathological lesions in, 11—12
immune response to stress in, 12
mixed-function oxygenase systems in, 9—10
monooxygenases in different species of, 84
Fixative solutions, composition of, 147
Fixing and staining, 112
Flatfish, 3
Flavobacterium, 95, 101—104, 108
Florida estuaries, 37
Flounder, see *Platichthys flesus*; Winter flounder
Flouranthene, 5
Flourene, 157, 164—171
Fore River, Maine, 36—40, 174
Formalin-sea water solution, 112, 147
Fossil fuels, as PAH source, 5
Fright reaction, 167—168
Frozen samples, 68, 71
Fuel oil
biochemical effects in fish, 164—171, 173
dosing with, 157
Fundulus, 71—72, 88
Fundulus heteroclitus (mummichog), 6, 10, 84
Fundulus similis (killifish), 30—35

G

Galeichthys felis (catfish), 39
Galveston Bay, Texas, 39—42
General adaptation syndrome (GAS), 10, 163, 168
Glucose
analysis in plasma and coelomic fluid, 159
in blue crabs exposed to pentachlorophenol, 163
correlation with cortisol, 170
in fish exposed to pollutants, 163—170, 175
in sandworms exposed to pollutants, 161
Glucose-6-phosphatase, 70, 81
β-Glucuronidase
assay for, 112—113
in *M. mercenaria*
digestive gland, 133, 136—138, 140—144
hemolymph, 125—129
Glutamic-oxaloacetic transaminase (GOT)
assay for, 114
in *M. mercenaria*
digestive gland, 138—140, 146
hemolymph, 132—137
Glutamic-pyruvic transaminase (GPT)
assay for, 114
in *M. mercenaria*
digestive gland, 138, 140, 145
hemolymph, 128—133
Glutathione
analysis in tissues, 160
depletion of, 87
in pollutant-exposed fish, 170—174
Glycera, 6, 8
Glycera antibacterial factor, 104—106
Glycera dibranchiata (worm), 94, 98, 103—108
β-Glycerophosphate, 112
Glycogen
analysis in tissues, 159
in blue crabs exposed to pentachlorophenol, 163
in fish exposed to pollutants, 163—164, 170, 175—177
in sandworms exposed to pollutants, 161, 174—175
Goldfish (*Carassius auratus*), 33—34

Green sunfish (*Lepomis cyanellus*), 31—32
Grouper, 37
Gulf Coast (Gulf of Mexico), 35—37, 39—40, 42—46

H

Handling stress, 168
Hard-shell clam, see *Mercenaria mercenaria*
Heavy metals, 4, 107
Helley's solution (Zenker-Formalin), 112, 147
Hemagglutinins
  determination of, 96
  in *Glycera dibranchiata*, 103—104
  in *Mercenaria mercenaria*, 100—101, 103
Hematoxylin and eosin, 112, 148—149
Hemolymph, clam, 121—133
Hemolysins
  determination of, 96
  in *Glycera dibranchiata*, 104
  in invertebrates, 100
  of *Mercenaria mercenaria*, 98—100, 103
Hexachlorobenzene (HCB)
  bioaccumulation
    in clams, 120—121
    in killifish, 30—32
  in clams, 117—123, 133—144
  dosing procedures with, 97—98
  in field studies, concentrations measured, 36
  invertebrate responses to, 100—103, 106—108
  as model pollutant, 5
  quantitation of, 24—30
Hexachlorobiphenyl, 86
α-Hexachlorocyclohexane, 36, 38—39
High pressure liquid chromatography (HPLC), 24, 82—83
Histocytochemistry, 112—113
Histology, 4
Histopathological analysis, 11—12, 176
Homeostasis, 2, 169, 176—177
Homogenization, of tissues, 69—70, 72
Hormesis, 135
Hormones, steroid, 10—11, 66
Humoral factors, 96—98
Hydrocarbons
  aromatic, 7, 87
  chlorinated, 4, 9
  as pollutants, 4, 8
Hypercholesterolemia, 10
Hyperglycemia
  in fish exposed to pollutants, 163—165, 170
  as index of pollutant stress, 3, 10
  in sandworms exposed to pollutants, 161, 163, 174
Hypothalamo-pituitary-interrenal axis, 166—170, 177

I

Ictalurus punctatus (catfish), 31
Immunochemical studies, 84
Immunocompetence
  Glycera sp. as test animal for, 8
  in invertebrates, 94—108
  of stressed animals, 12
Immune system, responses of, 4, 12
Indexes
  biological, selection of, 8—12
  biochemical, 140—142, 163, 174
  of pollution-induced damage in marine animals, 2—4
Indoleamines, 159
Induction, environmental, 84—89
Inhibitor control, in enzymatic assays, 112—113
Inhibitors, 77
Insecticides, organophosphate, 8
Insulin secretion, impaired, 170
Interferences, procedures for removing from samples, 27
Invertebrates
  histopathology in, 114—119
  immune response to stress in, 12
  immunocompetence in, 94—108
  mixed-function oxygenase systems in, 9
Ion exchange chromatography, 74—75, 86

J

Jellyfish, 39, 42

K

α-Ketoglutaric acid, 114
Killifish (*Fundulus similis*), 30—35
Kinetics, uptake and release, 8—9

L

Lepomis cyanellus (green sunfish), 31—32
Lepomis macrochirus (bluegill), 33
Libinia emarginata (spider crab), 56—61
Light-dark cycle, 115—117, 119, 121—122, 125—126, 139, 141
Lindane, 43—46
Lipids
  analysis of, 159
  in field study of fish, 175
  metabolism of, 170
  in pollutant-exposed sandworms, 174
Lipophilicity, of pollutants, 8—9, 11
Livers, embryo, 71—72
Liver weight/body weight ratio, and sufficiency of diet, 68
Lobsters, see also Spiny lobster, 37—38
Lower Fox River, Wisconsin, 35, 38, 46
Luidia clathrata (starfish), 39

Lymphotoxin system, 105
Lysozyme, 96—97, 105, 106, 108

M

Mammals
 cytochrome P-450 from, 87
 detoxification of foreign compounds in, 172
 general adaptation syndrome in, 10, 163, 168
 immune responses to stress in, 12
 mixed-function oxygenase system in, 10
Marker enzymes, 77
Mechanical trauma, of injection, 124
Membranes, cell, lipophilic pollutant interaction with, 11
Menippe mercenaria (stone crab), 56, 58
Mercenaria mercenaria (hard-shell clam)
 enzymes of
  digestive gland, 133-144
  hemolymph, 125—133
 in field studies, pollutant concentrations measured in, 38, 40
 histology of, normal, 114—115
 immunocompetence, 94—98
 immunological responses of, 12, 98—103, 107—108
 mixed-function oxygenase activity in, 10
 periodic changes in, 115
 pollutant dosing of, 97—98
 proteins of, hemolymph, 121—125
 as test animal, 6, 8
Mercury, 158, 170—171
Metabolic disturbances, secondary, 170
Metabolic transformation, 4, 9—10
Metabolites, carcinogenic, 9, 66—67
Metallothionein, 170, 173—174
Methylcholanthrene, 77, 84—86
Metyrapone, 77, 85
Mexico, 37—39
Microassay, for aryl hydrocarbon hydroxylase activity, 78—79
Micropogon undulatus, see Croaker
Microsomes
 contaminants in, 76
 crab, 52—53, 62—63
 preparation of, 52—53, 69—71, 73
 storage of, 71
Migration patterns, of test animals, 6
Mississippi Delta, 35—37, 40—43, 46
Mississippi estuaries, 37
Mississippi River, 41
Mitochondria, spectral analyses of, 76
Mixed-function oxygenase (MFO) system, 51—63
 in biotransformation, 9—11
 in crabs, 56—58
 induction of, 2, 4
 in polychaetes, 54—56
 steroid metabolism by, 12
 in worms exposed to crude oil, 7
Mollusca, 3, 6, 8
Mollusks
 mixed-function oxygenase system in, 9—10
 monitoring for pollutants, 3
 polycyclic aromatic hydrocarbon detection in, 5
 xenobiotic metabolism in, 8
Monooxygenases, 65—80
 in different fish species, 84
 environmental induction of, 84—89
 isolation from fish, 67—68
 subcellular distribution of, 69—70
 in winter flounder, sex differences in, 67—68
Morphological changes, stress-induced, 2
Morphometry, 115—117
Mortality, in pollutant-exposed fish, 163—165, 169
Mosquitofish, 39
Mugil cephalus, see Striped mullet
Mummichog (*Fundulus heteroclitus*), 6, 10, 84
Mussels, 41
"Mussel watch", 3
Mytilus, 3
Mytilus edulis, 11

N

NADH-cytochrome b_5 reductase, 77, 81
NADH-cytochrome c reductase, 81
NADPH-cytochrome c reductase, 9, 70, 81
NADPH-cytochrome P-450 reductase, 73, 77
Naphthalene
 biochemical effects in fish, 164, 167, 170—171
 dosing with, 157
α-Naphthoflavone (ANF), see also 5,6-Benzoflavone, 77, 82—83, 85—86
β-Naphthoflavone, see 7,8-Benzoflavone
Naphthol AS-B1 glucuronide, 113
Neanthes arenaceodentata, 7
Neanthes (Nereis) virens (clam worm or sandworm)
 cadmium in, 161—162
 collection and handling of, 28, 52, 156—157
 cytochromes P-450 and mixed-function oxygenases from, 54—55
 field studies of, 37—40, 174—175
 pentachlorophenol in, 161—162
 as test animal, 6—8
New Bedford Harbor, Massachusetts, 38, 40
New York Bight, 38, 175
Nonesuch River, Maine, 174
Norepinephrine, 159, 176
Norfolk, Virginia, 41
Nueces Estuary, Texas, 43—47

O

n-Octylamine, 62—63
Oil spills, 7
Ontario, Lake, 39, 41
Organic pollutants, see Pollutants, organic
Organochlorine residues, 35—40
Organophosphate insecticides, 8

Osmolality, 159, 163—164, 166, 170
Oyster (*Crassostrea virginica*)
  aryl hydrocarbon hdyroxylase induction in, 8
  benzo(a)pyrene hydroxylase in, 10
  field samples, pollutant concentrations in, 37, 41—42
  leukocytic infiltration in PCB-exposed, 11

P

Panulirus argus (spiny lobster), 59
Partition coefficients, octanol/water, 8
Pathology index, 176
Penaeus aztecus (brown shrimp), 42
Pentachlorophenol (PCP)
  bioaccumulation of, 32—36
  in clams, 117—119, 122—123, 133—144
  in crabs, 163
  derivatization of, 25—26
  dosing procedures with, 97—98, 157—158
  in field studies, concentrations measured, 36—46
  in fish, 164, 167—172
  invertebrate responses to, 100—103, 106—108
  as model pollutant, 5—6
  quantitation of, 24—30
  in sandworms, 161
Pesticides, 5, 43—47
Petrochemicals, 4
Petroleum, 3, 5, 7
pH, 163
Phagocytosis, 94—95, 107
Phenol, 12
Phenolphthalein glucuronic acid, 113
Phenols, chlorinated, 5—6
Phospholipids, 9
Phthalate esters, 35—41
Phyla, marine, 3, 6
Physical stresses, response to, 166—168, 172, 174
Physiological dysfunction, 169
Pimephales promelas (fathead minnow), 31
Pituitary-adrenocortical axis in fish, 3, 10—11
Platichthys flesus (flounder), 86
Plymouth Beach, Massachusetts, 175—176
Pollutants, organic
  accumulation in marine systems, 2, 8
  biological processes, interactions with, 12—13
  cell membrane, interactions with, 11
  chemicals chosen as models of, 3—6
  exposure studies of, 97—98, 156—174
  properties of, 8—9
Polychlorinated biphenyls (PCBs)
  field studies, concentrations measured in, 35—41, 43—46
  immune system effects of, 12
  in marine environment, 5
  in oysters, 11
  in worms, 54—55
Polychlorinated dibenzodioxins, 8
Polychloroaromatic hydrocarbons, 5
Polyclonal antibodies, 86
Polycyclic aromatic hydrocarbons (PAH), 5, 7
Polydora, 12, 108
Porgy, 7
Portland, Maine, 35—40, 43, 174—175
Postmitochondrial supernatant fluids, (''S-9''), 70
Proteins, of *M. mercenaria* hemolymph, 121—126, 140—142
Pseudopleuronectes americanus, see Winter flounder

R

Rainbow trout (*Salmo gairdneri*), 31, 34, 86—88
Raja erinacea, 10
Rangia cuneata (clam), 8
Rat, cytochrome P-450 from, 87
Reconstitutions, of purified cytochrome, P-450, 83
Recoveries, from spiked samples, 28—30
Rosette formation, 95—96, 107

S

D-Saccharic acid-1,4-lactone, 113
St. Croix estuary, Canada, 39
St. John's estuary, Canada, 39
Salmo gairdneri (rainbow trout), 31, 34, 86—88
Salmo trutta, 33
Salmon, Coho, 167, 169
Samples
  collection and handling of, 27—28
  extraction procedures for, 24—27
  frozen, 68, 71
  preparation for monooxygenase measurements, 68—72
Sampling sites, 27
Sand column dosing system, 28, 119—120, 158
Sand trout (*Cynoscion nothus*), 39
Sandworms, see also *Neanthes (Nereis) virens*, 156—158, 161
Sandy Hook-Raritan Bay area, 175
San Luis Pass, Texas, 40—43
Scup (*Stenotomus chrysops*)
  benzo(a)pyrene metabolites in, 88—89
  cytochrome P-450 from, 86
  enzymes, purification of, 72—76
  liver weight/body weight decline after capture, 68
  microsomes, preparation of, 73
  as test animals, 7
SDS-PAGE, 83, 114, 151—152
Sediments
  analytical procedures used for, 26—27, 29—30
  petroleum hydrocarbon degradation in, 7
  as PAH source, 5
  samples
    collection of, 27—28
    pollutant concentrations in, 35—37
  as sinks for contaminants, 6
Sepunculids, 100
Serotonin, 159, 176

Serratia marcescens, 97, 104—106
Sesarma cinerum (crab), 56—58
Sex differences, 67—68, 84
Shrimp, brown, 42
SKF 525-A, 77, 85
Sodium dodecyl sulfate, 114
Southern California, 37
Spade fish (*Chaetodipterus faber*), 39
Spectral analyses
  of cytochrome b_5, 76
  of cytochromes P-450, 53—56, 59—63, 76
Spider crab (*Libinia emarginata*), 58—61
Spiny lobster, 59
Staining solutions and procedures, 148—150
Starfish (*Luidia clathrata*), 39
Stenotomus chrysops, see Scup
Steroid hormones, 10—11, 66
Steroidogenesis, 10, 170
Steroids, 10, 12, 88, 170
Stone crab (*Menippe mercenaria*), 56, 58
Stress responses, 4, 10—11
  biochemical, 163—170, 176—177
  physical, 166—168, 172, 174
  primary, secondary, and tertiary, 177
Striped mullet (*Mugil cephalus*)
  bioaccumulation factor for pentachlorophenol in, 33
  biochemical responses to pollutants, 164—171
  collection and maintenance of, 156
  pollutant exposure of, 157—158
  as test animal, 6
Subcellular fractions, preparation of, 68—71
Substrates, 76—77, 112—114
Succinate cytochrome c reductase, 81
Sulfur and sulfur compounds, 27

T

Target organs, 3—4
L-Tartrate, 112
Temperature, of water, 175
Testosterone, 82, 88
Testosterone hydroxylase, 88
Texas estuaries, 37, 43—47
Thiols, see Acid-soluble thiols
Tidal cycle
  measurements in marine bivalves and, 115—119, 121—122, 125—127, 139, 141
  mummichog migration with, 6
Tilapia mossambicus, 164, 167
Tissue samples
  analytical procedures used for, 26, 29—30
  homogenization of, 69
Triglycerides, 159, 163
Trout, see also Rainbow trout; Sand trout, 33
Tumors, as index of chemical pollution, 3

U

Una pugnax (fiddler crab), 56—58

V

Vancouver Harbor, 41

W

Water samples, procedures for, 24—29
Winter flounder (*Pseudopleuronectes americanus*)
  biochemical response to pollutants in, 164, 171
  cytochrome P-450 and monooxygenase activities in, 67—68
  field studies of, 175—176
  induction in, significance of, 88
  as test animal, 6
Worms, polychaete, see also *Glycera dibranchiata; Glycera* sp.; *Neanthes (Nereis) virens*; Sandworms
  collection and maintenance of, 52
  cytochromes P-450 and mixed-function oxygenases from, 9—10, 54
  hemolysin, failure to induce in, 100
  stress response in, 11
  as test animals, 7—8

X

Xenobiotics, 8, 66—68

Y

Yaquina Bay, Oregon, 41
Yellow eel, 34